STUDENT STUDY GUIDE
for
UNDERSTANDING EARTH
Third Edition

FRANK PRESS
RAYMOND SIEVER

PETER L. KRESAN
University of Arizona, Tucson

REED MENCKE, PH.D.
University of Arizona, Tucson

ROBERT L. BINGHAM
University of Arizona, Tucson

W. H. FREEMAN AND COMPANY
NEW YORK

Supplements Editor: Bridget O'Lavin
Supplements Project Editor: Jodi Isman
Production Coordinator: Paul Rohloff
Composition & Text Design: Boleyn Baylor

ISBN: 0-7167-4122-9

Copyright © 2000 by W. H. Freeman and Company

No part of this book may be reproduced by any mechanical, photographic, or electronic process, or in the form of any phonographic recording, nor may it be stored in a retrieval system, transmitted, or otherwise copied for public or private use, without written permission from the publisher.

Printed in the United States of America

First Printing 2000

Contents

Preface

WELCOME TO GEOLOGY!

We live on a dynamic Earth! A fact we trust you will appreciate even more by the end of this semester. Although we do not wish to predict geological events, we will predict that your awareness of geological events and their significance will be increased in this course. We think you'll come to appreciate the following quote:

Civilization exists by geological consent . . .
subject to change without notice.
—*Will Durant*

We wish you a great semester studying geology. Jesse Jackson said, "It is attitude — not aptitude — that determines altitude." If your attitude is good and if you do the work assigned when it is assigned and not get behind, you will do very well in this course. We hope that your hard work combined with this Study Guide will lead to success in your geology course and inspire you to maintain a lifetime interest in serving as a good steward for our Earth.

This Study Guide is dedicated to your success!

Peter Kresan
Reed Mencke
Bob Bingham

Acknowledgments

The authors wish to thank the University Learning Center at the University of Arizona for considerable support during this project. Many materials in this Study Guide have been adapted from the University Learning Center Workshop materials. Particular thanks to Sylvia Mioduski for her continuing encouragement and understanding, to Stacey Hartman who developed some materials adapted for use in this guide, and to Nancy Frazier for review and comment on some chapters. We also thank Michelle Wallace and Carlotta Chernoff for their thoughtful reviews. Very special thanks to Boleyn Baylor and Teri Bingham who played key roles in the production of this guide—Bo for her invaluable and skillful design, layout, formatting, editing and proofing, and Teri for her generous assistance with graphics and production advice. Special thanks also to Bridget O'Lavin and W. H. Freeman and Company for the opportunity to work on this project and the freedom to be creative with its design and content. We are also thankful for the use of the outstanding illustrations produced for the text. Reed extends his greatest thanks to his partner in life Pat Mencke for supporting him through this project as she has through so many others.

PREVIEW OF THE STUDY GUIDE / UNDERSTANDING EARTH

This guide was designed to make your study of geology as effective and rewarding as possible. Effective learning is primarily a matter of being organized and systematic. The guide lays out a step-by-step method for studying geology. It will help you to do each of the following:

- Focus on the most important key ideas in each chapter.
- Master the material in each chapter thoroughly enough that you will be ready to move on to the next chapter.
- Study effectively. The study method is based on sound and tested principles of learning. Specific geology study strategies are woven throughout the guide, including *How to Take an Excellent Set of Notes* and *How to Prepare for Exams*. A final Appendix, *How to Study Geology*, brings all these strategies together into a short course.
- Be organized and efficient. Success in geology, or any other science, revolves around lecture. Study aids in the guide are laid out step-by-step (see Flow Chart: How to Study Geology) so you know what to do before lecture, during lecture, after lecture, and during exam preparation.

Before Lecture

Before lecture you should review the *Chapter Preview* questions.

The key to getting a good set of notes is to come to class with an overview of what will be covered already in mind. Overview means you know what geological processes will be explained and what key questions about these processes will be answered in the lecture. The method you use to gain an overview is called *Chapter Preview*. In each *Chapter Preview* we identify the three or four key questions the chapter will cover. Brief answers designed to help you start thinking about the material are supplied as well. You work with these questions and answers to insure that you arrive at lecture ready to listen and take a good set of notes.

During Lecture

During lecture your main task is to take an excellent set of notes.

With the basic questions in mind you have a head start on getting good notes. Note taking is a skill. You can get better at note taking if you practice. In early chapters we provide a *Note-Taking Checklist* of specific actions you can take to get more content into your notes in a usable form. Additional note-taking tips are provided for every text chapter: e.g., suggestions about what to listen for, or figures you need to review carefully before a particular lecture.

After Lecture

After each lecture we suggest you schedule two separate sessions: a brief *Note Review Session*, and a longer *Intensive Study Session* during which you achieve mastery of the content of that lecture.

Note Review Checklist. The purpose of note review is to clean up your notes and add important visual material. The end result will be notes that are useful for both learning the material and preparing for exams. For best results, note review should occur right after class while you can still remember the lecture vividly.

Intensive Study Session. The intensive study session is designed to insure you master the material. Just reading is a very inefficient way to learn. You learn best if you spend your study time answering questions. *Practice Exercises*, CD activities and a set of multiple choice questions are provided for each chapter of the text. Using these materials is the fastest way to learn and the best way to remember.

Exam Prep

Doing well on college exams is mostly about organization. The ideal is a systematic, orderly review during which you spend most of your time (about 70 percent) answering review questions. A *Chapter Summary* and *Practice Exam Questions* are provided for each chapter of the text.

Preparing for exams is a skill. You can improve your exam prep skill by practice. Tips for preparing for midterm and final exams, taking a multiple choice test, and using your personal learning style to advantage are built into the *Exam Prep* section of many chapters.

Appendix: How to Study Geology

All study tips and hints are rolled up into a single appendix. You can either go through this appendix in sequence as your own personal short course in study skills, or just use it as a reference.

How to Study Geology

During Lecture:

Take an Excellent Set of Notes
- Note Taking Check List

Before Lecture

Prepare for lecture: Arrive with an overview in mind.
- Chapter Preview
- Vital Information From Other Chapters
- Website and CD Preview

Right After Lecture

Note Review
Fill in what you missed. Add visual material
- Note Review Checklist

After Lecture:

Intensive Study Session
Master the key concepts.
- Website and CD Activities and Tools
- Practice Exercises and Study Questions

Exam Prep

Begin review one week before the exam
- Eight-Day Study Plan
- Chapter Summaries
- Practice Exam Questions
- Tips for Preparing for Geology Exams

Chapter 1
Building a Planet

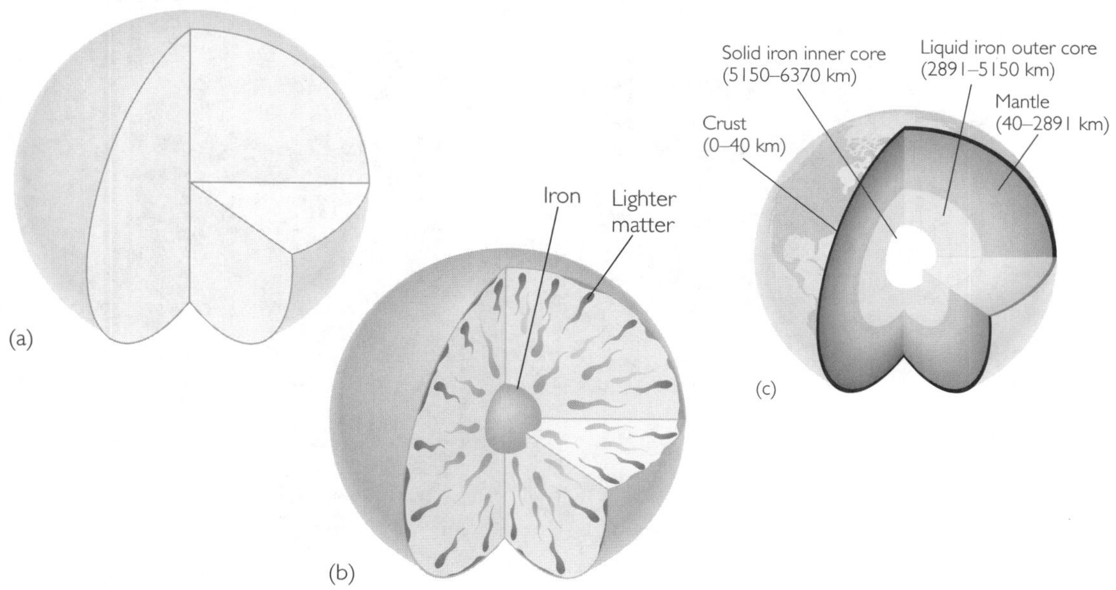

Figure 1.6 / Early Earth (a) was probably a homogeneous mixture with no continents or oceans. In the process of differentiation, iron sank to the center and light material floated upward to form a crust (b). As a result, Earth is a zoned planet (c) with a dense iron core, crust of lower density rock, and a residual mantle between them.

BEFORE LECTURE

CHAPTER PREVIEW

- **How do geologists study Earth?**
  Brief answer: Using the scientific method.

- **How did our solar system form?**
  Brief answer: It accreted from gas and dust about 4.5 billion years ago.

- **How did the Earth form and change over time?**
  Brief answer: The Earth's core, mantle, crust, oceans and atmosphere evolved as the interior of the planet heated up, melted and differentiated.

- **What are the basic elements of plate tectonics?**
  Hint: See *Plate Tectonics: A Unifying Theory for Geologic Science* in Chapter 1.

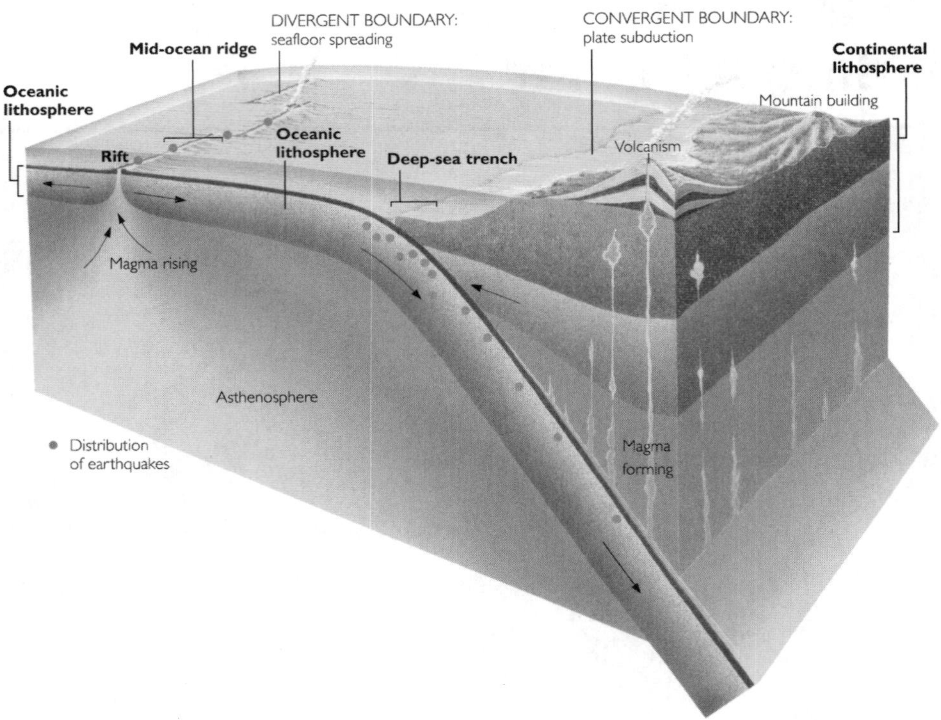

Figure 1.16 / The basic elements of plate tectonics are characterized here. The oceanic lithosphere grows at the spreading center by magmatic and volcanic processes and is consumed by subduction at the convergent boundary. Many geologic processes occur at the convergent plate boundary—subduction of lithosphere, the generation and upwelling of magma, volcanism, mountain building, and earthquakes. This illustration is not drawn to scale. The horizontal distance has been compressed and the vertical dimension exaggerated to show detail.

Civilization exists by geological consent. . . . subject to change without notice.
—*Will Durant*

DURING LECTURE

This will be your very first lecture for this course. Get off to a good start. Arrive ten minutes early. Find the best seat in the house: close to the front of the room, where you can hear the lecture and see the slides. If there is a projector or other equipment blocking your view, change locations.

OK, you've got a good seat. You have ten minutes before lecture begins. What should you do with it?

First, motivate yourself to want to listen to this lecture. Open your text to Chapter 1. Thumb through the chapter just looking at the pictures. Look for stuff that interests you. Ask yourself what you would like to know about this chapter. What would you ask your teacher if this were a one-on-one tutorial?

Second, prepare your mind for learning. A master football player warms up by running, stretching, and passing the football. A master learner warms up by focusing attention on what will be covered during lecture. Spend a minute or two looking over the Chapter Preview questions. Try to anticipate how these questions will be answered in lecture. If you have time, read the *Chapter Summary* for Chapter 1. Read over some of the questions.

LEARNING TIP
Do a "learning warm-up" before every lecture. Arrive 10 minutes early. "Psych" up. Prepare your mind for learning. Anticipate questions that will be answered in lecture.

The man who is afraid of asking is ashamed of learning.
—Danish proverb

AFTER LECTURE

Review Notes

The perfect time to review your notes is right after lecture while the material is still fresh in your mind. Review to be sure you got all the key points and wrote them down in a form that will be readable later.

Meet Press and Siever: How to Use Your Geology Textbook

As you begin your study of geology, first take a minute or two to get acquainted with your text. First of all, why do you think it was written? Here is what your text authors Frank Press and Raymond Siever have to say about why they wrote this book:

"Geology fascinates and excites us. We wrote *Understanding Earth* to help you discover for yourselves how interesting geology is in its own right and how important an understanding of geology has become for making decisions of public policy. What can we do to protect people and property from natural disasters such as volcanoes, earthquakes, and landslides? How can we use the resources of Earth—coal and oil, minerals, water, and air—in ways that minimize damage to the environment? In the end, understanding Earth helps us understand how to preserve life on Earth."

Press and Siever add that people tend to enjoy what they do well. They designed their text to help you do well in your geology course. Many aids to learning are built into the text.

Clues and Tools: "What's Important?"

One of the toughest things about taking an introductory course is that you don't know the subject matter! That means that not only do you have a lot to learn but you have a lot to learn about *how* to learn it. Where should you focus your attention? What skills and concepts should receive the bulk of your study energy? Aids to help you see what material is important are built into every page of *Understanding Earth*, but you have to know where to look for these aids and how to use them. Here is a short list of learning aids in your new text and a few preliminary thoughts about how to use them to master geology:

Chapter Outline

Press and Siever begin each chapter with an outline of what will be covered. To use this tool you need to approach it like a detective: looking actively for clues. Look at the outline for Chapter 1. It probably won't surprise you that the first section in a geology text would be "The Scientific Method." But look further. Would you expect a section about: "The Origin of Our System of Planets"? Why do you suppose a geology text would need to talk about how the planets are formed? If you don't know, read with that question in mind.

Chapter Summary

At the end of each chapter a brief summary emphasizes the most important ideas of the chapter.

> **TIME-SAVER TIP**
> **Before reading a chapter of the text read the *Chapter Summary* first, referring as you do so to the *Chapter Outline*. This will provide an overview, or organizer, that will greatly accelerate your reading and understanding of material.**

Key Terms

Key terms within the text are printed in bold. You can further speed up your reading time by being vigilant for key terms in bold. When you see such a bolded term you know it is the most important concept in that paragraph. Focus on understanding that concept. Read with purpose.

Pictures

Geology is a very visual science. Pictures in the text are essential. Use the pictures as a "virtual field trip" to help you learn what particular rocks and formations look like. Refer to these as you read.

> **STUDY TIP**
> **Having a tough time getting yourself started with a study session? Use the pictures to motivate yourself. Start a chapter study session by scanning the pictures and captions to find material that interests you. Begin your reading with that material. This approach works like fire building. You generate a small spark of interest, then fan it by bringing in new material**

Figures (Flow Charts and Tables)

Figures are even more important than pictures. Flow charts such as the Rock Cycle in Chapter 3 present the key concepts of each chapter. Pay careful attention to the arrows in flow charts and ask yourself what drives the process. If you are a visual learner you may even want to study the figures and sketches before you start to read. Then read text on an as-needed basis to clarify the figure.

Color

Color is used in the figures to provide you with clues about how each process works and what is involved. Look at the Rock Cycle (Figure 3.1. in Chapter 3). What color is used to describe processes that utilize heat? Water? Cooling? Such cues can support learning at a subliminal level of awareness, particularly for visual learners. But they work even better if you pay close attention to them.

Clues to Rock Texture

Clues to rock texture are also cleverly built into the text. Look at Figure 3.1 again. Why do you think igneous rocks are depicted as "dots and specks of varying size"? Why are sediments and sedimentary rock depicted with "horizontal markings"? Why is metamorphic rock depicted with a series of "distorted and folded bars"? You will see as you study the next several chapters that each of these textures is descriptive not only of how the rocks look but of the processes that produce them. Pay close attention to these visual learning clues. They will help you learn geology.

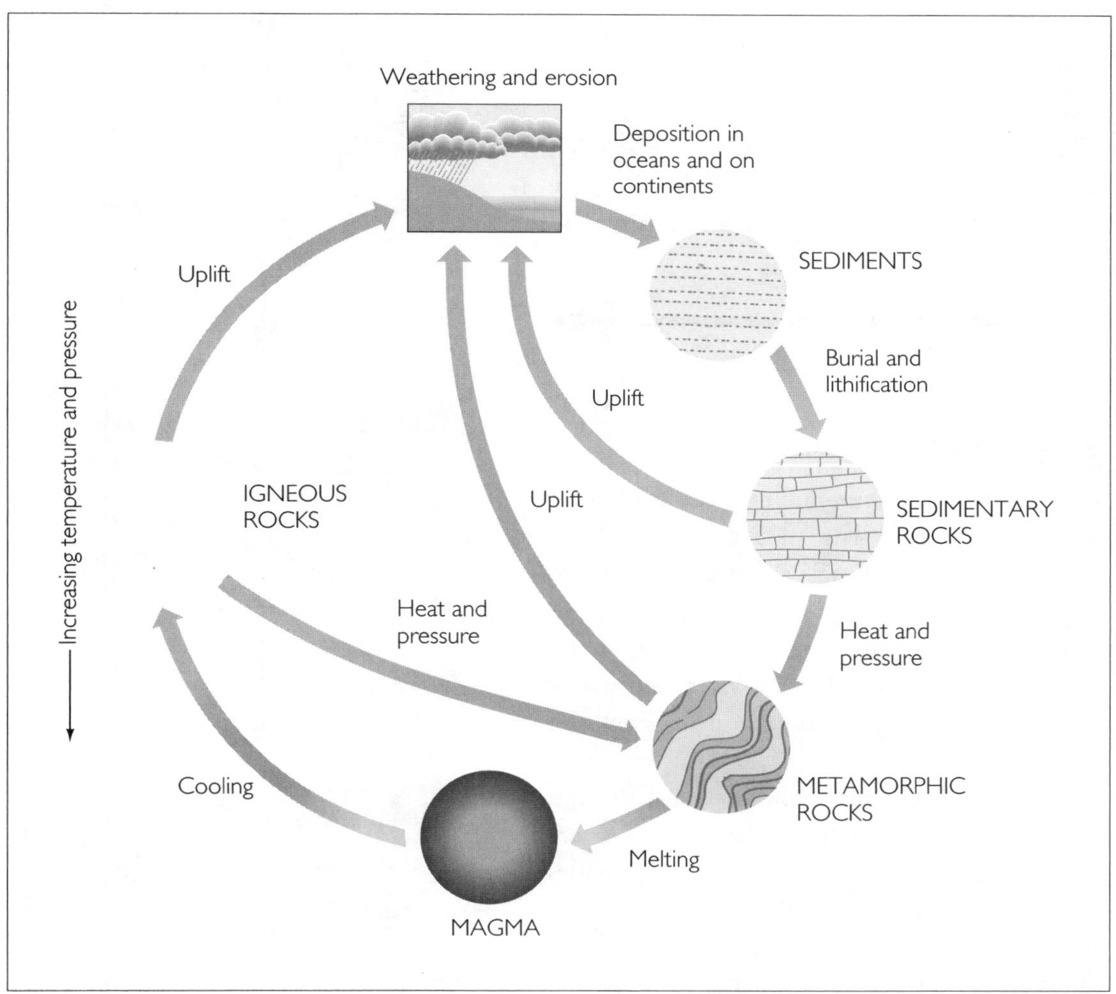

Figure 3.1 / The Rock Cycle figure makes great use of patterns and colors.

Website and CD Activities and Tools

http://www.whfreeman.com/presssiever

- **Website Q & A Practice Multiple Choice Questions:**

  At the Website complete the *Q & A*. Pay particular attention to the explanations for answers. Flashcards at the Website will help you learn new terms. The *Interactive Exercise: Three Types of Plate Boundaries* at the Website is a good review of the basic elements of plate tectonics.

PRACTICE EXERCISES AND STUDY QUESTIONS

(Answers and explanations are at the end of this chapter.)

Exercise 1: The Evolving Early Earth

Fill in the blanks in the flow chart (right) which characterizes a popular hypothesis for the early history of the Earth after the sun and planets had formed.

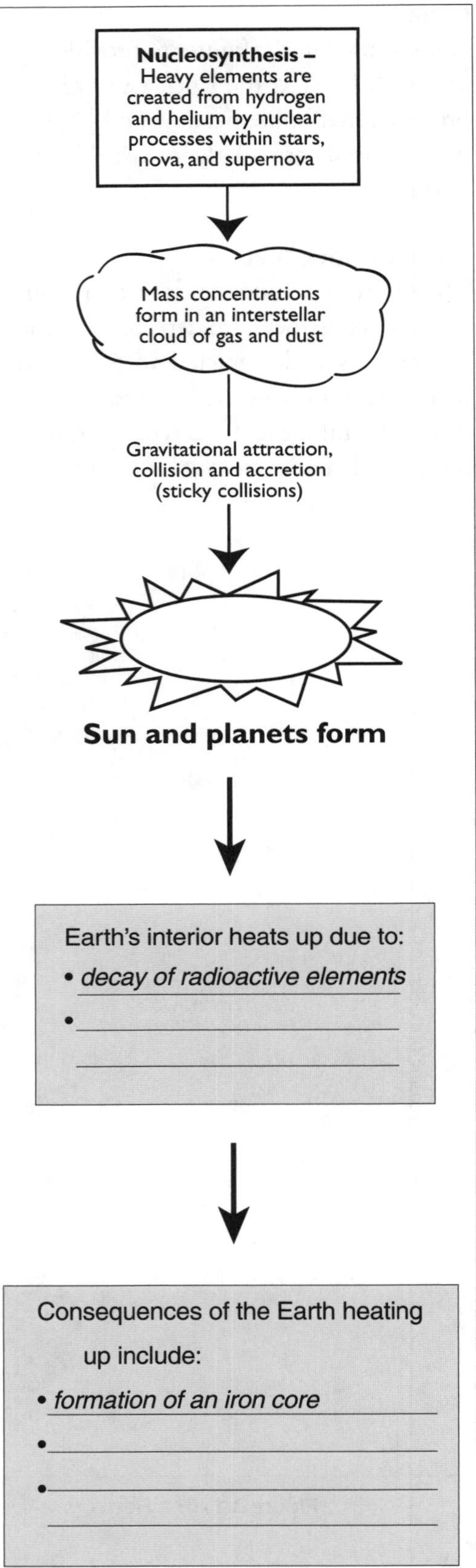

EXAM PREP

Materials in this section are most useful during your preparation for midterm and final examinations. For optimal performance, midterm preparation should begin about eight days prior to the exam (see Appendix: *How to Study Geology* for details). The basic idea is a systematic review of material divided into short study sessions.

The following *Chapter Summary* and *Practice Exam* should simplify review. Read the *Chapter Summary* to begin your review session. It provides a helpful brief overview that should get you back into the material. Then try the *Practice Exam*. Take it just as you would a midterm: time yourself and keep your book closed until you have finished. Next, score your exam to see how you stand in regard to mastery of this chapter. Finally, and most important of all: review any item that you missed. Identify and correct the misconception that made you miss the item.

CHAPTER 1 SUMMARY

Website and CD Activities and Tools

http://www.whfreeman.com/presssiever

At the Website in the *Geology Essays* section, the essay *Planetary Geology* will provide you with interesting information on meteorites and our solar system.

- Field and lab observations, experiments, and the human creative process help geoscientists to formulate hypotheses (models) for how the Earth works and its history. A hypothesis is a tentative explanation which can help focus attention on plausible features and relationships of a working model. If a hypothesis is eventually confirmed by a large body of data, it may be elevated to a theory. Theories are abandoned when subsequent investigations show them to be false. Confidence grows in those theories that withstand repeated tests and are able to predict the results of new experiments.

- Our solar system probably formed when a cloud of interstellar gas and dust condensed about 4.5 billion years ago. The planets vary in chemical composition in accordance with their distance from the Sun and with their size.

- Earth probably grew by accretion of colliding chunks of matter. Very early after the Earth formed, it is thought that our Moon formed from material ejected from the Earth by the impact of a giant meteorite.

- Heat generated from the Moon-forming impact and the decay of radioactive elements probably caused much of the Earth to melt. Melting allowed iron and other dense matter to sink towards the Earth's center and form the core. Lower density (lighter) matter floated upward to form the mantle and crust. Release of trapped gases (mostly water) from within the Earth gave rise to the oceans and an early atmosphere. In this way, the Earth was transformed into a differentiated planet with chemically distinct zones: an iron core; a mantle of mostly magnesium, iron, silicon, and oxygen; and a crust rich in oxygen, silicon, aluminum, calcium, potassium, sodium, and radioactive elements.

- As the Earth cooled, an outer relatively rigid shell called the lithosphere formed. Dynamic processes driven by heat transfer, density differences, and gravity broke the outer shell into plates which move around the Earth at a rate of centimeters per year. The boundaries between divergent and convergent plates are zones of intense volcanism, earthquake activity, and mountain building. The oceanic lithosphere is produced at divergent boundaries and recycled back into the mantle at convergent boundaries. Major earthquakes occur along transform fault boundaries.

PRACTICE EXAM QUESTIONS FOR CHAPTER 1

(Answers and explanations are at the end of this chapter.)

1. Of the following statements regarding the scientific method, which one is true?
   A. The outcome of scientific experiments cannot be predicted by a hypothesis.
   B. A hypothesis that has withstood many scientific tests is called a theory.
   C. For a hypothesis to be accepted, it must be agreed upon by more than one scientist.
   D. After a theory is proven to be true, it may not be discarded.

2. During the formation of our solar system, what was the process that caused dust and condensing material to accrete into planetesimals?
   A. Nuclear fusion
   B. Rapid spin of the proto-sun
   C. Heating of gases
   D. Gravitational attraction and material collisions

3. A major source of internal heat still in the Earth today is
   A. Ocean tides
   B. Radioactivity
   C. Solar energy
   D. Volcanoes

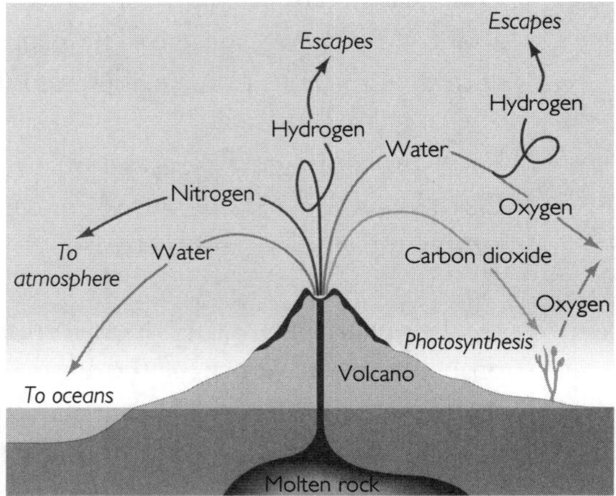

Figure 1.8 / The evolution of the Earth's atmosphere.

4. Which of the following processes is thought responsible for the formation of the present Earth atmosphere?
   A. Chemical breakdown of minerals
   B. Capture of oxygen and nitrogen from space
   C. Both degassing of Earth's interior and oxygen-liberation by plants
   D. Degassing of the Earth's interior

5. How old is the Earth estimated to be, based on radiometric dates of lunar rock samples and meteorites?
   A. 45 million years old
   B. 4.5 billion years old
   C. 10 billion years old
   D. 100 billion years old

6. Geochemists have determined that the differences in elemental distribution in the present Earth's crust, mantle, and core—in contrast to the initial solar nebula—may be due in part to
   A. inhomogeneities within the initial interstellar dust cloud
   B. early melting and differentiation of materials according to density
   C. cosmic ray bombardment over 4.5 billion years
   D. nuclear synthesis within our sun

7. The process most important to ocean floor building is
   A. volcanism
   B. deposition of sediment derived from the land
   C. metamorphism
   D. precipitation of carbonate rocks

8. The rates of horizontal movements on a continental scale of the Earth's crust occur on the order of
   A. several millimeters per year
   B. several centimeters per year
   C. several meters per year
   D. several kilometers per year

9. Along a transform plate boundary like the San Andreas Fault the two plates
   A. move apart to create a widening rift valley
   B. are being consumed by subduction
   C. are being forced together so as to produce a mountain system
   D. slip horizontally past each other

10. Of the following, which is an example of a divergent tectonic plate boundary?
    A. Mid-Atlantic Ridge
    B. San Andreas Fault
    C. Mississippi River
    D. Japanese Trench

11. The geologic principal that describes the idea that processes operating today worked much the same over all of geologic time is known as the
    A. principal of past parallelisms
    B. principal of humanitarianism
    C. principal of uniformitarianism
    D. principal of equality

TEST TAKING TIP

When taking a test, be alert to questions that give away the answers to other questions. Example: Question 12 provides the answer to Question 11.

12. "The present is the key to the past" is a phrase that states the principle of
    A. the hydrologic cycle
    B. the rock cycle
    C. uniformitarianism
    D. the scientific method

13. Earth's lithosphere can be characterized as
    A. having an average thickness of about 100 km
    B. including the crust and upper mantle
    C. being solid and above the asthenosphere
    D. all of the above

14. Press and Siever describe the origin and evolution of the early Earth as
    A. a proven fact
    B. a theory
    C. a hypothesis
    D. pure guesswork

15. What valuable purpose does a scientific hypothesis (i.e., scientific model) serve?
    A. It reveals the absolute truth.
    B. It provides a basis for testable predictions.
    C. It can never be changed.
    D. It is a complete representation of the real world.

ANSWERS AND EXPLANATIONS FOR EXERCISES AND QUESTIONS

Exercise 1: The Evolving Early Earth

Earth's interior heats up due to:
- *decay of radioactive elements*
- Moon-forming impact and other collisions

↓

Consequences of the Earth heating up include:
- *formation of an iron core*
- differentiation of the mantle and crust
- release of gases from the interior to form the oceans and atmosphere

1. B. A hypothesis is a tentative explanation which can help focus attention on plausible features and relationships of a working model. If a hypothesis is eventually confirmed by a large body of data, it may be elevated to a theory. Theories are abandoned when subsequent investigations show them to be false. Confidence grows in those theories that withstand repeated tests and are able to predict the results of new experiments.
2. D. Gravitational attraction caused the dust and condensing material to collide and clump together (sticky collisions). Refer to Figure 1.2.
3. B. The heat generated by the decay of radioactive elements continues to heat the Earth today. Along with meteoric impacts, it is thought to have partially melted the early Earth.
4. C. Volcanic degassing accounts for all major gases in the Earth's atmosphere except oxygen. Oxygen is a byproduct of photosynthesis by algae and plants (refer to Figure 1.8).
5. B. Figure 1.5 provides a time line for the early history of the Earth.
6. B. The Earth's core, mantle, and crust are thought to have formed when much of the Earth melted. Melting allowed materials within the earth to segregate (differentiate) according to their density—heavier matter sinks towards the center (refer to Figure 1.6).
7. A. Basaltic volcanism along divergent plate boundaries generates ocean crust (refer to Figure 1.16).
8. B. Typical plate movement today is between 2 to 17 centimeters per year.
9. D. Plates slip past each other along transform boundaries, like the San Andreas Fault.
10. A. Figures 1.12 and 1.14 are good references for this question.
11. C. Uniformitarianism is an historic concept important in the history of geology.
12. C. See the discussion in *Modern Theory and Practice of Geology.*
13. D. The lithosphere is typically characterized by all three.
14. C. A hypothesis is a tentative explanation which can help focus attention on plausible features and relationships of a working model. If a hypothesis is eventually confirmed by a large body of data, it may be elevated to a theory.
15. B. Confidence grows for those hypotheses that withstand repeated tests and are able to predict the results of new experiments.

It is in the stars.
The stars above us, govern our condition.
—*W. Shakespeare*
King Lear *Act IV, scene III*

Chapter 2
Minerals: Building Blocks of Rocks

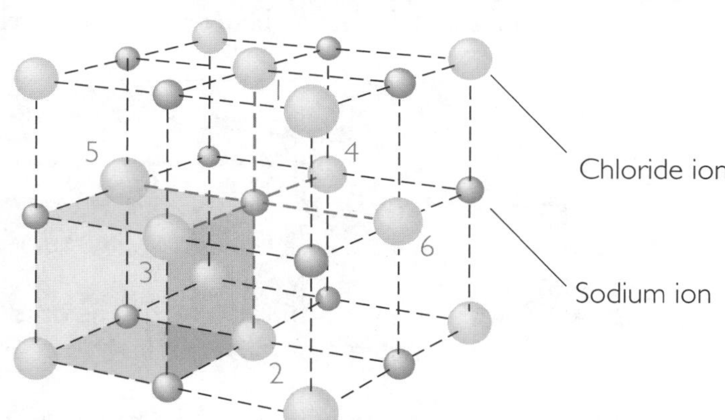

Figure 2.9 / Structure of sodium chloride. The dashed lines between the ions show the cubic geometry of this mineral; they do not represent chemical bonds.

BEFORE LECTURE

Pre-Lecture Chapter Preview

Why preview? Introductory courses can be difficult. There are lots of new terms, new ideas, and new skills during a first semester geology course. But which ideas are "important"? How do you focus your effort?

Picture a house under construction. Imagine further that the contractor was very careless and neglected to construct whole sections of the frame. Finishing those areas lacking a frame would be impossible. It's pretty difficult to tack siding onto thin air! This metaphor describes your geology lecture. You are the contractor. You need to arrive at class with an overview of the lecture already in mind. An overview means you have already identified which geological processes will be explained and what key questions the lecture will answer. Lacking a frame of key questions is like building a house with no supporting structure. You will have nothing on which to hang the lecturer's main points (information). Without the main points the details are meaningless. As the lecture progresses you are likely to feel increasingly confused and bored. By the time the lecturer gets to the most important material you may be completely lost. Research suggests that the average note taker gets less than 15 percent of what is covered in the last segment of a lecture into their notes.

Where do you get the overview? The answer is surprisingly simple. You just need to spend a few minutes before lecture previewing the chapter that will be covered. Previewing is the method by which you generate and master a framework for listening. Here's how to do it:

Three Question Preview Method

Step 1: Skim the text. Look for the three or four most important questions the material will cover.

Step 2: List the questions. Limit yourself to just three or four so you will be sure to remember your framework during lecture. Be sure you identify the questions that are most important.

Step 3: Answer the questions. Spend the rest of your pre-lecture study time finding preliminary answers in your text. Read just enough to get a general idea of the answer. Don't allow yourself to get lost in details.

The chapter on Minerals will provide your first opportunity to try out this method. So, before you go to that lecture be sure to spend some time previewing Chapter 2. We have made it easier by identifying the three key questions on Minerals for you. Use the following *Chapter Preview* questions as a guide. These questions constitute the basic framework for the topic of Minerals. Working with the preview questions before lecture and committing them to memory should help you both understand the lecture better and get an excellent set of notes.

CHAPTER PREVIEW

- **What are minerals?**
  Brief answer: a naturally occurring, inorganic, crystalline solid.
- **How are atoms combined to form the crystal structure of minerals?**
  Hint: See structures for silicate minerals in Figures 2.16 and 2.17.
- **What is the atomic structure of minerals?**
  Brief answer: For silicate minerals it is various arrangements of silica tetrahedra.
- **What are the major rock-forming minerals and their physical properties?**
  Brief answer: The silicates and carbonates. Diagnostic properties include: hardness, cleavage, luster, color, density and the shape of the crystal.

Vital Information from Other Chapters

- Differentiation of Earth into crust, mantle, and core in Chapter 1.

Minerals are the alphabet and rocks are the words.
—*Tom Dilley*

DURING LECTURE

Do a learning warm up. Arrive early. Look for pictures in the text that catch your interest. Look over the preview questions and keep them handy for easy reference.

Your basic goal during lecture is to get a good set of notes. Those notes should contain in-depth answers to the three preview questions. To avoid getting lost in details, keep the big picture in mind: Chapter 2 explains how rocks are built out of minerals and minerals are built out of atoms.

Your lecturer may show slides of mineral crystals and may pass around specimens of particular minerals. You will get more out of these if you sit close to the front row where you can see the sample rocks as he discusses them. Remember to focus carefully on clues he provides for recognizing a particular sample in the field. As you listen try to formulate at least one good question you can ask during the lecture.

AFTER LECTURE

Review Notes

The perfect time to review your notes is right after lecture while the material is still fresh in your mind. Review to be sure you got all the key points and wrote them down in a form that will be readable later.

Intensive Study Session

You learn geology much as you would build a house. Before each lecture construct a frame of questions. During lecture attach details and ideas to the frame. After lecture master those ideas during an intensive study session. Now you have constructed the first story. But geology is a skyscraper with 23 floors (one for each chapter). Each chapter supports those above it. If you don't completely master this chapter the next will be more difficult.

Schedule at least one hour after each lecture for intensive study. Devote this time to mastering key concepts. Mastery is not gained by just reading the text. Mastery occurs as the result of asking yourself questions (and answering them). The following *CD Activities*, *Practice Exercises* and *Study Questions* are designed to help you reach mastery level quickly. The greater the number of these exercises you can work into your study schedule early in the course, the easier subsequent chapters will be.

Everything you do in life is worth infinite care and infinite effort
—*J.D. McDonald*

Website and CD Activities and Tools

http://www.whfreeman.com/presssiever

- Website Q & A Practice Multiple Choice Questions:
  At the Website complete the *Q & A* and do the Interactive Exercises called *Elements of Common Minerals* and *The Periodic Table.*

Tools and Resources

- Textbook: Appendix 3: Properties of the Most Common Minerals of the Earth's Crust
- *Interactive Tools and Exercises* at the Website: Mineral Databank
- *Flashcards* for Chapter 2 at the Website.

Conceptual Flow Chart illustrating the factors that influence the physical properties of mineral

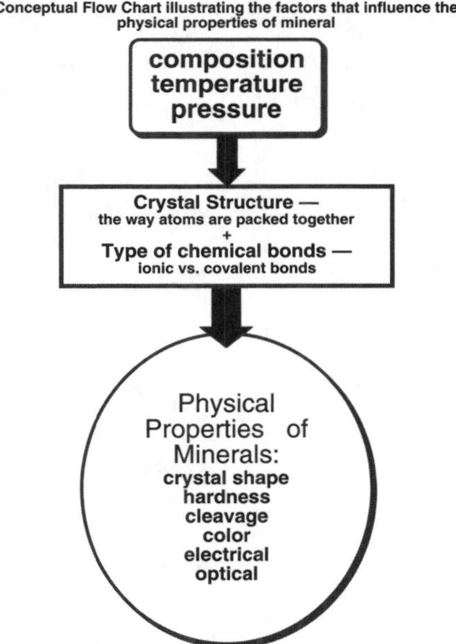

"Crystal" comes from the Greek word Krustallos, which means ice.
Is ice — water ice — a mineral?

PRACTICE EXERCISES AND STUDY QUESTIONS

(Answers and explanations are at the end of this chapter.)

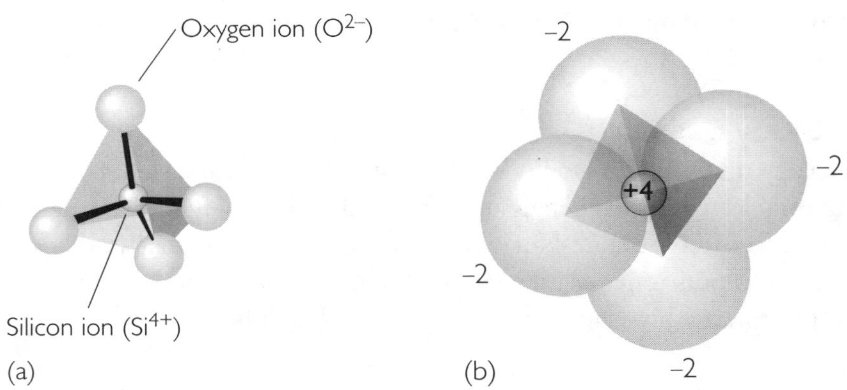

Figure 2.16 / Silicate Ion. (a) A model showing the structure of the silicon-oxygen molecule. In the center of this model is one silicon ion carrying a positive charge of 4. It is surrounded by four oxygen ions, each carrying a minus charge of 2. After bonding, the resulting silicate ion carries a negative charge of 4 (4 minus 8 equals – 4). (b) A more realistic model of the silicate ion with the atoms drawn to scale and filling space.

Exercise 1: Crystal Structures of Some Common Silicate Minerals

Three crystal structures of silicate minerals are illustrated below.

- Fill-in-the-blanks with the name for each major silicate structure (single chain, double chain, sheet, framework, or isolated tetrahedra);
- Give an example of a mineral with that crystal structure; and
- Using the symbols given in the key to cleavage properties, characterize the cleavage properties for minerals with this crystal structure.

Hint: Refer to Figure 2.17 and Table 2.2

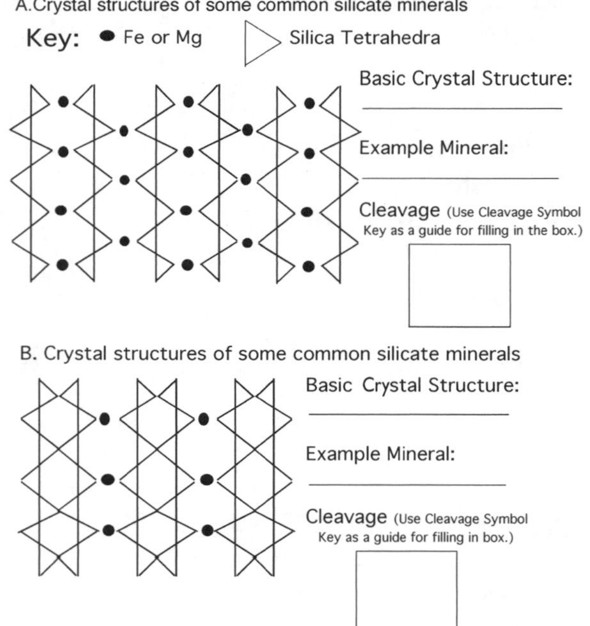

C. Crystal Structure of some Silicate Minerals

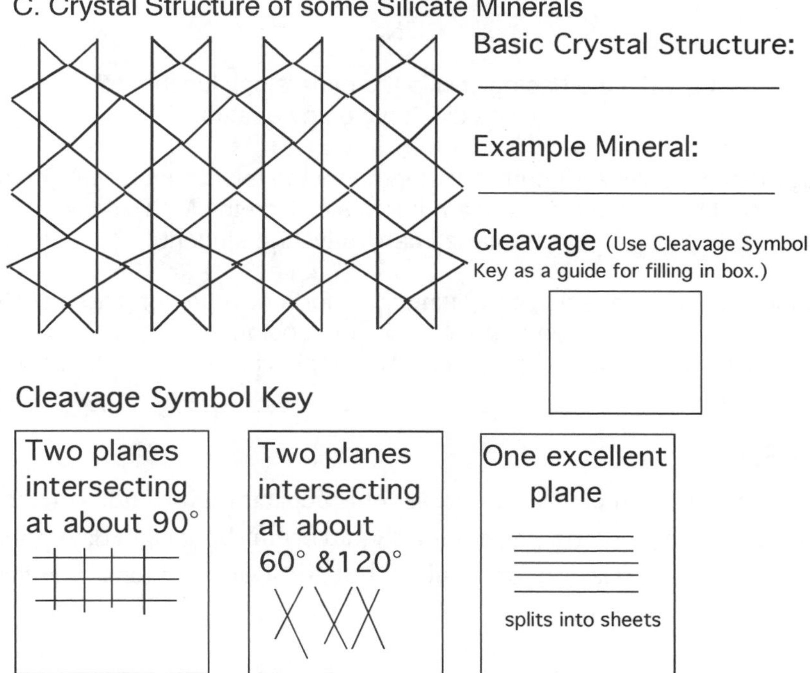

Basic Crystal Structure:

____

Example Mineral:

____

Cleavage (Use Cleavage Symbol Key as a guide for filling in box.)

Cleavage Symbol Key

Two planes intersecting at about 90°	Two planes intersecting at about 60° &120°	One excellent plane splits into sheets

If you think the mineral will exhibit no cleavage, please write this inside the box.

Exercise 2: Major Mineral Classes

Fill-in-the-blank with the name of the mineral class to which each mineral belongs.
The common classes of minerals include: silicates, carbonates, sulfates, sulfides, oxides, native elements, and halides. Hint: Refer to Table 2.1 in textbook and the *Mineral Databank* at the website.

1. hornblende *silicate* (hint: an amphibole)
2. beryl ______________
3. clay minerals ______________
4. calcite ______________
5. gypsum ______________
6. quartz ______________
7. diamond ______________
8. dolomite ______________
9. plagioclase ______________ (hint: a feldspar)
10. garnet ______________
11. halite ______________
12. hematite ______________
13. graphite *native element*
14. galena ______________
15. muscovite ______________ (hint: a mica)
16. anhydrite ______________
17. augite ______________ (hint: a pyroxene)
18. pyrite ______________
19. malachite ______________
20. orthoclase ______________ (hint: typically a tan to pink feldspar)
21. silver ______________
22. albite ______________ (hint: typically a white feldspar)
23. olivine ______________
24. biotite ______________ (hint: a black mica)

PUNS

Why did the geologist skip the mineralogy luncheon?
She didn't have any apatite.

The geology teacher asks a student if he can name the felsic mineral with crystal faces, conchoidal fracture, and a hardness of 7 on the Moh's scale.
"Of quartz I can," says the student.

So, the geologist is at the eye doctor, getting fitted for a new pair of glasses. "Do you ever see double?" asks the doctor.
"Only when I look through calcite samples in lab," says the geologist.

EXAM PREP

Materials in this section are most useful during your preparation for midterm and final exams. For optimal performance, midterm preparation should begin about eight days prior to the exam (see Appendix: *How to Study Geology* for details). The basic idea is a systematic review of material divided into short study sessions.

The following *Chapter Summary* and *Practice Exam* should simplify review. Read the chapter summary to begin your review session. It provides a helpful, brief overview that should get you back into the material. Then try the *Practice Exam*. Take it just as you would a midterm: time yourself and keep your book closed until you have finished. Next, score your exam to see how you stand in regard to mastery of this chapter. Finally, and most important of all: review any item that you missed. Identify and correct the misconception that made you miss the item.

CHAPTER 2 SUMMARY

- Minerals are naturally occurring inorganic solids with a specific crystal structure and chemical composition.
- Minerals form when atoms or ions chemically bond and come together in an orderly, three-dimensional geometric array—a crystal structure.
- The strength of the chemical bonds and the crystalline structure determines many of the physical properties, e.g., hardness, cleavage, of minerals.
- Silicate minerals are the most abundant class of minerals in the Earth's crust and mantle. There are three important groups of silicates:
  - ferromagnesium silicates, e.g. olivine and pyroxene - common in the mantle
  - feldspar and quartz - common in the crust
  - clay mineral - commonly produced by chemical weathering
- Other common mineral classes include carbonates, oxides, sulfates, sulfides, halides and native metals.
- Why study minerals?

  One can deduce the composition of rocks because minerals have a definite composition.

  One can infer the environment in which the rock formed because most minerals are stable only under certain conditions of temperature and pressure.

  For fun and profit. Gemstones are treasured for their beauty, color and rarity. Many minerals are used in industrial processes and manufacturing.

PRACTICE EXAM QUESTIONS FOR CHAPTER 2

(Answers and explanations are at the end of this chapter.)

Exercise 1: Identifying Minerals Using Their Physical Properties

Give the mineral name for each of the mineral samples described below.
Hint: Refer to *Appendix 3* in the textbook and the *Mineral Databank* at the Website.

Mineral Sample A

This mineral is colorless with a silvery shine. It separates into thin sheets or flakes that can bend without breaking. It is within a rock that contains large crystals of quartz and feldspar. Hint: Mineral is named after Moscow where it was commonly used as a substitute for window glass.

Mineral name is ____________________

Mineral Sample B

This mineral sample is pale yellow and occurs as a vein within a rock fracture. Many small cubic crystals of this mineral are exposed in the vein when the rock is broken apart along the fracture. The powdered form (streak) of the mineral is black and has a sulfur-like smell. Hint: A common name for this mineral is "fool's gold".

Mineral name is ____________________

Mineral Sample C

This is on display at a mineral exhibition both as a mineral specimen and as cut and polished pieces in jewelry. It consists of beautiful concentric bands of various shades of green. A knife blade can easily powder its surface but my fingernail can not scratch it. Acid will react with the mineral, especially if powdered. The specimen is labeled as coming from a copper mine in Bisbee, Arizona. Hint: See the Website's *Photo Gallery*, under *Additional Resources* for Chapter 2.

Mineral name is ____________________

Mineral Sample D

This tan to pink sample occurs as many 5 to 7 centimeter crystals surrounded by flakes of mica and grains of white to clear quartz. The tan to pink crystals are prismatic — rectangular-boxed shaped. A knife blade may or may not scratch the mineral. Hint: It is one of the most common minerals in the Earth's continental crust. A semi-precious variety, called moonstone, is used in jewelry.

Mineral name is ____________________

Mineral Sample E

Exhibiting excellent cleavage in three directions, this mineral breaks into rhombohedral-shaped pieces. Samples vary in color from clear, to white, to tan. A fingernail will not scratch the surface, but a knife blade easily powders the surface. Weak acid readily reacts with the surface of the mineral. Hint: Marine organisms typically produce this mineral from seawater to generate their skeletons (sea shells).

Mineral name is ____________________

Mineral Sample F

This sample was found as a series of thin layers with other layers of mud and silt. My fingernail could easily scratch its surface and form a white powder. Water dissolves the powdered form but the sample does not react with acid. Samples commonly exhibit a splintery aspect but will not separate into individual fibers like asbestos. The sample does easily break (cleave) along one plane but will not form the thin sheets typical of mica. Hint: This mineral is a major constituent of plaster of Paris and wall board used in buildings and houses.

Mineral name is ______________________

1. The bonds between Na and Cl in halite are strongly ionic. In ion form, Cl has seven electrons in its outer shell and Na has one. After these two elements bond, Cl has ___ electrons in its outer shell and Na has ___ in its outer shell.
   A. six, two
   B. four, four
   C. one, seven
   D. eight, eight

2. _____________ and ___________ are examples of minerals with an identical chemical composition but different crystal structures.
   A. calcite, dolomite
   B. hematite, magnetite
   C. pyrite, gypsum
   D. graphite, diamond

3. The term geologists use for a naturally occurring aggregate of minerals is
   A. element
   B. rock
   D. compound
   E. crystal

4. When a substance is made of atoms that are arranged in a fixed, orderly and repeating pattern, it is said to be
   A. amorphous
   B. glassy
   C. crystalline
   D. liquid

5. Which of the following statements is NOT true of minerals?
   A. They are crystalline.
   B. They possess a definite chemical composition.
   C. They are naturally occurring.
   D. They are organic.

6. Which of the following objects is not a mineral?
   A. native copper
   B. glass
   C. diamond
   D. water ice

7. The most common mineral group on Earth is
   A. carbonates
   B. silicates
   C. oxides
   D. halides

8. The structure of feldspars, such as orthoclase and plagioclase, consists of
   A. double chains of silica tetrahedra
   B. three dimensional framework of silica tetrahedra
   C. single chains of silica tetrahedra
   D. isolated silica tetrahedra

9. Micas are known for their tendency to split apart into thin sheets. Their silicate crystal structure is
   A. isolated silica tetrahedra
   B. chains of silica tetrahedra
   C. sheets of silica tetrahedra
   D. A framework arrangement of silica tetrahedra

10. Which of the following does NOT belong to the carbonate class of minerals?
    A. calcite
    B. anhydrite
    C. dolomite
    D. aragonite

11. Which of the following minerals has a sheet silicate structure?
    A. clay minerals
    B. feldspar
    D. amphibole
    E. graphite

12. Amphiboles and pyroxenes are mineral groups that belong to the class of minerals called
    A. chain silicates
    B. framework silicates
    C. sulfide minerals
    D. sheet silicates

13. Which of the following statements is NOT true about quartz and calcite?
    A. Calcite has excellent cleavage; quartz has poor to no cleavage.
    B. Quartz is a carbonate mineral.
    C. Both are often colorless.
    D. Quartz has hardness about 7; calcite about 3.

14. The characteristic of certain minerals to break along planes of weakness is called
    A. crystal symmetry
    B. cleavage
    C. hardness
    D. luster

15. Graphite and diamond are two very different minerals. Why are the physical properties of diamond and graphite so different even though they are both made from pure carbon?
    A. The chemical bonds between the carbon atoms and their crystal structures are significantly different.
    B. Graphite has strikingly different physical properties from diamond because of impurities within its crystal structure, whereas diamond is pure carbon.
    C. Actually, graphite and diamond are not minerals because they are made from carbon and all matter made from carbon is organic. Their physical properties are different because they were produced by very different living organisms.
    D. None of the above.

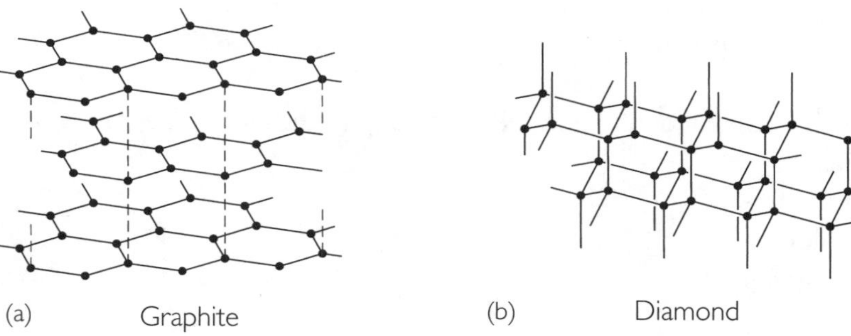

Figure 2.15 / The crystal structures of (a) graphite and (b) diamond.

16. Chemical bonding along cleavage planes within the crystal structure of a mineral are typically
    A. more covalent
    B. more ionic
    C. more magnetic
    D. involving electron-sharing between atoms

17. Iron and magnesium ions are similar in size and both have a +2 positive charge. Therefore, we would expect
    A. iron and magnesium to bond easily
    B. iron and magnesium to share electrons
    C. iron and magnesium to form minerals with different crystal structures
    D. iron and magnesium to substitute for each other within the crystalline structure of minerals

Hint: Refer to Figure 2.14 and the section in the textbook entitled *Cation Substitution: Same Crystal Structure, Different Chemical Composition.*

18. The chemical bonds between carbon atoms within diamond are predominately
    A. covalent
    B. ionic
    C. metallic
    D. nuclear

ANSWERS AND EXPLANATIONS FOR EXERCISES AND QUESTIONS

After Lecture

Exercise 1: Crystal Structures of Some Common Silicate Minerals

A. single chains of silica tetrahedra
    pyroxene (augite)
    two good cleavage planes intersecting at about 90°

B. double chains of silica tetrahedra
    amphibole (hornblende)
    two good cleavage planes intersecting at about 60° and 120°

C. sheets of silica tetrahedra
    mica (muscovite)
    Excellent cleavage in one direction

Exercise 2: Major Mineral Classes

1. *silicate*
2. beryllium silicate
3. sheet silicate
4. calcium carbonate
5. hydrated calcium sulfate
6. framework silicate
7. native element (carbon)
8. calcium/magnesium carbonate
9. framework silicate
10. silicate
11. sodium halide
12. iron oxide
13. *native element (carbon)*
14. lead sulfide
15. sheet silicate
16. calcium sulfate
17. chain silicate
18. iron sulfide
19. copper carbonate
20. framework silicate
21. native element (silver)
22. framework silicate
23. silicate (single tetrahedra)
24. sheet silicate

Exam Prep

Exercise 1: Identification of Minerals Based on Their Physical Properties

Mineral A: muscovite
Mineral B: pyrite
Mineral C: malachite
Mineral D: orthoclase
Mineral E: calcite
Mineral F: gypsum

1. A. Chloride is an anion and easily loses an electron to sodium which is a cation.
2. D. Diamond and graphite are both composed of pure carbon but have significantly different crystal structures. Refer to Figure 2.15 in the textbook.
3. B. A rock is an aggregate of one or more minerals.
4. C. All minerals are crystalline solids.
5. D. By definition, minerals are inorganic. Are graphite and diamond minerals? Yes. They are both made from pure carbon—a common element in organic material. However, even though they are made of carbon, graphite and diamond are not produced by biological processes. Carbon is a common, naturally occurring element. Graphite is typically found in metamorphic rocks and diamond originates in the Earth's mantle.
6. B. Glass is an amorphous material that lacks a crystal structure. Native copper, diamond and water ice all fit the definition of a mineral.
7. B. Silicate minerals are the most common mineral group in the Earth's crust and mantle. The Earth's core is thought to consist mostly of an iron-nickel alloy.
8. B. Refer to Figure 2.17 and 2.18.
9. C. Refer to Figure 2.17 and 2.18.
10. B. Refer to Table 2.1 and Appendix 3.
11. A. Clay minerals have a sheet silicate structure.
12. A. Refer to Figure 2.17 and Table 2.2.
13. B. Quartz is a silicate and calcite is a carbonate.
14. B. Cleavage is the tendency for minerals to break along planes of weaker chemical bonds within their crystal structure.
15. A. The physical characteristics of a mineral are determined by its composition, the nature of the chemical bonds and the crystalline structure. Although diamond and graphite are both pure carbon, their crystal structures and the chemical bonds within the crystal structure are significantly different.
16. B. Chemical bonds typically exhibit a mixture of ionic and covalent characteristics. Bonds with a more covalent character are stronger and the bond length is shorter. Bonds with a more ionic character are weaker and the atoms tend to be farther apart.
17. D. Cations of similar sizes and charges tend to substitute for one another and to form compounds having the same crystal structure but differing chemical composition. Cation substitution is common in silicate minerals.
18. A. Chemical bonds between carbon atoms within diamond are predominately covalent. Compared to ionic bonds, covalent bond length is shorter and covalent bond strength is higher.

Chapter 3
Rocks: Record of Geologic Processes

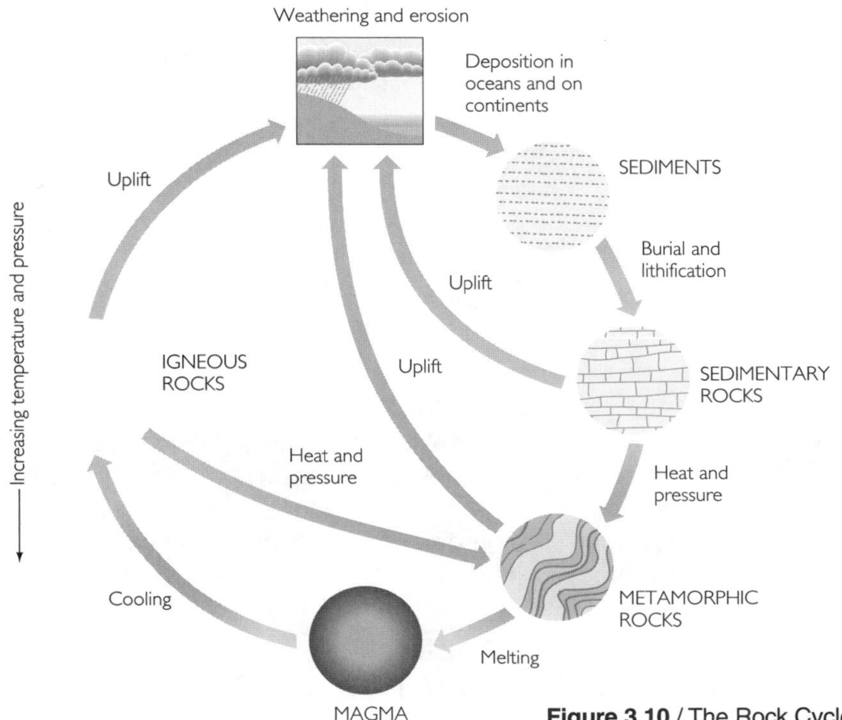

Figure 3.10 / The Rock Cycle

BEFORE LECTURE

Before you attend the lecture be sure to spend some time previewing Chapter 3. Use the following *Chapter Preview* questions as a guide. They constitute the basic framework for this chapter. Working with the preview questions before lecture and committing them to memory should result in your understanding the lecture much better and getting a better set of notes. Need a reminder about previewing? See the Appendix: *How to Study Geology*. This appendix integrates all study strategy suggestions from this guide into one essay.

CHAPTER PREVIEW

- **What determines the properties of rocks?**
  Brief answer: Mineral content and texture.
- **What are the three types of rocks and how are they formed?**
  Brief answer: igneous, sedimentary, and metamorphic rock.
- **How does the rock cycle describe rock formation?**
  Hint: See Figure 3.10.

MAJOR ROCK TYPE	IGNEOUS	SEDIMENTARY	METAMORPHIC
Source of materials	Melting of rocks in hot, deep crust and upper mantle.	Weathering and erosion of rocks exposed at the surface.	Rocks under high temperatures and pressures in deep crust and upper mantle.
Rock-forming process	Crystallization (solidification of melt).	Sedimentation, burial, and lithification.	Shearing and new minerals recrystallize in the solid state.

Figure 3.1 / The minerals and textures of the three major rock types.

STUDY TIP

Recommendation: Commit the *Rock Cycle* to memory. It provides a brief overview, a backbone outline, of the entire course. Subsequent chapters will develop the *Rock Cycle* in more detail. Refer back to the *Rock Cycle* frequently to see how chapters fit together. Don't allow yourself to become lost in a "forest of details".

Vital Information from Other Chapters

- Review the section, *Plate Tectonics: a Unifying Theory for Geological Science* in Chapter 1.

Website and CD Preview

http://www.whfreeman.com/presssiever

- On the Website complete the *Rock Cycle Interactive Geology Exercise* for Chapter 2.

DURING LECTURE

Your basic goal during lecture is to get a good set of notes. Those notes should contain in-depth answers to the *Chapter Preview Questions*. It may be helpful to keep a copy of the *Chapter Preview Questions* in front of you or, better yet, develop preliminary answers to the questions and commit them to memory before the lecture.

To avoid getting lost in details, keep the big picture in mind: Chapter 3 explains the three basic types of rocks and how these are formed by the processes of the Rock Cycle. You may want to keep a copy of the Rock Cycle figure handy during the lecture so you can refer to it.

NOTE-TAKING CHECKLIST

- ✔ Organize your notes in a 3-ring binder so that you can easily reorganize them.
- ✔ Save space in your notes to add important visual material (flow charts, simple sketches, comparison charts) after class. To make this easy employ a double page note taking format (take notes on the right-hand page; save the left-facing page as a "sketch page").
- ✔ Date each day's notes so you can find material later.
- ✔ Take notes in a format that makes the main topics and concepts easy to identify. Some students accomplish this by taking notes in outline format, but many other approaches are possible. Visual learners sometimes find it helpful to highlight main points (after class) with a standard color. Another good approach would be to use the questions we provide under *BEFORE LECTURE/Chapter Preview* as headers. These can be added during class or during your after-class Review Session.
- ✔ Indicate areas where you need to do follow-up work with your text, instructor or tutor with a question mark in the margin.
- ✔ Indicate possible test questions by putting a "TQ" (Test Question) in the left margin where you can easily see it.
- ✔ Write assignments in the left column where you can easily find them. After class enter the due date in your personal calendar.

The important thing is not to stop questioning.
—Albert Einstein

AFTER LECTURE

Intensive Study Session

Schedule at least one hour after lecture for intensive study. Use this time to master the key concepts about types of rocks produced by the Rock Cycle. Begin by trying the *Practice Exercises* and *Study Questions*. See if you can answer the questions with your book closed. Then check your answers. Read the text to clear up points you feel confused about. Then go on to Q & A on the Website.

Website and CD Activities and Tools

http://www.whfreeman.com/presssiever

- **Website Q & A Practice Multiple Choice Questions:**

  At the Website complete the *Q & A*. Pay particular attention to the explanations for answers. Flashcards at the Website will help you learn new terms. The *Rock Cycle Tool* in the *Interactive Geology Tools* section of the Website provides an excellent tour through the rock cycle and how it works in the context of plate tectonics. Use it for review.

PRACTICE EXERCISES AND STUDY QUESTIONS

(Answers and explanations are at the end of this chapter.)

Exercise 1: Website Rock Cycle

http://www.whfreeman.com/presssiever

Chapter 3 of your textbook and the *Rock Cycle Tool* for Chapter 3 in the *Interactive Tools & Exercises* section of the Website are your sources of information for this exercise.

A. What are the three plate tectonic settings for the generation of magma?

____

____

____

B. Cooling rates of igneous rocks

Given the two types of igneous rocks, intrusive and extrusive, fill-in the table with their characteristic cooling rates (fast vs. slow) and textures (fine grained vs. coarse grained).

TYPES OF IGNEOUS ROCKS	COOLING RATES	TEXTURES
Extrusive		
Intrusive		

C. There are three major agents for the transport and deposition of sediments on land. They are:

____

____

____

D. What are the two processes that transform loose sediments into sedimentary rock?

Hint: more than one word can be used to describe these processes. Read the appropriate *Rock Cycle* section carefully.

____

____

E. What are the two main types of sedimentary rocks and what are they made out of?

Sedimentary Rock Type	What are they made out of?

F. What are the two major conditions that result in metamorphic rocks?

____

____

G. Are metamorphic rocks formed by melting? __________ (Yes/No)

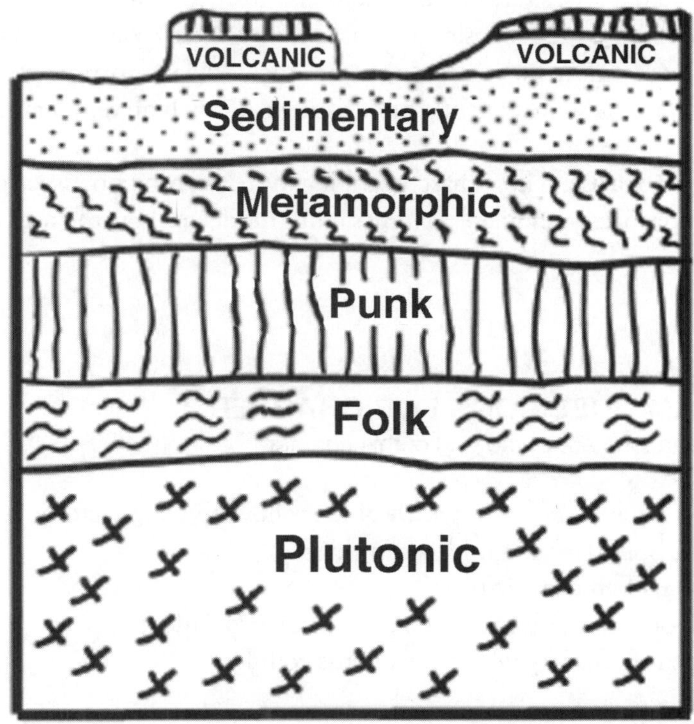

Layers of Rock – Volcanic/Sed/Meta/Punk/Folk/Plutonic Rock

EXAM PREP

Save the exercises in this section for the week before your midterm. Use them during a systematic review. Complete recommendations for conducting such a review are in the Appendix: *How to Study Geology*.

When you are ready to review Chapter 3, begin by reading the *Chapter Summary*. It provides an overview that should help you get back into the material. Then try the *Practice Exam*. Take it just like you would the midterm. Then score it to see how you stand on mastery of this chapter. Most importantly, review any items you missed. Identify and correct any misconceptions that led to incorrect answers

CHAPTER 3 SUMMARY

- The properties of rocks and the rock's name are determined by mineral content (the kinds and proportions of minerals that make up the rock) and texture (the size, shapes, and spatial arrangement of its crystals or grains).

- There are three major rock types. **Igneous rocks** solidify from molten liquid (magma). Crystal size within igneous rocks is largely determined by the cooling rate of the magma body. **Sedimentary rocks** are made of sediments formed from the weathering and erosion of any pre-existing rock. Deposition, burial and diagenesis (compaction and cementation) transform loose sediments into sedimentary rocks. **Metamorphic rocks** are formed by an alteration in the solid state of a pre-existing rock by high temperatures and pressure.

- The *Rock Cycle* is a useful flow chart describing the relationships between geologic processes and the major rock types. Plate tectonics is our model for how the rock cycle operates on Earth today.

PRACTICE EXAM QUESTIONS FOR CHAPTER 3

(Answers and explanations are at the end of this chapter.)

1. Uplift is most likely to occur in which of the following plate-tectonic settings?
   A. divergent boundary at a mid-ocean ridge
   B. subsidence of an oceanic plate
   C. hot spot of mantle plume in the interior of a continent
   D. convergent boundary where continents collide

2. Which of the following rocks form from molten material cooling and solidifying within the Earth's crust?
   A. volcanic
   B. plutonic
   C. sedimentary
   D. metamorphic

3. Molten rock within the Earth's crust is called
   A. silica
   B. lava
   C. magma
   D. mica

4. Coarse-grained igneous rocks, such as granite, are exposed today at the Earth's surface due to
   A. uplift and erosion
   B. quickly cooled lavas erupting from ancient volcanoes
   C. silicate minerals precipitated from rainwater
   D. all of the above

ANSWERS AND EXPLANATIONS FOR EXERCISES AND QUESTIONS

After Lecture

Exercise 1: The Rock Cycle

A. Plate tectonic settings for magma generation are

divergent boundaries
convergent boundaries
hot spots/mantle plumes

B. Igneous Rocks

TYPES OF IGNEOUS ROCKS	COOLING RATES	TEXTURES
Extrusive	fast cooling	fine-grained
Intrusive	slow cooling	coarse-grained

C. Agents for the transport and deposition of sediments are:

water
wind
ice

D. Two processes that transform loose sediments into rock are

burial (or compaction)
cementation (cementing or lithification)

E. The two main types of sedimentary rocks and their constituents are

Sedimentary Rock Type	What are they made out of?
clastic	rock and mineral fragments
chemical/biochemical	weathering

F. Metamorphism occurs at elevated heat and pressure.

G. No. The rock is not melted during metamorphism, although a minor amount of melt "sweat" may be generated during high grade metamorphism. Igneous rocks are formed from the solidification of melts (magmas).

Exam Prep

1. D. The highest mountains (greatest uplift) on Earth occurs along convergent plate margins, e.g. Himalayas and the Andes.
2. B. Plutonic rocks solidify from melts, called magmas. Refer to the *Rock Cycle Tool* at the Website and *The Rock Cycle* section in Chapter 3 of your textbook.
3. C. Most magmas are generated from the melting of silicate rocks within the Earth's crust and mantle. On rare occasion, a magma composed of carbonates erupts on the Earth's surface. On Io, the moon of Jupiter, sulfur magmas erupt from about ten active volcanoes. Io's eruptions can shoot fountains of sulfur compounds 360 km high.
4. A. An igneous rock with a coarse texture, where individual mineral grains (crystals) are visible without magnification, forms when the rock crystallized slowly beneath the Earth's surface. Solidification of a magma body may take tens to hundreds of thousands of years within the crust and millions of years more to cool after completely crystallized. To expose a coarse grain igneous (plutonic) rock at the Earth's surface requires significant uplift and erosion of the rocks that once sat on top.

Chapter 4
Igneous Rocks: Solids from Melts

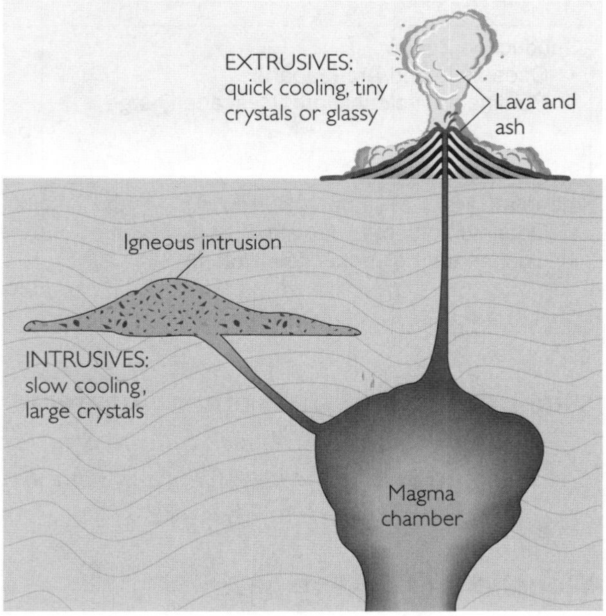

Figure 3.2 / Intrusive vs. extrusive igneous rocks. Differences in crystal grain sizes due to differences in cooling rate distinguishes intrusive from extrusive igneous rocks. Intrusive (plutonic) rocks cool slowly and therefore are typically coarse grained. Extrusive (volcanic) rocks cool quickly and therefore are typically finer grained.

BEFORE LECTURE

Before you go to lecture be sure to spend some time previewing Chapter 4. We have made it easier by identifying the three key questions on Igneous Rocks for you (see *Chapter Preview* questions). These questions constitute the basic framework for understanding the chapter. Working with these questions before lecture and committing them to memory should help you understand the lecture better and get an excellent set of notes. Need a refresher on previewing? See Appendix: *How to Study Geology*.

CHAPTER PREVIEW

- **How are igneous rocks classified?**
  Brief answer: composition and texture (cooling history).
- **How and where do magmas form?**
  Hint: Figure 4.8 in textbook provides an overview.
- **How does magmatic differentiation account for the great variety of igneous rocks?**
  Hint: Figures 4.9, 4.10, 4.11 and 4.12 will help introduce you to how minerals can become segregated within a magma chamber or lava.
- **What are the forms of intrusive and extrusive igneous rocks?**
  Hint: Figure 4.13 shows the basic extrusive and intrusive igneous rock bodies.
- **How do igneous rocks relate to plate tectonics?**
  Brief answer: Magmas tend to form at divergent and convergent plate boundaries and hotspots. Refer to *Tectonic Activity, Rock Composition and Types of Magmas* sections of Chapter 4 for more details.

PLATE TECTONIC SETTING (EXAMPLES)	MAGMA TYPE (EXAMPLE ROCK TYPE)
Spreading Centers: • oceanic ridge (Mid-Atlantic Ridge) • continental rift (East Africa Rift)	• mafic (basalt) • mafic to felsic (silicic) - more variable because some continental crust may melt
Subduction Zones: • Oceanic Island Arc (Japan) • Continental volcanic arc (Cascade Range,	• mafic to intermediate • mafic to felsic (silicic) - more variable because some continental crust may melt
Intraplate Mantle Plumes ("hotspots"): • Oceanic hotspots (Hawaii) • Continental hotspots (Yellowstone)	• mafic (basalt) • mafic to felsic (silicic) - more variable because some continental crust may melt

Vital Information from other Chapters

- Review Table 2.2: *Major Silicate Structures* in Chapter 2 and *The Rock Cycle and Plate Tectonics* and *The Rock Cycle* in Chapter 3.

Website and CD Preview

http://www.whfreeman.com/presssiever

- Review *The Rock Cycle Tool* from Chapter 3 on the Website. Pay particular attention to information on melting, magmas, cooling, and igneous rocks.

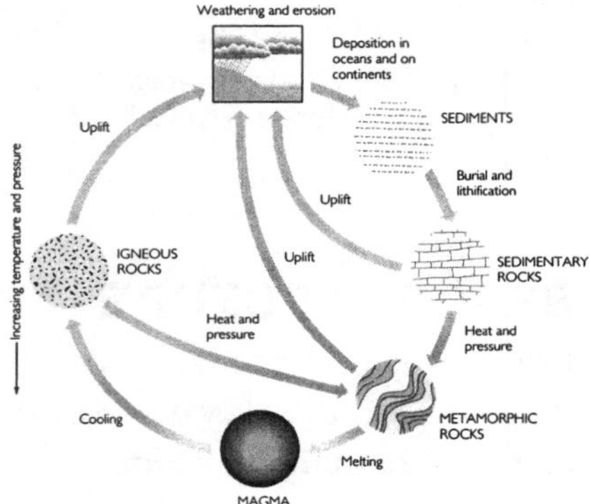

Figure 3.10 / The Rock Cycle.

DURING LECTURE

Your basic goal during lecture is to get a good set of notes. By the end of lecture those notes should contain helpful and detailed answers to the *Preview Questions.*

Big Picture for Chapter 4

To avoid getting lost in details keep the big picture in mind: Chapter 4 explains how plate tectonics drives the formation of magma and how igneous rocks of varying composition and texture are produced at particular plate locations. Example: fine-textured volcanic rock of basaltic composition would be produced at a diverging ocean plate (see Figure 4.8).

Key Figures for Lecture

This lecture may get a little technical. Two key figures from the chapter will help you stay on track. Keep copies of Table 4.2 and Figure 4.11 handy during lecture so you can refer to them. Table 4.2 explains how physical properties like melting temperature correlate to the composition (mafic to felsic) of igneous rocks. Figure 4.11 describes Bowen's Reaction Series.

Ask Questions

This chapter is challenging and you will certainly need to ask questions.

NOTE TAKING TIP

Sit close enough to the lecturer that you can see the texture of samples. Your lecturer may show slides of igneous rock formations and may pass around specimens of igneous rocks. You will get more out of these if you sit close to the front row where you can see the sample rocks as they are discussed. Remember to focus carefully on any clues your lecturer may provide for recognizing a particular sample in the field. Focus in particular on the texture of samples and learn to see the difference between a fine-textured volcanic rock and a coarse-textured plutonic rock.

AFTER LECTURE

The perfect time to review your notes is right after lecture, while the material is still fresh in your mind. Review to be sure you got all the key points and wrote them down in a form that will be readable later. Here's a checklist that will help you check your notes.

NOTE REVIEW CHECKLIST

✔ All notes legible? (Rewrite so they read easily.)

✔ Important points clearly identified? You should now have headers in your notes that tie to each of the questions in the *BEFORE LECTURE/Chapter Preview.*

✔ Holes (missing material) filled in from memory?

✔ Areas where you don't remember what was said marked for a follow-up session with your instructor, tutor, or study partner?

✔ Possible test questions indicated in the margin (TQ)?

✔ Additional visual material. *Suggestions for Chapter 4: Be sure to copy Table 4.2 (explaining mafic to felsic composition). You might also want to add Figure 4.8 to help you understand how plate movement influences the formation of magma, and your own version of Figure 4.11 (Bowen's Reaction Series).*

✔ Reworked notes into a form that is efficient for your learning style?

✔ Write (or draw) a brief "big picture" overview of this lecture? *Suggestions for Chapter 4: The material in Chapter 4 is very visual. Try to summarize the major concepts of the chapter in a visual summary. Hint: Base your summary on Table 4.2 and Figures 4.8 and 4.11.*

Intensive Study Session

You learn geology much as you would build a house. Before each lecture, construct a frame of questions. During lecture attach details and ideas to the frame. After lecture, master those ideas during an intensive study session. Now you have constructed the first story. But geology is a skyscraper with 23 floors (one for each chapter). Each chapter supports those above it. If you don't completely master this chapter, the next will be more difficult.

Schedule at least one hour after each lecture for intensive study. Devote this time to mastering key concepts. Mastery is not gained by just reading the text. Mastery occurs as the result of asking yourself questions (and answering them). The following *CD Activities*, *Practice Exercises* and *Study Questions* are designed to help you reach mastery level quickly. The greater the number of these exercises you can work into your study schedule early in the course, the easier subsequent chapters will be.

Website and CD Activities and Tools

http://www.whfreeman.com/presssiever

- **Website Q & A Practice Multiple Choice Questions:**

  At the Website complete the *Q & A*. Pay particular attention to the explanations for answers. Flashcards at the Website will help you learn new terms. In the *Interactive Tools and Exercises* section of the Website, work on the *Bowen's Reaction Diagram* and *Igneous Rock Properties* exercises. Hint: Table 4.2 and Figure 4.11 are very useful.

PRACTICE EXERCISES AND STUDY QUESTIONS

(Answers and explanations are at the end of this chapter.)

Table 4.2 Changes in some Major Chemical Elements from Felsic to Mafic Rocks

	Felsic	Intermediate		Mafic
Coarse-Grained (intrusive)	Granite	Granodiorite	Diorite	Gabbro
Fine-Grained (extrusive)	Rhyolite	Dacite	Andesite	Basalt

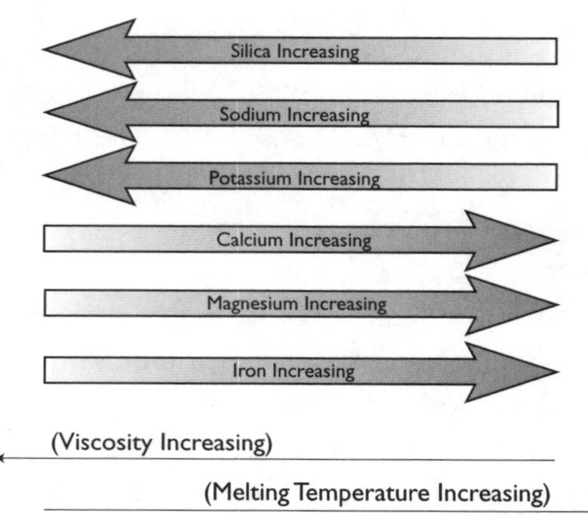

Exercise 1: Igneous Rock Textures

In the four boxes below, sketch the igneous rock texture that is consistent with the origin of the rock as described next to each box. Fill in each blank with the appropriate texture term from the list below. Figures 3.2 and 4.1 are helpful.

Texture terms:

Fine-grained (aphanitic)
Intermediate grain sizes - visual grains but not very coarse grained
Coarse-grained (phaneritic)
Mixture of coarse and fine grains (porphyritic)

A. Draw the texture of an igneous rock from a pluton solidified at depth within the crust.

phaneritic
name of texture

B. Draw the texture of an igneous rock from a shallow magma body, like a dike or sill.

____
name of texture

C. Draw the texture of a lava erupted from a magma chamber after some minerals had begun to crystallize.

____
name of texture

D. Draw the texture of a lava flow that erupted before the magma chamber underwent any crystallization.

____
name of texture

Exercise 2: Distribution of Igneous Rocks within the Earth

Complete this table by filling in the blanks using terms from the lists below. Refer to Tables 4.1 and 4.2 in your textbook. Some answers are provided. Note that the Earth's core is not included in this table because it is thought to be composed mostly of iron and nickel and not silicate minerals.

Major Layer Within the Earth	Example Igneous Rock	General Compositional Group	General Chemical Composition
Continental crust (for continental crust there are two appropriate answers)	Granite		
		Intermediate	
Ocean Crust			
Mantle			Less Si, Na, K More Fe, Mg, Ca

Igneous Rocks

granite/rhyolite
basalt/gabbro
andesite/diorite
peridotite
granodiorite/dacit

Compositional Group

ultramafic
felsic
intermediate
mafic

General chemical compositions of igneous rocks

More (higher amounts) Si, Na, K
More (higher amounts) Fe, Mg, Ca
Less (lower amounts) Si, Na, K
Less (lower amounts) Fe, Mg, Ca
Intermediate

1. A common igneous rock in the continental crust is
   A. Fe and Ni
   B. granite
   C. olivine
   D. basalt

2. Granite is mainly made up of
   A. quartz, orthoclase (K-feldspar), Na-plagioclase, and biotite
   B. quartz, Ca-plagioclase, Na-plagioclase, and amphibole
   C. quartz, pyroxene, and muscovite
   D. quartz, orthoclase (K-feldspar), Ca-plagioclase, and olivine

3. What rock has the same mineralogy as granite, but has a fine-grained texture? Hint: Refer to Table 4.2.
   A. andesite
   B. basalt
   C. obsidian
   D. rhyolite

4. Which of the following minerals is common in basalt?
   A. pyroxene
   B. quartz
   C. muscovite
   D. Na-plagioclase

5. Which of the following rocks contains the most silica?
   A. basalt
   B. rhyolite
   C. fissure eruptions
   D. dacite

6. In the field or in hand-sized specimens, intrusive and extrusive igneous rocks are distinguished by which characteristic?
   A. composition
   B. color
   C. porphyritic versus non-porphyritic texture
   D. grain size

7. Which of the following pairs of intrusive and extrusive rocks are made from the same minerals, i.e. they have the same chemical composition?
   A. gabbro and basalt
   B. diorite and basalt
   C. granite and andesite
   D. gabbro and rhyolite

EXAM PREP

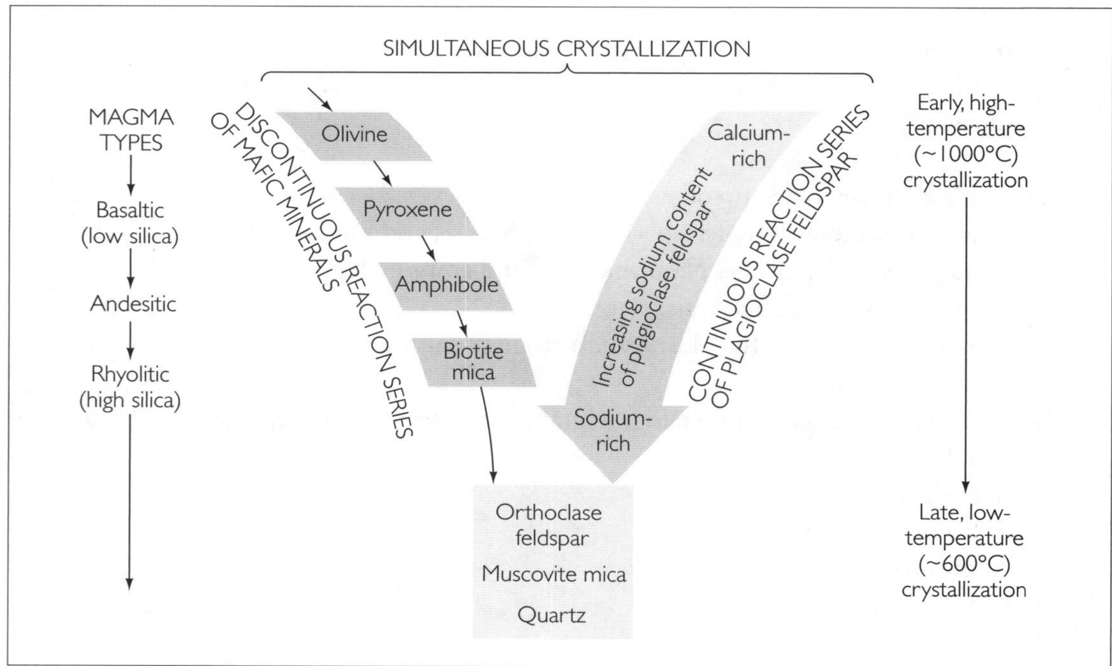

Figure 4.11 / Bowen's Reaction Series

CHAPTER 4 SUMMARY

- Igneous rocks can be divided into two broad textural classes: (1) the coarsely crystalline rocks, which are intrusive (plutonic) and therefore cooled slowly, and (2) the finely crystalline rocks, which are extrusive (volcanic) and cooled rapidly. Within each of these broad textural classes, the rocks are subdivided according to their composition. General compositional classes of igneous rocks are felsic, intermediate, mafic and ultramafic, in decreasing silica and increasing iron and magnesium content.

- The lower crust and upper mantle are typical places where physical conditions induce rock to melt. Temperature, pressure, rock composition and the presence of water all affect the melting temperature of the rock.

How to Melt a Rock – the Generation of Magma

- Increase temperature. Not all minerals melt at the same temperature (refer to Figure 4.11, Bowen's Reaction Series), so the mineral composition of the rock affects the melting temperature. Felsic rocks with higher silica content melt at lower temperatures than mafic rocks which contain less silica and more iron / magnesium.

- Lower the confining pressure. A reduction in pressure can induce a hot rock to melt. A reduction in confining pressure on the hot upper mantle is thought to generate the basaltic magmas which intrude into the oceanic ridge system to form ocean crust (refer to Figure 4.18).

- Add water. The presence of water in a rock can lower its melting temperatures up to a few hundred degrees. Water released from rocks subducting into the mantle along convergent plate boundaries may be an important factor in magma generation, especially at convergent plate boundaries.

- There are an amazing variety of igneous rocks on Earth. Two processes help to explain how the composition of igneous rocks can be so variable: **partial melting** and **fractional crystallization.**

  Most rocks probably only partially melt. Then, due to differences in density and mobility, the melt can collect into larger bodies. Since iron, magnesium, and calcium are concentrated in silicate minerals with high melting temperatures and silica, sodium, and potassium are concentrated in silicate minerals with low melting temperatures, the degree to which the rock melts will influence the bulk chemical composition of the melt and remaining solids. For example, as more of the rock melts, the melt typically becomes enriched in iron, magnesium and calcium because minerals with high melting temperatures and high iron content contribute increasingly more to the melt.

  Fractional crystallization simply involves the separation of a fraction of the early-formed crystals from the melt. Segregation of the melt and solids from each other may be a result of the melt migrating upwards, leaving the solids behind or the settling of early-formed, iron-rich, heavy minerals, leaving a pool of magma depleted in iron and enriched in silica relative to the bulk composition of the original magma body. In this way, fractional crystallization can enhance compositional differences between the parent magma and the rock that eventually crystallizes from the magma.

  The Bowen's Reaction Series in Figure 4.11 is a flow chart describing how the very general bulk composition of a magma can change as the magma solidifies or as a rock melts.

- Names are given to igneous rock bodies based on their size and shape. Figure 4.13 summarizes the common igneous rock bodies, such as batholith, pluton, dike and sill.

- In the context of plate tectonics, mafic magmas are thought to be generated by partial melting of the upper mantle along divergent plate boundaries and beneath hotspots. Felsic and intermediate magmas are commonly associated with convergent plate boundaries and are thought to be generated by partial melting induced by the release of water from subducting slabs of crust.

Website and CD Activities and Tools
http://www.whfreeman.com/presssiever
The photomicrographs in the *Photo Gallery* at the Website and also on the CD can be very helpful in understanding the differences in grain-size for many rocks discussed in the following questions. The *Photo Gallery* is found under *Additional Resources* on the CD, and as a menu category for any chapter at the Website.

PRACTICE EXAM QUESTIONS FOR CHAPTER 4

(Answers and explanations are at the end of this chapter.)

Exercise 1

Circle the answers that correctly complete the following statements:

A. The atomic structure of the earliest formed silicate minerals in a magma tend to be MORE / LESS complex than the crystalline structures of minerals formed during later stages in the solidification of the magma.

B. During the solidification of a magma, the minerals with the highest silica content will crystallize FIRST / LAST.

C. In the last stages of solidification of a magma, the remaining silicate melt will contain MORE / LESS silica than the original melt.

Hint: You do not need to know the actual bulk composition (mafic, intermediate, felsic) of the magma to answer these questions.

Exercise 2: Partial Melting and Magma Composition

Circle the answers that correctly complete the following statements:

Compared to the bulk composition of the rock, the minerals with the lower melting temperature are:

A. HIGHER / LOWER in the Bowen's Reaction Series

B. DEPLETED / ENRICHED in Mg, Fe, and Ca

C. DEPLETED / ENRICHED in Si, Na, K

Hint: Refer to Figure 4.11, Bowen's Reaction Series.

Exercise 3: Magma Generation at Subduction Zones

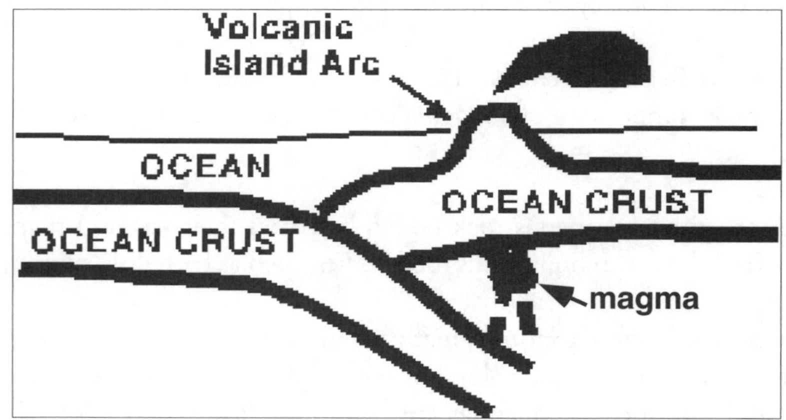

Circle the answers that correctly complete the following statements:

Compared to the basaltic ocean crust, the magma generated by partial melting of the subducting slab of ocean crust will:

A. have MORE / LESS silica

B. have MORE / LESS iron and magnesium

C. have MORE / LESS sodium and potassium

D. be MORE / LESS mafic

1. Only ferromagnesium minerals are found in which of the following lists of minerals?
   A. pyroxene, hornblende, K-feldspar, biotite
   B. plagioclase, biotite, pyroxene, clay
   C. quartz, muscovite, biotite, plagioclase
   D. biotite, pyroxene, olivine, hornblende

2. How does a rising magma make space for itself as it moves through the solid crust?
   A. by breaking off large blocks of rock that sink into the magma chamber
   B. by wedging open the overlying rock
   C. by melting surrounding rocks
   D. all of the above

3. How would you distinguish a lava flow from a sill exposed at the Earth's surface?
   A. Sills tend to be coarser-grained than lava flows because they cool slower.
   B. Their chemical compositions (that is, the minerals present) would be very different.
   C. Sills tend to be finer-grained due to slower rates of crystallization.
   D. Lava flows are coarser-grained due to very rapid rates of cooling.

4. Igneous rock names are based on
   A. texture and composition
   B. fine-grained and coarse-grained textures
   C. intrusive and extrusive
   D. where the magma chamber erupts

5. The terms intrusive and extrusive describe what important characteristics of igneous rocks?
   A. fine-grained and coarse-grained textures
   B. texture and composition
   C. phenocrysts and groundmass
   D. mineralogy and structure

6. Following Bowen's Reaction Series, the later, lower-temperature fractions of liquid magma become progressively
   A. depleted in silica
   B. enriched in silica
   C. enriched in magnesium
   D. depleted in potassium

7. Which of the following minerals are the earliest, highest-temperature minerals to crystallize in Bowen's Reaction Series?
   A. quartz and feldspar
   B. plagioclase and amphibole
   C. plagioclase and olivine
   D. chert and mica

8. Which set of conditions will result in basalt melting at the lowest temperature?
   A. dry basalt at low pressure
   B. dry basalt at high pressure
   C. wet basalt at low pressure
   D. wet basalt at high pressure

9. The formation of granitic batholiths occurs
   A. within the ocean crust
   B. within the continental crust
   C. along spreading centers in the ocean
   D. under hot spots

10. Considering the following minerals, which pairs would you predict NOT to be found together in the same igneous rock?
   A. K-feldspar and biotite
   B. Na-plagioclase and muscovite
   C. quartz and Na-plagioclase
   D. quartz and olivine

Hint: Refer to the Bowen's Reaction Series, Figure 4.11.

11. According to Bowen's Reaction Series, one would expect a zoned plagioclase crystal to
   A. have a sheet-like crystal structure
   B. have a center richer in calcium and concentric layers progressively richer in sodium
   C. be richer in iron than in magnesium
   D. have a center richer in sodium and concentric layers progressively richer in calcium

Hint: The photomicrograph on your CD and at the Website's *Photo Gallery* (slide 4.10) offers a great view of a zoned plagioclase.

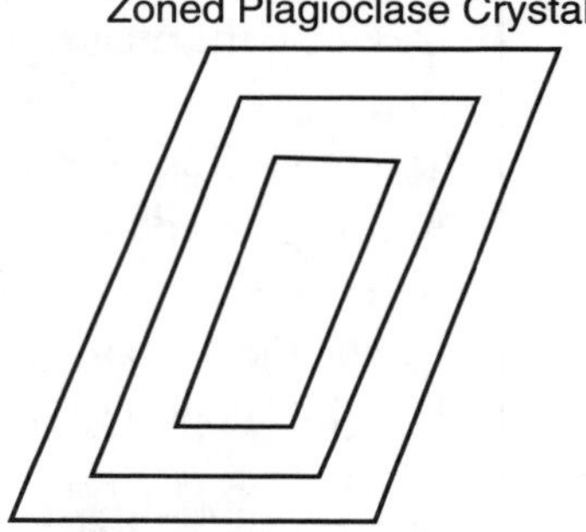

12. If lava flows on the slopes of a volcano are derived from one large magma chamber which crystallizes slowly and feeds eruptions over a period of many thousands of years, how would you predict the gross composition of the lava to change as the lava flows become younger? Hint: Use Bowen's Reaction Series and review the textbook section *Magmatic Differentiation*.
   A. Younger lava flows would become progressively enriched in iron and become more mafic.
   B. Younger lava flows would become progressively enriched in silica and become more felsic.
   C. Lava flows would be the same composition since they all came from the same magma chamber.
   D. Lava flows would alternate in composition.

13. The rock type of most batholiths found in the continental crust is
    A. gabbro
    B. obsidian
    C. granite
    D. basalt

14. The source for most mafic magmas is thought to be
    A. partial melting of felsic and intermediate rocks in the upper continental crust
    B. partial melting of ultramafic rocks within the upper mantle
    C. melting of preexisting granites and sediments
    D. directly from the molten core of the Earth

15. The production of basalt can be achieved by the partial melting of
    A. gabbro
    B. ultramafic rocks
    C. a mixture of gabbro and oceanic sediments
    D. rhyolite

16. An igneous rock made of a mixture of both coarse and fine grain minerals is called porphyritic and is formed by
    A. rapid cooling followed by a period of slow cooling
    B. slow cooling followed by a period of rapid cooling
    C. very slow cooling of a water-rich magma
    D. very rapid cooling in the presence of water

17. Andesites—intermediate magmas—are typically associated with
    A. divergent plate margins, like the mid-oceanic ridges
    B. fractures in the crust that allow magmas from the upper mantle to rise to the surface
    C. hotspots, like Hawaii
    D. convergent plate margins, like the western edge of South America—the Andes

ANSWERS AND EXPLANATIONS FOR EXERCISES AND QUESTIONS

After Lecture

Exercise 1: Igneous Rock Textures

Texture term/sketch description

A. phaneritic / coarse-grained, large interlocking crystals

B. phaneritic / visible crystals but not as coarse grained as sample A

C. porphyritic / mixed cooling history has large and very small crystals

D. aphanitic / fine grained – crystals may not be visible without magnification, looking like tiny dots in a sketch

Exercise 2: Distribution of Igneous Rocks within the Earth

Major Layer Within the Earth	Example Igneous Rock	General Compositional Group	General Chemical Composition
Continental crust (for continental crust there are two appropriate answers)	granite		more Si, Na, K less Fe, Mg, Ca
	andesite / diorite	intermediate	intermediate
Ocean Crust	basalt / gabbro	mafic	more Fe, Mg, Ca less Si, Na, K
Mantle	peridotite	ultramafic	less Si, Na, K more Fe, Mg, Ca

1. B. Continental crust is primarily felsic (granitic) and intermediate (dioritic) in composition. Iron and nickel are thought to be the primary constituents in the Earth's core. The mineral olivine is thought to be abundant in the mantle and also occurs in basalts, which make up the ocean crust.
2. A. Refer to Table 4.1 and Figure 4.11, Bowen's Reaction Series. The Bowen's Reaction Series is a very useful general guide to the general mineralogical composition of igneous rocks.
3. D. Rhyolite is an aphanitic volcanic rock and granite is a phaneritic plutonic rock. So, their cooling histories are different and therefore their textures are different. However, they have the same general composition (felsic).
4. A. Pyroxenes like augite are a common mineral in basalts. They are a ferromagnesium mineral with a single chain silicate crystal structure.
5. B. Rhyolite contains the most silica of the rocks listed. Fissure eruptions may be composed of a great variety of lavas but are typically basaltic.
6. D. The distinction between intrusive (plutonic) and extrusive (volcanic) rocks is solely grain size which is typically a consequence of cooling rate.
7. A. Gabbro is the intrusive equivalent of basalt, which is extrusive.

Exam Prep

Exercise 1

A. LESS. The presence of iron and magnesium in the magma greatly influences the complexity of the silicate structure because they act to "poison" the polymerization of the silica tetrahedra, preventing more complex silica tetrahedra crystalline structures. Since much of the iron and magnesium is incorporated into the early-formed crystals, as illustrated in Bowen's Reaction Series, minerals that crystallized later in the history of the solidification of the magma tend to be depleted in iron and magnesium, enriched in silica, and have more complex silicate structures. Refer to Figure 4.11 and Table 2.2, Chapter 2.

B. LAST. Refer to the explanation for A. and Figure 4.11, Bowen's Reaction Series.

C. MORE. Refer to the explanation for A. and Figure 4.11, Bowen's Reaction Series.

Exercise 2

A. LOWER. Refer to Figure 4.11, Bowen's Reaction Series.

B. DEPLETED. Much of the iron and magnesium in a magma is incorporated into the early formed crystals, as illustrated in Bowen's Reaction Series. Minerals that crystallized later in the history of the solidification of the magma tend to be depleted in iron and magnesium and enriched in silica.

C. ENRICHED. Refer to the explanation for B and Figure 4.11, Bowen's Reaction Series.

Exercise 3

A. MORE. As illustrated by the Bowen's Reaction Series, silicate minerals with more silica content have lower melting temperatures. So, a partial melt will be enriched in silica relative to the igneous rock from which it was generated.

B. LESS. Silicate minerals rich in iron and magnesium have higher melting points and are the last to melt compared to minerals lower on the Bowen's Reaction Series. Therefore, a magma generated from a partial melt will be enriched in silica and depleted in iron and magnesium relative to the bulk composition of the original rock from which the melt was generated.

C. MORE. Table 4.2 and Figure 4.11 are very helpful. Na and K concentrate in minerals with greater amounts of silica. So, they are enriched in minerals lower on the Bowen's Reaction Series and have lower melting temperatures compared to minerals with more iron, magnesium and calcium.

D. LESS. A partial melt will always have more silica and less iron/magnesium than the parent rock from which it is generated. Therefore, it will be less mafic.

1. D. Table 4.1 is a good reference for this question.
2. D. The textbook discusses the subject of magma generation and movement in a variety of places in Chapter 4, including the sections *The Formation of Magma Chambers* and *Forms of Magmatic Intrusions*.
3. A. Refer to textbook section *From Laboratory to Field Observation: The Palisades Intrusion* and *Forms of Magmatic Intrusions*. Slides 4.2 and 4.4 in the *Photo Gallery* at the Website and on the CD are good photomicrographs of diabase and basalt rocks that reveal these textural differences very well.
4. A. Many students mistakenly choose B or C as the answer. However, grain size alone is not the only basis for the classification and naming of igneous rocks. Composition (mineral content) is the other criteria for the classification of igneous rocks.
5. A. Intrusive (plutonic) rocks are typically coarser grained due to slower cooling. Extrusive (volcanic) rocks are finer grained due to quicker cooling rates. Cooling rates are not the only factors that influence the grain size of igneous rocks. Rapid changes in pressure within a magma chamber can induce rapid crystallization and therefore fine-grained crystals within an intrusive rock. High water content in residual fluids from a solidifying magma can enhance crystal growth and size.
6. B. The early formed magma from a partial melt or the residual melt in a solidifying magma chamber are enriched in silica relative to the parent material.
7. C. Olivine and calcium-rich plagioclase are the first minerals to crystallize from a magma with a mafic composition. Refer to Figure 4.11 in textbook.
8. C. Temperature is not the only factor to influence the melting point of solids. Lower pressure and an increase in water content typically lower the melting temperature of silicate minerals.

9. B. A batholith is a very large body of igneous rock. The continental crust is felsic to intermediate in bulk composition. Therefore, granitic batholiths occur within the continental crust. In rare circumstances, due to partial melting and fractional crystallization, felsic igneous rocks can be generated within the mafic ocean crust, like in Iceland. However, this is very unusual.
10. D. Quartz and olivine do not crystallize from the same magma body. If there was enough iron in the original magma, the silica is consumed in the formation of the ferromagnesium minerals and plagioclase feldspars. The magma completely solidifies before pure quartz can crystallize. A parent magma with enough silica to generate quartz will not contain enough iron and magnesium to generate olivine. Therefore, the two minerals are mutually exclusive.
11. B. Although Ca ion (+2) and Na ion (+1) are almost identical in size (refer to Figure 2.14), calcium-rich plagioclase crystallizes earlier and at higher temperatures than sodium-rich plagioclase, according to Bowen's Reaction Series. As temperature decreases and the magma becomes depleted in calcium, sodium takes the place of calcium in the crystal structure of plagioclase. The substitution of sodium for calcium within the crystal structure is analogous to changing some furniture in a house. The whole house doesn't change (get torn down and rebuilt), just some details within the house change. Zoned plagioclase crystals are relatively common.
12. B. If no new magma is ejected into the magma chamber, the composition of the magma may change over time because of fractional crystallization and the segregation of the earlier-formed iron-rich minerals by settling from the melt. The remaining magma becomes progressively enriched in silica and depleted in iron and magnesium.
13. C. Refer to question 9.
14. B. Refer to sections on *Current Theories of Magmatic Differentiation* and *Igneous Activity and Plate Tectonics* in the textbook. Keep in mind that the partial melt from an igneous rock will always be enriched in silica relative to the parent rock. D is incorrect because the Earth's core is thought to be composed mostly of metallic iron and nickel, not silicate minerals.
15. B. Basalt is a mafic rock. This question is essentially the same as question 14.
16. B. Slow cooling typically produces larger crystals and rapid cooling produces finer crystals. The presence of water, which facilitates ion mobility, enhances the rate of crystal growth and crystal size. So, an igneous rock with a mixture of coarse and fine grained minerals formed under conditions where the cooling history was mixed. Both volcanic and plutonic rocks can exhibit porphyritic texture. A porphyritic volcanic rock erupts from a magma that has begun to crystallize—crystals already exist within the magma. Early-formed crystals are literally carried to the surface by the remaining melt, which cools quickly upon erupting on the Earth's surface. A porphyritic texture in a plutonic rock may be a result of its cooling history or changes in pressure and other conditions within the magma chamber.
17. D. Subduction at convergent plate margins produces large amounts of andesites. In fact, andesites are named after the Andes Mountains in South America because they are very abundant there. Magmas generated within the upper mantle, like at divergent plate boundaries and hotspots, are typically mafic (refer to Figure 4.8 in textbook).

Chapter 5
Volcanism

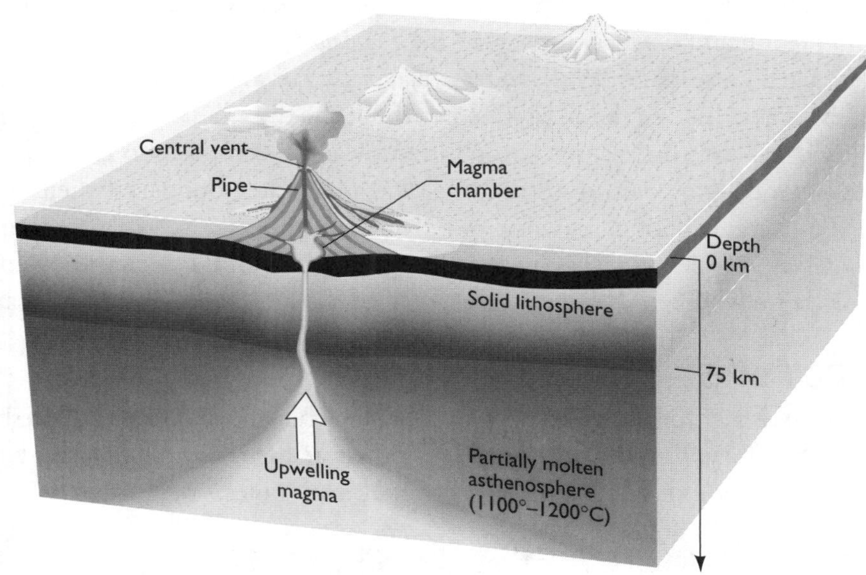

Figure 5.1 / Volcanism is a surface expression of magmatism at depth.

BEFORE LECTURE

Before you go to lecture be sure to spend some time previewing Chapter 4. We have made it easier by identifying the three key questions on Volcanism for you (see *Chapter Preview* questions). These questions constitute the basic framework for understanding the chapter. Working with the *Chapter Preview* questions before lecture and committing them to memory should help you understand the lecture better and get an excellent set of notes. Need a refresher on previewing? See Appendix: *How to Study Geology*.

CHAPTER PREVIEW

- **Why does volcanism occur?**
  Brief answer: Volcanism is the surface expression of magmatism at depth.
- **What are the three major lava types and how do they relate to eruptive style and volcanic landforms?**
  Brief answer: mafic (basalt), intermediate (andesite), and felsic (rhyolite).
- **How is volcanism related to plate tectonics?**
  Brief answer: Volcanism is concentrated at convergent and divergent plate boundaries and hotspots.

Vital Information from other Chapters

- Review the following sections in Chapter 4: *Extrusive Igneous Rocks*, *Chemical and Mineral Composition*, *Tectonic Acitivty*, *Rock Composition*, and *Types of Magmas*.

DURING LECTURE

Your basic goal during lecture is to get a good set of notes. By the end of lecture those notes should contain helpful and detailed answers to the *Preview Questions*. To avoid getting lost in details, keep the big picture in mind: Chapter 5 explains the varying eruptive styles of volcanoes. Keep in mind what you learned in the last chapter about the three basic types of lava. Eruptive style of a volcano depends on magma composition. Magma composition depends on plate tectonics: that is, whether a volcano formed at a diverging plate boundary (where two plates spread apart), a converging boundary (where one plate dives under another), or in the middle of a plate (where it passes over a mantle plume—hot spot). Much of this was already covered in Chapter 4. Volcanism will be easy if you review and understand the basic ideas of Chapter 4.

AFTER LECTURE

The perfect time to review your notes is right after lecture, while the material is still fresh in your mind. Review to be sure you got all the key points and wrote them down in a form that will be readable later. Here's a checklist that will help you check your notes.

NOTE REVIEW CHECKLIST

✔ All notes legible? (Rewrite so they read easily.)

✔ Important points clearly identified? You should now have headers in your notes that tie to each of the questions in the *BEFORE LECTURE/Chapter Preview.*

✔ Holes (missing material) filled in from memory?

✔ Areas where you don't remember what was said marked for a follow-up session with your instructor, tutor, or study partner?

✔ Possible test questions indicated in the margin (TQ)?

✔ Additional visual material. *Suggestions for Chapter 5: Draw to compare. Add some simple sketches that will help you remember the different landforms discussed in this chapter. Example: compare the slope of a shield volcano with that of a composite volcano by drawing a simple sketch of each. Simple means you only have to show how steep and pointed each volcano type is. So, you just draw or trace the outline of the slope. No fancy artwork, just a steep inverted "v" for one and a flattened inverted "v" for the other. Show how much relative area each is likely to cover (which spreads out over many miles and which is confined to a relatively small area). You will find that sketching provides a good check on how well you understand the differences.*

✔ Reworked notes into a form that is efficient for your learning style?

✔ Created a brief "big picture" overview of this lecture (using a sketch or written outline)? *Hint: Do Exercise 1 in the Practice Exercises and Study Questions and you will have a perfect summary of Chapter 5.*

Intensive Study Session

As usual, this section contains some *Practice Exercises* and *Study Questions* to help you master the material on volcanism, but before we turn to that let's take a look at text marking and annotating, two skills that can greatly reduce the amount of time you spend rereading text material.

How to Mark and Annotate Your Text

Marking your text as you read is a vital part of learning the material. Good marking is good thinking. First, you think about what is important. Then you mark it. Press and Siever have already marked key terms in bold letters. To this bold lettering you will want to add underlines, numbers, and annotations in the margin. Here are some brief suggestions for how and when to use each kind of mark

USEFUL MARKINGS FOR	**GEOLOGY: HOW-TO TIPS**
Underline important points. Underline to highlight the key ideas.	(1) Read before you mark. To avoid overusing underlining, set your pencil down on the table as you begin to read a paragraph. Don't let yourself pick up the pencil until you have read the entire paragraph. Then underline the key point. (2) Be selective. Underline brief, but meaningful phrases or key words that will trigger your memory. Mark just enough so you can review without re-reading the paragraph. (3) Use color marking if it helps you remember.
Annotate the margin. Write a brief summary statement of **key geologic processes** in your own words in the text margin. Use these annotations to facilitate exam review.	(1) Use your own words. Putting ideas into your own words is a powerful learning strategy. (2) Be neat. This takes time. But it will pay off later when you review, because your annotation will be easily perceived. (3) Organize your annotations into categories. Grouping ideas into categories or bulleted lists makes them easier to remember. (4) Use annotations during exam review. Avoid rereading the text word for word. You don't have time.
Other useful text marks:	(1) **Circled numbers** in the margin indicate sequences such as the processes of the Rock Cycle. (2) **Question mark (?)** Put a question mark in the margin to remind yourself to ask your instructor about a point you do not understand. (3) **Asterisks ***** Use asterisks to mark ideas of special importance. Use asterisks sparingly. Save them for the two or three most vital ideas in the entire chapter.

Website and CD Activities and Tools

http://www.whfreeman.com/presssiever

- **Website Q & A Practice Multiple Choice Questions:**
  At the Website complete the *Q & A*. Pay particular attention to the explanations for answers. Flashcards will help you learn new terms. *Volcano Plumbing* and *Volcanism and Plate Boundaries* in *Interactive Exercises* at the Website are useful reviews of key illustrations, Figures 5.1 and Figure 5.30. *The Interactive Tool, Important Volcanoes of the US and Canada* at the Website is an excellent overview of the eruption history for volcanoes that you, family, or friends may live near.

Check out the following videos from the *Additional Resources* section of your CD:
Eruption of Kilauea • *Eruption of Kilauea: Explosive Phase* •*Mount Shasta: Composite Volcano*

PRACTICE EXERCISES AND STUDY QUESTIONS

(Answers and explanations are at the end of this chapter.)

Exercise 1: Lava Types

This table will provide you with a useful study guide for much of Chapter 5. The finished table makes an ideal summary of the chapter that should be very useful when you return to this unit in preparation for your midterm exam. Fill in the blanks with the typical characteristics of each lava type. Keep in mind that different lavas exhibit a range of properties and behaviors. Give the best answer that generally characterizes each. Some answers have been provided as guidelines. Bullets mark information to fill in.

LAVA TYPES	Basalt (mafic)	Andesite (intermediate)	Rhyolite (felsic)
PROPERTIES eruption temperature silica content gas content viscosity typical flow velocity typical flow length typical flow thickness	 • • low, up to a few —% low-fluid magma 0.7 to 30 m/minute 10 to 160 km 5 to 15 m	 intermediate intermediate variable intermediate 9 m/day 8 km 30 m	 • high (70%) high (up to 15%) • less than 9 m/day less than 1.5 km 200 m
ERUPTION STYLES	typically not very explosive	•	•
DEPOSITS	Flood basalt • • • •	Lava flow Dome Pyroclastic flow – tuff and welded tuff	Obsidian dome • •
LANDFORMS	• • Cinder cone Small caldera	Composite volcano Summit crater Caldera Cinder cone	• • •
Association with **PLATE TECTONICS**	•	•	•
HAZARDS	• •	Lava flow Pyroclastic / ash flow Explosive blast Hot gases Mudflow	Explosive blast Hot gases • •

> The longest word in the English language is supposedly **pneumonoultramicroscopicsilicovolcanoconios** — a lung disease caused by breathing in particles of volcanic matter or a similar fine dust.

1. What feature of obsidian (volcanic glass) makes it so unusual?
   A. It is both plutonic and volcanic.
   B. It contains no silica.
   C. It contains no minerals.
   E. It is very coarse-grained.

2. Which extrusive rocks contain the most silica?
   A. andesite
   B. rhyolite
   C. granite
   D. basalt

3. Lavas are almost always fine-grained rocks because
   A. the type of minerals they are made up of do not form large crystals
   B. very little water is in the extruded magma
   C. the lava crystallizes under pressure too low for large crystals to grow
   D. they cool too rapidly for large crystals to grow

4. During volcanic eruptions, the most common gas released is
   A. hydrogen
   B. water
   C. nitrogen
   D. carbon dioxide

5. A cloud of superheated steam and hot ash produced during a volcanic eruption, and moving rapidly parallel to the ground, is called
   A. a welded tuff
   B. a pyroclastic flow
   C. volcanic bombs
   D. a cinder flow

6. A good example of shield volcanoes can be found in the
   A. volcanoes of northern California, Washington and Oregon
   B. Hawaiian island volcanoes
   C. Caribbean islands, such as Mt. Pelée on Martinique
   D. volcanoes found along western South America

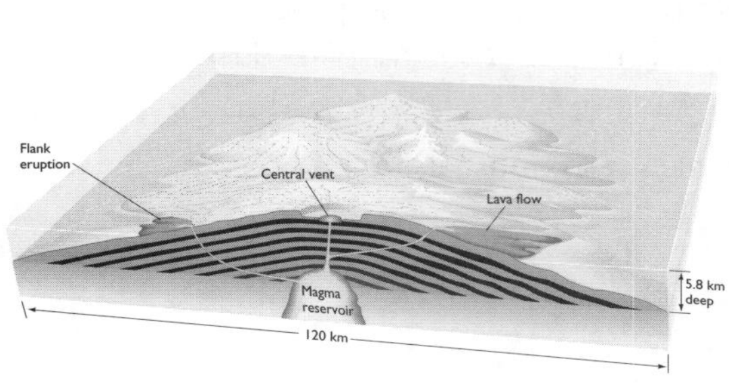

Figure 5.10 / Shield Volcano

Central vent filled from previous eruption
Radiating dikes
Pyroclastic layers
Lava flows

Figure 5.14 / Composite Volcano

7. Composite volcanoes (stratovolcanoes) are largely composed of
   A. basalt lava flows and basaltic cinders
   B. pillow and pahoehoe lava flows
   C. rhyolitic and intermediate lavas and pyroclastic flows
   D. dikes and sills

EXAM PREP

Materials in this section are most useful during your preparation for midterm and final exams. For optimal performance, midterm preparation should begin about eight days before the exam (see Appendix: *How to Study Geology* for details). The basic idea is a systematic review of material divided into short study sessions.

If you have used regular intensive study sessions to master the material, now you get a pay back! Review for your exam will proceed smoothly and take far less time.

The following *Chapter Summary* and *Practice Exam* should simplify review still further. Read the *Chapter Summary* to begin your session. It provides a helpful overview that should get you back into the material. Next, try the *Practice Exam*. Take it just as you would a midterm to see how you stand in regard to mastery of this chapter. After you answer the questions, score them. Finally, and most important of all: review any question that you missed. Identify and correct the misconception that resulted in missing the question.

CHAPTER 5 SUMMARY

- Volcanism occurs when molten rock inside the Earth rises buoyantly to the surface because it is less dense than the surrounding rock. Volcanism is a surface expression of magma generation within the Earth.
- Silicate lavas can be classified into three major types, felsic (rhyolite), intermediate (andesite), and mafic (basalt), respectively based on decreasing amount of silica and increasing amounts of iron and magnesium.
- Eruption styles, volcanic deposits, landforms, and potential hazards are strongly linked to the chemical composition and gas content of the lava (refer to *Exercise 1*). Because basaltic lavas are relatively fluid and dry, they typically exhibit less explosive eruptions

and erupt as lava flows. Rhyolite lavas are very viscous and usually wet. Therefore, they typically erupt very explosively as pyroclastic flows or form domes.

- There is a strong connection between major types of volcanism and crustal plate boundaries. Basaltic lavas occur at divergent plate boundaries and hot spots. The ocean crust is created by basaltic volcanism at the ocean ridge system. Basalt is thought to be generated by partial melting of the ultramafic upper mantle. Basaltic, andesitic, and rhyolitic lavas erupt at convergent zones. The lavas generated along any particular convergent zone will depend in large part on what rocks are being subducted and melted within the overriding crust.
- There are both benefits and hazards associated with volcanism. Geothermal heat is growing in importance for electric energy generation. Earth's oceans and atmosphere are thought to have condensed from volcanic degassing of our planet's interior. Volcanic dust and gases can impact global climate. Volcanic eruptions and associated mudflows can have disastrous impacts on a region and its people. Important ore-forming processes occur when groundwater circulates around the magma chamber and through hot rock beneath a volcano or seawater circulates through ocean-floor rifts.

Website and CD Activities and Tools

http://www.whfreeman.com/presssiever

Under *Additional Resources* on your CD, review the following animations: *Shield Volcano and Caldera Formation*, *Composite Volcano*, and *Cinder Cone*. On the CD, in the *Additional Resources* section for Chapter 5, the *Photo Gallery* has some great views of volcanoes and volcanic features. The *Photo Gallery* is also at the Website.

PRACTICE EXAM QUESTIONS FOR CHAPTER 5

(Answers and explanations are at the end of this chapter.)

1. Of the following states, which is essentially all volcanic rock?
   A. Alaska
   B. Hawaii
   C. Oregon
   D. California

2. Compared to basalt, rhyolite lava flows are very thick and tend to form domes because rhyolite lava
   A. contains less gas than basalt lava and is therefore more fluid
   B. is richer in silica than basalt lava and is therefore less fluid
   C. cools more quickly than basalt lava
   D. is less dense than basalt lava

3. On a recent three-day hike up a gently sloping mountain, your friends describe to you features they encountered. They frequently crossed lava flows and fissures, and occasionally they had to detour around large cinder cones. Based on your friends' description, you tell them they were hiking on a
   A. caldera
   B. composite volcano
   C. gabbro pluton
   D. shield volcano

4. Composite volcanoes are commonly associated with which tectonic setting?
   A. passive continental margins and deep ocean basins
   B. convergent plate boundaries, such as those of the circum-Pacific
   C. ocean spreading centers, such as Iceland
   D. transform faults, such as the San Andreas fault

5. Typically, explosive volcanic eruptions are associated with
   A. basalt lavas
   B. shield volcanoes
   C. magmas that are poor in both silica and dissolved gases
   D. magmas high in both silica and dissolved gases

6. Shield volcanoes and composite volcanoes differ in shape based on
   A. the composition of the magma that forms each
   B. the particular part of the ocean that each forms in
   C. the latitude at which each forms
   D. factors that are completely unknown to us at present
   Hint: Refer to Figures 5.10 and 5.14.

7. Calderas usually result
   A. due to molten material from the core coming very near the Earth's surface at a thin point in the crust
   B. after a steam explosion, when magma comes into contact with abundant underground water
   C. when an eruption literally blows the top of a volcano off
   D. after large volumes of magma erupt leaving a void in the magma chamber into which the superstructure of a volcano can collapse
   Hint: Refer to Figure 5.16.

8. Your friends have described to you an eruption that took place at an undisclosed location. The lava they described merely flowed out of a fissure and spread rapidly over a large area. You would inform that person that the rock type being formed would most likely be
   A. granite
   B. andesite
   C. basalt
   D. rhyolite

9. You have been informed that an explosive volcanic eruption has taken place at an undisclosed location and that a huge nuée ardente (pyroclastic flow) flowed off the steep-sided volcano. You could respond that the magma type is likely to be
   A. basaltic
   B. rhyolitic
   C. ultramafic
   D. gabbroic

10. Where are andesitic volcanoes located?
    A. where diverging plate boundaries occur
    B. at transform boundaries
    C. along the mid-oceanic ridge crest
    D. along converging plate boundaries

11. A large resort located on a beautiful lake, and known for its hot springs, is experiencing a drop in business due to publicity about the recent swarm of small earthquakes in the area. The lake is actually located within a caldera, and beautiful rock towers and spires of weathered volcanic tuff are found all along the edge of the lake. As the director of the resort you're concerned about the change in business and the potential risk to your guests. What should you do?
    A. You cannot be worried because you know a volcano can't blow up on you.
    B. You know that earthquake swarms can be a precursor to volcanic eruptions, and that very explosive eruptions have happened at this place in the recent geologic past, so you decide that the resort should close until the situation is safe.
    C. Advertise the resort as the best place to see beautiful basalt lava fountains.
    D. Earthquakes have occurred occasionally along a nearby known fault, and there has been no historic volcanic eruptions, so you're not concerned.

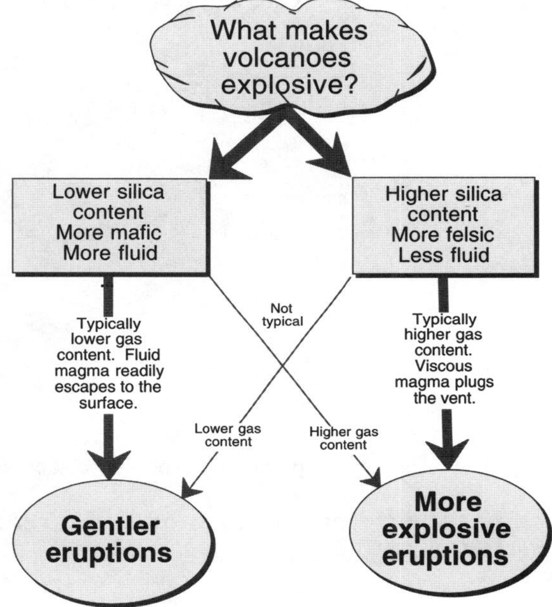

12. Large volcanoes can potentially impact global climate when they erupt because they release
    A. geothermal heat
    B. volcanic dust, sulfur, and carbon dioxide gas
    C. lahars and lava flows
    D. nitrogen and argon gases

13. The largest crystals found in a lava flow can be found
    A. near the top surface of the flow
    B. in the center of the flow
    C. near the bottom surface of the flow
    D. the crystals would have the same grain size throughout

14. Which of the following volcanic deposits can be formed from felsic lava?
    A. pahoehoe
    B. flood basalt
    C. volcanic dome
    D. shield volcano

15. Which of the following statements is true?
    A. Mafic rocks are richer in silica than felsic rocks.
    B. Mafic rocks crystallize at higher temperatures than felsic rocks.
    C. Mafic rocks are more viscous than felsic rocks.
    D. Mafic rocks tend to be lighter in color than felsic rocks.

16. If lava flows of progressively younger ages all erupted from a single large magma chamber, how would you expect their composition to have systematically changed?
    A. The lava flows will be progressively enriched in iron as they get younger.
    B. The lava flows will be progressively enriched in silica as they get younger.
    C. The lava flows will be progressively more mafic as they get younger.
    D. The lava flows will be progressively more fluid as they get younger.

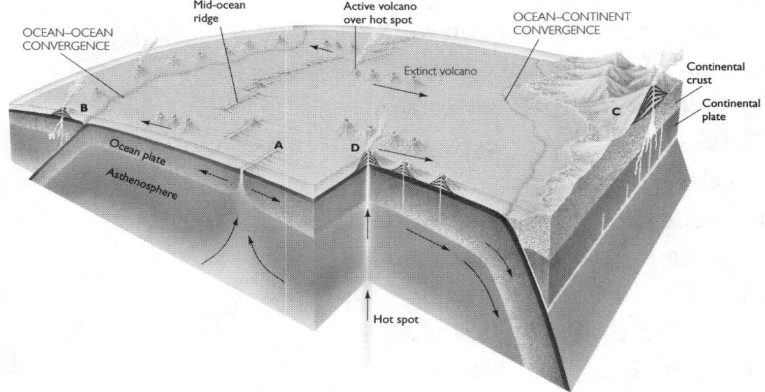

Figure 5.30 / The association of volcanism with plate tectonics.

17. Which type of lava is most likely to erupt at plate tectonic setting "A"? Refer to Figure 5.30 above.
    A. basalt
    B. andesite
    C. rhyolite
    D. diorite

18. Which type of lava is most likely to erupt at plate tectonic setting "B"? Refer to Figure 5.30 above.
    A. basalt
    B. andesite
    C. rhyolite
    D. diorite

19. Which of the following volcanoes formed at a plate tectonic setting similar to "C"? Refer to Figure 5.30 above.
    A. Hekla, Iceland
    B. Mauna Loa, Hawaii, USA
    C. Yellowstone, USA
    D. Mount St. Helens, USA

20. Which type of lava is most likely to erupt at location "D" in Figure 5.30 above? Note: Location "D" is not associated with a crustal plate boundary.
    A. basalt
    B. andesite
    C. rhyolite
    D. diorite

ANSWERS AND EXPLANATIONS FOR EXERCISES AND QUESTIONS

After Lecture

Exercise 1: Lava Types

This table will provide you with a very useful study guide for much of Chapter 5.

LAVA TYPES	**Basalt (mafic)**	**Andesite (intermediate)**	**Rhyolite (felsic)**
PROPERTIES eruption temperature silica content gas content viscosity typical flow velocity typical flow length typical flow thickness	 1000 to 1200° C low (50%) low, up to a few —% low-fluid magma 0.7 to 30 m/minute 10 to 160 km 5 to 15 m	 intermediate intermediate variable intermediate 9 m/day 8 km 30 m	 800 to 1000° C high (70%) high (up to 15%) typically high less than 9 m/day less than 1.5 km 200 m
ERUPTION STYLES	typically not very explosive	explosive	typically very explosive
DEPOSITS	Flood basalt Fissure flow Pahoehoe and aa flows Pillow lava Cinder	Lava flow Dome Pyroclastic flow – tuff and welded tuff	Lava dome Pyroclastic flow – tuff and welded tuff
LANDFORMS	Shield volcano Lava plateau Cinder cone Small caldera	Composite volcano Summit crater Caldera Cinder cone	Composite volcano Large caldera Summit crater
Association with **PLATE TECTONICS**	Divergent boundaries Hotspots	Convergent boundaries	Convergent boundaries
HAZARDS	Lava flow Explosive in contact Hot gases	Lava flow Pyroclastic / ash flow Explosive blast Hot gases Mudflow	Pyroclastic / ash Explosive blast Hot gases Mudflows (lahars)

1. C. Being a (volcanic) glass, obsidian lacks a crystalline structure. Therefore, by definition it cannot contain minerals. Since obsidian forms from the very quick cooling of lava, it is not plutonic and not coarse-grained. Even under a microscope crystals are not visible. X-ray analysis of volcanic glass reveals that it is an amorphous solid with no crystalline structure. Obsidian contains from 50 to 77% silica. Because obsidian lacks a crystalline structure it fractures into conchoidal chips and holds a sharp edge, which made it a preferred material for arrowheads and spear points.
2. B. Figure 4.6 and Table 4.2 provide basic compositional trends for igneous rocks. Rhyolite contains up to about 77% silica. Basalt is mafic with about 50% silica. Fissure eruptions are typically basaltic in composition because basalt is a fluid lava, compared to andesite and rhyolite, and can therefore erupt along long, narrow fissures. Andesite is intermediate in composition.
3. D. Cooling rate determines the size of the mineral grains within lava flows. If minerals already exist in the magma chamber before an eruption occurs, the lava will erupt with larger crystals floating in it and solidify into a volcanic rock with a mixture of grain-sizes, known as a porphyry.
4. B. Water is the main constituent of volcanic gas (75 to 95%), followed by carbon dioxide, sulfur dioxide, and traces of nitrogen, hydrogen, carbon monoxide, sulfur, and chlorine.
5. B. Your textbook describes *Pyroclastic Deposits* and *Pyroclastic Flows*.
6. B. Hawaii is a classic example of a shield volcano (refer to Figure 5.10). Composite volcanoes occur in the Cascade Range of Oregon (Crater Lake) and Washington and in the Andes of South America.
7. C. Refer to Figure 5.14 in the textbook.

Exam Prep

1. B. The Hawaiian Islands are a chain of volcanoes generated as the Pacific plate moves over a hot spot in the mantle.
2. B. The higher silica content of rhyolitic lavas results in a much higher viscosity which favors very thick lava flows or domes (refer to Figure 5.11).
3. D. Refer to Figure 5.10, 5.14 and 5.16 for a comparison of shield, composite and caldera type volcanoes. A gabbro pluton is an intrusive rock body and not relevant to the question.
4. B. Refer to the textbook section entitled *The Global Pattern of Volcanism*.
5. D. Magmas rich in silica and dissolved gases cause the most explosive volcanic eruptions. Increasing silica content progressively increases the viscosity of the magma. More viscous magmas are more likely to plug the throat of a volcano until pressure builds up high enough to cause an explosive eruption.
6. A. Shield volcanoes are formed dominantly by basaltic lavas which are more fluid and therefore tend to spread out. Composite volcanoes are dominantly constructed from intermediate and felsic lava flows, domes and pyroclastic flows. Intermediate and felsic lavas are more viscous and form thicker flows than basalt.
7. D. Refer to Figure 5.16 in your textbook.
8. C. Refer to Figure 5.21 and the section on *Fissure Eruptions* in your textbook.
9. B. Refer to Figure 5.9 and the section on *Pyroclastic Deposits, Pyroclastic Flows and Composite Volcanoes* in your textbook.

10. D. Refer to Figures 5.28 and 5.30 to review the association of volcanism with plate tectonics.
11. B. Refer to *Reducing the Risks of Hazardous Volcanoes* in your textbook. The Interactive Tool, *Important Volcanoes of the United States and Canada,* at the Website is an excellent overview of the eruption history for volcanoes that you, family, or friends may live near or visit someday.
12. B. Refer to *Earth System Interactions: Volcanism and Climatic Change* in your textbook.
13. B. The center of a lava flow will cool the slowest because it is insulated by the surrounding margins of the flow. In basaltic flows containing olivine, sometimes the larger olivine crystals will concentrate near the bottom of the flow due to settling because they are iron-rich and dense (heavy).
14. C. Refer to Figure 5.11 in your textbook.
15. B. Refer to Table 4.2 and Figure 4.11. Review Bowen's Reaction Series, which is a very useful flow chart relating the sequence of silicate mineral formation in a magma to magma composition and temperature.
16. B. A review of *Magmatic Differentiation* in Chapter 4 would be useful. If a volcano is fed by one large magma chamber without any additional injections of magma from below, then the remaining melt in the magma chamber becomes progressively enriched in silica and depleted in iron and magnesium as the magma solidifies. Younger lava flows are enriched in silica relative to flows that erupted when more of the magma was molten. Having a higher silica content and a lower iron content, younger lava flows will be less mafic and more viscous (less fluid).
17. A. Basaltic eruptions occur along the oceanic ridge system and form the ocean crust.
18. B. Andesitic volcanoes are commonly associated with convergent plate boundaries. Partial melting of the subducting slab or materials above the subducting slab typically generates intermediate magmas.
19. D. Location C in Figure 5.30 is a convergent plate boundary where oceanic crust is subducting beneath continental crust. One example of where this process is happening today is beneath the Cascade Range of volcanoes that extend from northern California through Washington. Mount Saint Helens is within the Cascades.
20. A. Location D in Figure 5.30 is a hotspot. Hotspots are surface expressions of magma plumes coming up from the ultramafic mantle. These plumes are basaltic in composition. The composition of hotspot volcanism may become more felsic if the magma plume rises through continental crust like under Yellowstone. Hotspots like Hawaii and Yellowstone occur within the middle of crustal plates. Hotspots are also located along divergent plate boundaries like in Iceland.

Chapter 6
Weathering and Erosion

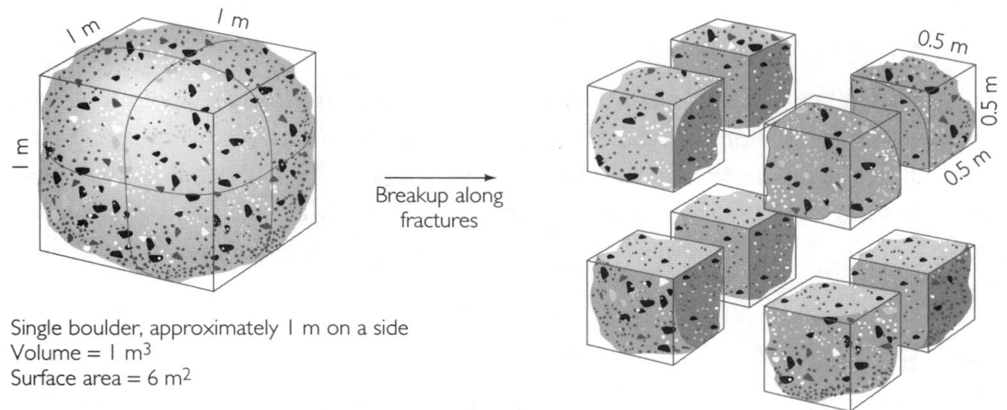

Figure 6.5 / As a rock mass breaks into smaller pieces, increasingly more surface becomes available for the chemical reactions of weathering.

BEFORE LECTURE

Before you attend lecture be sure to spend some time previewing the chapter. For an efficient preview use the questions below.

CHAPTER PREVIEW

- **How does weathering fit in the rock cycle?**
  Hint: see Figure 3.10.

- **How does chemical weathering work?**
  Brief answer: Water, oxygen, acids, and physical weathering facilitate chemical weathering reactions which alter and break down minerals to form new minerals, oxides, salts, and release silica.

- **How does physical weathering work?**
  Brief answer: Physical weathering breaks rocks into fragments. Processes which aid physical weathering include chemical weathering, frost wedging, and plant roots.

- **How do soils form as products of weathering?**
  Brief answer: Soils form within rock materials due to chemical weathering. Soil formation is influenced by the composition of the rock, stability of the weathering surface (topography), time, and most importantly climate. Life processes and their by-products are also important factors in soil formation and soil type.

Vital Information from other Chapters

- Rock-Forming Minerals: Chapter 2.
- The Rock Cycle: Chapter 3.

When we try to pick up anything by itself,
we find it entwined with everything else in the universe.
—John Muir

DURING LECTURE

Learning Warm-Up. Much that we consider beautiful in nature is the product of some sort of a weathering process. Just before lecture, spend five minutes scanning pictures in the text that show beautiful products created by weathering. Ask yourself, "What colors and shapes are associated with various weathering processes?" in this chapter. Make a list. As you take notes during this lecture, be sure to get details on specific chemical weathering processes such as oxidation and dissolution.

AFTER LECTURE

The perfect time to review your notes is right after lecture. If you wait even one day most (80%) of what you heard will have disappeared from memory.

NOTE REVIEW CHECKLIST

✔ All notes legible? (Rewrite so they read easily.)

✔ Important points clearly identified? You should now have headers in your notes that tie to each of the questions in the *BEFORE LECTURE/Chapter Preview.*

✔ Holes (missing material) filled in from memory?

✔ Areas where you don't remember what was said marked for a follow-up session with your instructor, tutor, or study partner?

✔ Possible test questions indicated in the margin (TQ)?

✔ Additional visual material. *Suggestion for Chapter 6:There are a lot of chemical and physical weathering processes in Chapter 6. It may be useful to make a list of processes that you can look at as a group all in one place in your notebook. Use a 2-column format for this. Your list of processes in the left column, details you need to remember in a second column to the right.*

✔ Reworked notes into a form that is efficient for your learning style? Are you a kinesthetic learner? If so you may find it helpful to rewrite your notes as questions that you will answer as you review.

✔ Created a brief "big picture" overview of this lecture (using a sketch or written outline)? *Suggestion for Chapter 6: Try reworking Figure 6.15 into a version that helps you remember. Re-state processes in words you will remember easily. Simplify the figure. Experiment with removing terms and boxes and omitting details. The idea is to wind up with a figure you will remember perfectly.*

Intensive Study Session

In this session you need to master the distinctions between chemical and physical weathering. You should also spend some time working with specific chemical weathering processes, particularly:

- Feldspar ⟶ weathering to clay.
- Oxidation of ferromagnesium minerals ⟶ to form iron oxides.
- Dissolution of calcite.

STUDY TIP

If you feel confident you understand these distinctions already, then go directly to the study questions. Take them like a real test. Don't look at answers until you are finished. Focus the remainder of the session on items you missed. First, read the brief explanation for each answer provided in the answer key. Follow up by studying text material that explains what you missed.

Website and CD Activities and Tools

http://www.whfreeman.com/presssiever

- **Website Q & A Practice Multiple Choice Questions:**

  At the Website complete the *Q & A*. Pay particular attention to the explanations for answers. Flashcards at the Website will help you learn new terms. The Interactive Exercise, *Mineral Stability*, at the Website will help you review the susceptibility of common minerals to chemical weathering. Note that the susceptibility of common silicate minerals to chemical weathering is closely related to their silicate crystal structure. Quartz, a framework silicate at the bottom of the Bowen's Reaction Series (Figure 4.14), is very stable on the Earth's surface. Whereas olivine, with a isolated silica tetrahedra crystal structure, is very susceptible to chemical weathering.

PRACTICE EXERCISES AND STUDY QUESTIONS

(Answers and explanations are at the end of this chapter.)

1. Of the following, which is NOT involved in the process of chemical weathering?
   A. water
   B. oxygen
   C. carbon dioxide
   D. nitrogen

2. The following products all result from chemical weathering EXCEPT
   A. feldspar
   B. iron oxides
   C. silica in solution
   D. clay minerals

3. Which of the following minerals does NOT chemically weather into a clay mineral?
   A. muscovite
   B. K-feldspar
   C. pyroxene
   D. quartz

4. Of the following materials, which one would make the longest lasting tombstone?
   A. limestone
   B. shale
   C. granite
   D. sandstone cemented with calcium carbonate

5. An example of chemical weathering is
   A. rusty streaks on a rock wall
   B. angular blocks of rock rubble in the mountains
   C. pot-holes in pavement
   D. rocks wedged apart by tree roots

6. Although water is an important agent of chemical weathering in its own right, it becomes more effective if small amounts of carbonic acid are present. Carbonic acid is formed when
   A. carbon from coal beds or from graphite deposits is pulverized along a fault or fracture and then added to water
   B. carbon dioxide from the atmosphere or from organic decomposition is added to water
   C. sulfur from coal-fired power plants is added to water
   D. water comes in contact with the calcite in a limestone layer

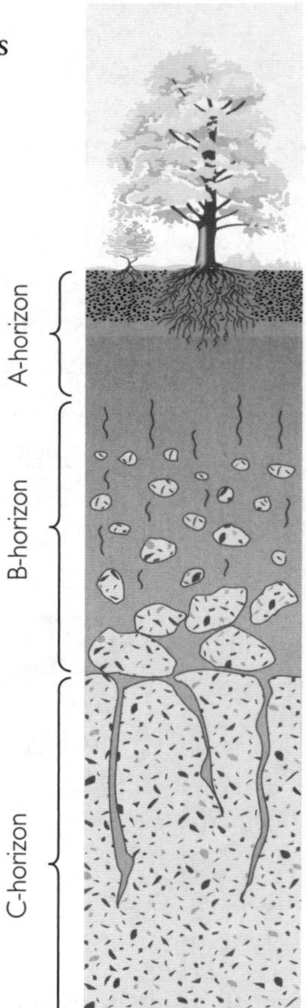

Figure 6.17 / Soil Profile: The thickness and distinctiveness of the soil profile depends on climate, parent rock material, and the length of time the soil has been forming. The transition from one horizon to another is normally indistinct.

7. The potential for chemical weathering can be greatly enhanced by physical weathering because physical weathering
   A. increases the surface area of the rock particles
   B. increases the availability of chemical agents
   C. increases drainage and thereby reduces contact with water
   D. increases the size of the rock particles

8. Limestone and other carbonate rocks weather relatively fast in a ____________ climate due to __________________ weathering reaction.
   A. dry / oxidation
   B. dry / hydrolysis
   C. wet / physical
   D. wet / dissolution promoted by carbonic acid

9. Georgia soils, along with other warm, humid regions, are deep red in color. This color is due to
   A. iron oxides
   B. quartz
   C. feldspar
   D. clay minerals

10. Clay minerals, like kaolinite, are a product of ____________________ weathering of __________________ minerals and are a raw material for ______________________.
   A. chemical / silicate / pottery
   B. physical / sulfate / cement
   C. chemical / sulfide / asphalt
   D. physical / carbonate / fertilizers

11. Soluble mineral leaching in a soil occurs which soil horizon?
   A. A-horizon
   B. B-horizon
   C. C-horizon
   D. all of the above

12. Which of the following climatic regions experiences the most rapid chemical weathering?
   A. hot, low precipitation
   B. extreme cold, low precipitation
   C. hot, high precipitation
   D. extreme cold, high precipitation

13. Which soil horizon tends to accumulate the most clay minerals and/or salts (such as calcium carbonate)?
   A. A-horizon
   B. B-horizon
   C. C-horizon
   D. R-horizon

14. Which of the following statements accurately describe a pedocal soil?
    A. They are typically enriched in aluminum in the B-horizon compared to laterite soils.
    B. They are typically enriched in calcium carbonate.
    C. They are typically enriched in iron and aluminum in the B-horizon compared to pedalfers.
    D. They are strongly leached soils and typically lack any significant amount of soluble material.

15. Generally, pedocal soils form in what climatic environments?
    A. cool climates, in heavily forested areas
    B. warm climates, with high precipitation
    C. warm climates, with low precipitation
    D. tropical climates, with extended periods of weathering

Exam Prep

Materials in this section are most useful during your preparation for midterm and final examinations. For optimal performance, midterm preparation should begin about eight days prior to the exam. The basic idea is a systematic review of material divided into short study sessions.

TIPS FOR PREPARING FOR GEOLOGY EXAMS

✔ Use the clues your instructor has provided in lecture about what is important. Even when a department agrees on a common core of material (rare) each instructor carves out a course that is unique, has a particular character or flavor, and distinct areas of emphasis. Your instructor is the ultimate guide in regard to the question "What is important?".

✔ Be sure you know the format of the exam. Multiple choice? True false? Essay? Thought problems? What?

✔ Review your notes for material marked TQ (Test Question).

✔ Ask your instructor if exams are available from the previous semester. Review these to check the format of questions, see what areas of content are stressed, and what types of problem solving are included. Don't make the mistake of assuming the same questions will be asked this semester.

✔ Be sure to attend review sessions if these are offered. Attending a review session will raise your midterm score.

✔ If your class has tutors, preceptors, supplemental instruction leaders or other peer helpers who have taken the class, ask for their suggestions about preparing for the midterm.

✔ Once you are clear about the nature of the midterm exam, begin your review. Conduct review in an orderly systematic manner that insures focused review of all the important material. The *Eight-Day Study Plan* (Appendix: *How to Study Geology*) is a good model for orderly review.

The following *Chapter Summary* and *Practice Exam* should simplify review still further. Read the *Chapter Summary* to begin your session. It provides a helpful overview that should get you back into the material. Next, try the *Practice Exam.* Take it just as you would a midterm to see how you stand in regard to mastery of this chapter. After you answer the questions, score them. Finally, and most important of all: review any question that you missed. Identify and correct the misconception that resulted in missing the item.

CHAPTER 6 SUMMARY

- Physical weathering breaks rock into smaller pieces and chemical weathering alters and dissolves the minerals within the rock. The principal factors that influence weathering are the composition of the parent rock, topography (stability of the land surface), and climate. Typically there is a positive feedback between physical and chemical weathering where one enhances the other if conditions are favorable. A good example of this positive feedback is soil formation. As physical and chemical weathering proceeds to alter a stable surface of rock material, a soil forms. The formation of soil promotes weathering by increasing the availability of moisture and producing acidic chemical conditions. Soils also promote plant growth and other organisms which aid both physical and chemical weathering.

- How chemical weathering works is well illustrated by three examples. First, the chemical weathering of feldspars, which are the most abundant silicate mineral in the Earth's crust, illustrates how water with the help of carbonic acid can transform feldspars into clay minerals and dissolve silica and salts (cations). Second, the reaction of oxygen with the iron in ferromagnesium minerals like pyroxene illustrates oxidation. Third, the reaction of calcite and other carbonate minerals that make up limestone exemplifies the role naturally acidic water plays in dissolving rock.

- Physical weathering involves a variety of processes that break rock into fragments. Physical weathering is promoted by chemical weathering which weakens grain boundaries within the rock. Physical weathering also promotes chemical weathering by increasing the surface area of the broken rock fragments. Frost wedging, crystallization of minerals like salts, and life processes play a major role in breaking rock apart.

- Soils are a product of chemical weathering of rock that has remained in place for a period of time. Soil formation is most effected by climate. The composition of the parent rock, life processes and topography are also important factors in soil formation. The three most general types of soils are pedocals (arid climates), pedalfers (temperate climates), and laterites (tropical climates). Soils, water, and the air we breathe are the three most basic natural resources.

Man can live without gold but not without salt.
—Falvius Magnus Cassiodorus, Roman politician of the Fifth Century A.D.

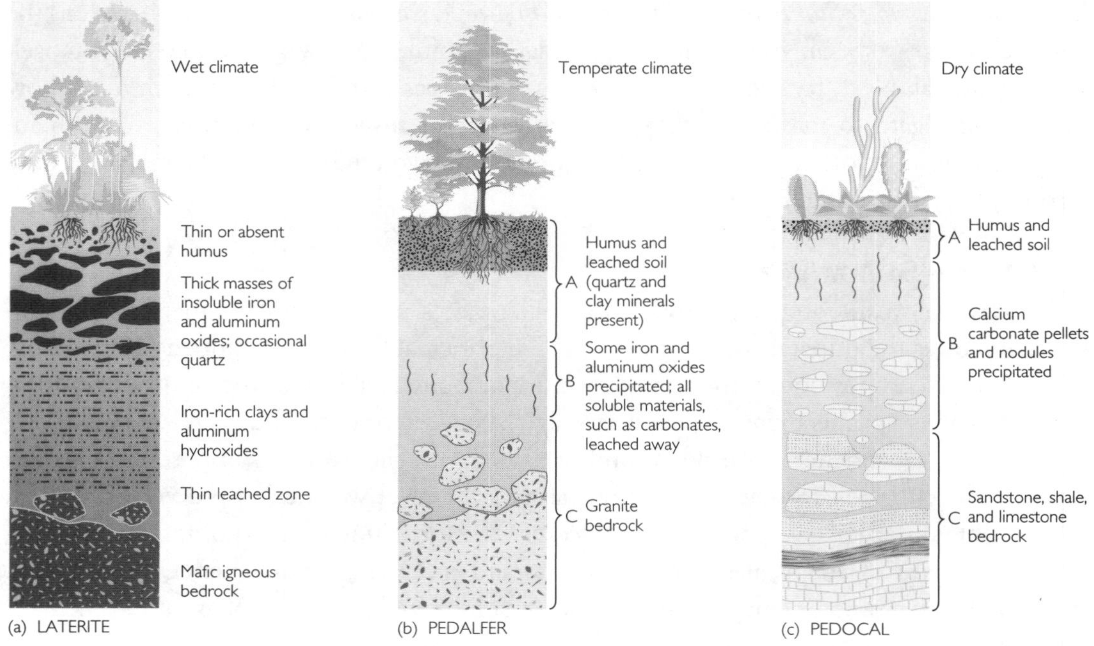

Figure 6.18 / Major soil groups. (a) A laterite soil typical of a tropical region is characterized by residual iron and aluminum oxides. (b) Pedalfer soils are characteristic of a temperate region like the eastern United States and are rich in clay minerals. (c) Pedocal soils form in arid or semiarid region like the southwestern United States and are enriched in calcium carbonate.

PRACTICE EXAM QUESTIONS FOR CHAPTER 6

(Answers and explanations are at the end of this chapter.)

Exercise 1: Physical and Chemical Weathering

Fill in the blanks in the flow chart.

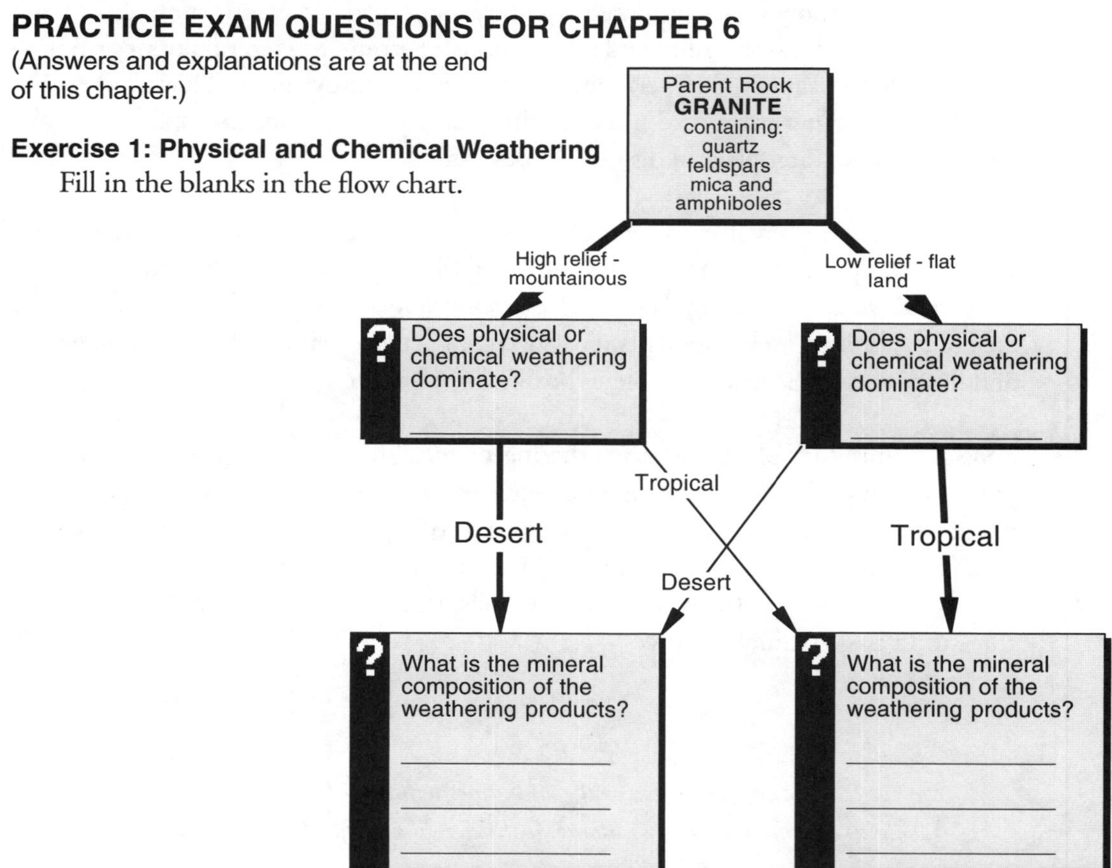

Exercise 2: Soil Types

Fill in the blanks in the table below.

CLIMATE	SOIL TYPE	SOIL CHARACTERISTICS	AGRICULTURAL POTENTIAL
Desert (warm and dry)	pedocal		
Temperature (moderate)		variable - clay rich	
Tropical (warm and wet)			Lush vegetation is supported by an organic-rich A-horizon. Crops can be grown for only a few years before nutrients are depleted. Source of bauxite, an aluminum ore.

Exercise 3: Soil Formation in Different Regions

Given the parent rock and location, briefly characterize the major soil type, e.g. pedocal, pedalfer, or laterite, and the diagnostic characteristic of each.

A. Soil on a quartz sandstone in semi-arid Kansas.

Major soil type: *loose sand / sand dunes - minimum soil development.*

Characteristic(s): *The sandstone will break apart into quartz sand grains and very little else.*

B. Soil on a granite in semi-arid Kansas.

Major soil type: ______

Characteristic(s): ______

C. Soil on a granite in warm and humid (temperate) Georgia.

Major soil type: ______

Characteristic(s): ______

D. Soil on a granite in semi-arid southern Arizona.

Major soil type: ______

Characteristic(s): ______

1. Soil production is often described as a "positive feedback process". Why?
   A. Carbon dioxide in rainwater is used up by organisms, so weathering of underlying rock is impeded.
   B. Rainwater becomes more acidic as it percolates through the soil, so weathering of underlying rock is promoted.
   C. Once a layer of soil is formed, the underlying rock is protected from further weathering.
   D. Plant growth reduces the potential for weathering and therefore of soil development.

2. Of the following minerals, the one most rapidly altered by chemical weathering would be
   A. mica (sheet silicate), such as biotite
   B. amphibole (double chain silicate), such as hornblende
   C. pyroxene (single chain silicate), such as augite
   D. isolated silica tetrahedra mineral, such as olivine

3. The climate with the fastest rate of chemical weathering has
   A. low temperatures and low rainfall
   B. high temperatures and high rainfall
   C. high temperatures and low rainfall
   D. low rates of mechanical weathering

4. The potential for chemical weathering is _______________ due to acid rain.
   A. neutralized
   B. not effected
   C. increased
   D. decreased

5. In the eastern United States, the widespread, good agricultural soil is called a
   A. pedalfer
   B. bauxite
   D. laterite
   E. pedocal

6. The generally great thickness of soils in tropical regions is due to
   A. spheroidal weathering of rock outcrop
   B. rapid chemical weathering enhanced by high rainfall and high temperature
   C. pressure release
   D. action from the large variety of bacteria

7. Which of the following processes generally result in a lateritic soil?
   A. intense compaction of pedocal soils to form brick-hard material
   B. intense chemical weathering
   C. moderate to slight weathering
   D. moderate weathering of very resistant lithologies

8. What happens to the quartz sand grains, chemically, in a calcite-cemented sandstone that is undergoing moderate chemical weathering?
   A. They combine with water.
   B. They dissolve.
   C. They oxidize.
   D. Virtually nothing – they become grains of quartz sand.

9. Rock material that is ____________________ tends to result in the most fertile soils.
   A. not weathered at all
   B. very weakly weathered
   C. moderately weathered
   D. intensely weathered

10. Which of the following naturally occurring materials could provide the most weather-resistant facing for buildings?
   A. granite
   B. halite
   C. limestone
   D. olivine basalt

Hint: Refer to Figure 6.1 and 6.5.

11. Physical and chemical weathering in the warm, wet climates of the Earth's surface will alter an exposed granite to
   A. quartz and feldspar sand
   B. olivine sand
   C. iron-rich soil
   D. quartz sand and clay

12. Even when weathering is intense, the A-horizon of a soil may take ________ of years to form.
   A. tens
   B. hundreds
   C. thousands
   D. millions

13. In the days of the Pharaohs of Egypt a cherished status symbol was the obelisk, a stone column decorated with hieroglyphs (designs carved into the stone – usually sandstone). In 1879 the obelisk of Thothmes III from the temple of Heliopolis, Egypt, was moved to Central Park in New York City. Within about 60 years the hieroglyphs were barely visible on the obelisk, while its counterpart still standing in Egypt has remained in nearly perfect condition in the desert sun for almost 4,000 years. Why did the stone obelisk deteriorate "so quickly" when it was moved to New York City?

____

____

____

ANSWERS AND EXPLANATIONS FOR EXERCISES AND QUESTIONS

After Lecture

1. D. Oxygen is the principal chemical agent for oxidation reactions. Water is the universal solvent and carbon dioxide combines with water to form carbonic acid. All three play an important role in chemical weathering. Nitrogen gas occurs as a molecule of two atoms of nitrogen. The nitrogen molecule is relatively non-reactive.
2. A. Silica in solution, iron oxides and clay minerals are all products of chemical weathering. Feldspar is a framework silicate mineral which crystallized from a magma.
3. D. The weathering products of many silicate minerals includes clay. However, quartz is pure silicon dioxide and is relatively resistant to weathering. Where chemical weathering is intense, as in the tropics, even quartz will dissolve.
4. C. Limestone and the calcium carbonate cement of the sandstone are very susceptible to dissolution reactions in wet climates, but are more resistant to weathering in dry climates. Shale is a soft sedimentary rock made from compacted clay and typically weathers rapidly. A silica cemented quartz sandstone might be as resistant to weathering as granite.
5. A. Rusty streaks represent the oxidation of iron-bearing minerals. The other answers are examples of physical weathering.
6. B. Rainfall is naturally acidic due to carbonic acid which forms when carbon dioxide from the atmosphere dissolves in water. Water passing through organic-rich soils dissolves additional carbon dioxide from the decay of organic matter and becomes more acidic.
7. A. Refer to Figure 6.5 in your textbook.
8. D. Table 6.1 shows that calcite, the major mineral in limestones is very soluble in wet conditions. Also refer to the section in your textbook entitled, *Fast Weathering: Dissolution of Carbonates.*
9. A. Refer to Figure 6.8.
10. A. Refer to *From Feldspar to Kaolinite Clay* in your textbook.
11. A. The A-horizon is defined as the zone of leaching.
12. C. Chemical weathering is most intense in hot and wet regions.
13. B. The B-horizon is defined as the zone of accumulation.
14. B. Refer to Figure 6.18 (c).
15. C. Pedocals are soils rich in calcium carbonate. Calcium carbonate is relatively soluble in water but it will accumulate in soils where evaporation rates are high like in arid and semi-arid regions. As the soil dries out, the calcium carbonate is precipitated.

Exam Prep

Exercise 1: Physical and Chemical Weathering

Refer to Figure 6.16 for another, expanded view of this flow chart.

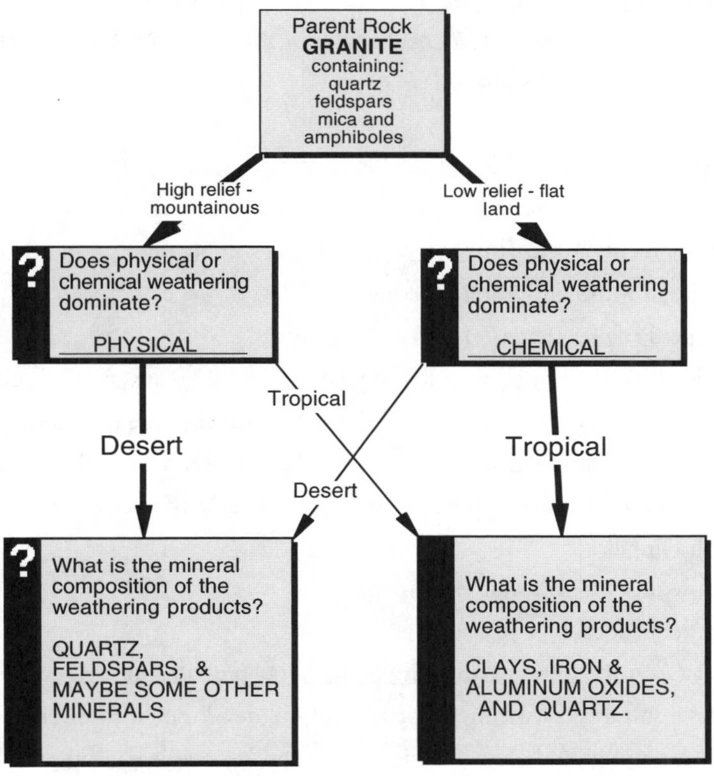

Exercise 2: Soil Types

CLIMATE	SOIL TYPE	SOIL CHARACTERISTICS	AGRICULTURAL POTENTIAL
Desert (warm and dry)	pedocal	Rich in calcium carbonate in the B-horizon.	Higher levels of salts within the soil may compromise the growth of some crops
Temperature (moderate)	pedalfer	variable - clay rich	Pedalfers make good agricultural soils.
Tropical (warm and wet)	laterite	Intensely leached soil rich in iron and aluminum oxides.	Lush vegetation is supported by an organic-rich A-horizon. Crops can be grown for only a few years before nutrients are depleted. Source of bauxite, an aluminum ore.

Exercise 3: Soil Formation in Different Regions

Given the parent rock and location, briefly characterize the major soil type, e.g. pedocal, pedalfer, or laterite and the diagnostic characteristic of each.

A. Soil on a quartz sandstone in semi-arid Kansas.

Major soil type: *loose sand / sand dunes - minimum soil development.*

Characteristic(s): *The sandstone will break apart into quartz sand grains and very little else.*

B. Soil on a granite in semi-arid Kansas.

Major soil type: Pedalfer

Characteristic(s): Clay and oxide rich; lacking accumulation of calcium carbonate. Granite contains quartz, feldspars, some mica and amphiboles. In wetter environments all these minerals but quartz will weather into clay minerals and oxides.

C. Soil on a granite in warm and humid Georgia.

Major soil type: Pedalfer (Laterite)

Characteristic(s): You might be tempted to give laterite for an answer since Georgia has a warm and wet climate. Most of the soils in Georgia are probably between a laterite and a pedalfer in characteristics. They will be clay and iron oxide rich, but tend to be nutrient poor because of the relatively intense leaching due to the high rainfall.

D. Soil on a granite in semi-arid southern Arizona.

Major soil type: Pedocal

Characteristic(s): A semi-arid to arid climate with low rainfall and high evaporation rates result in slow weathering rates and soil development. Lower rainfall results in less plant growth, less organic matter, and less acidic conditions which further slows weathering rates. Also soluble salts like calcium carbonate will tend to accumulate in the soil.

1. B. Soil gases can be rich in carbon dioxide released by the decay of organic matter and life processes. Water percolating down through organic-rich soils becomes more acidic as it dissolves the carbon dioxide in soil gas. Increased acidity of the water enhances chemical weathering reactions.
2. D. Relative to other silicate minerals, olivine is very susceptible to chemical weathering. Its simple (less polymerized) silicate crystal structure of single tetrahedra bonded with cations of iron and/or magnesium is easily attacked by chemical agents. Refer to Table 6.2.
3. B. Chemical weathering is most intense in warm and wet regions.
4. C. Pollution and gases from volcanic eruptions combine with rainfall to enhance its acidity. Before air pollution controls, rainfall in some regions of the United States was as acid as lemon juice. Acids are powerful chemical agents. Chemical weathering is greatly enhanced by increasing the acidity of water.

5. A. Pedalfers are good agricultural soils. Pedocals can also be farmed. Mineral nutrients in laterite soils are stored in organic-rich A-horizon and can quickly be depleted by farming. Bauxite is aluminum ore and lacks most mineral nutrients necessary for plant growth.
6 B. Climate is the most important factor controlling chemical weathering and soil development.
7. B. Lateritic soils form in tropical regions where chemical weathering is intense.
8. D. As the carbonate cement dissolves, the quartz sand grains will weather from the rock surface, but weathering will have little impact on the quartz grains themselves. Quartz is a very stable mineral on the Earth's surface.
9. C. Soil fertility depends on the availability of mineral nutrients which are released by chemical weathering from rock forming minerals. When chemical weathering is very slow, nutrients remain tied up in the silicate minerals within the rock and are not easily extracted by plant roots. In regions where chemical weathering is intense, most of the mineral nutrients are washed out of the soil. Therefore, fertile soils form where weathering occurs at moderate rates. The parent rock is another important influence on soil fertility. Weak soils developed on a quartz sandstone are likely to be nutrient poor because quartz is pure silicon dioxide and lacks vital mineral nutrients, like potassium, calcium, iron, magnesium, for plant growth.
10. A. Refer to Table 6.1 and 6.2.
11. D. Granite is composed mostly of feldspars and quartz. Clay minerals are a major product from the chemical weathering of feldspar. Quartz is very resistant to weathering.
12. C. Soils typically take thousands of years to develop.
13. The sandstone obelisk deteriorated so quickly after being moved from Egypt to New York City because the climate in New York is significantly wetter than in Egypt. Commonly cemented by calcium carbonate, the quartz grains in sandstone are released from the rock as naturally acidic rain waters dissolve the calcium carbonate cement. Other factors that may have contributed to the enhanced weathering of the obelisk include acid rain from air pollution and frost wedging, since freeze/thaw conditions are common in New York during the winter.

Chapter 7
Sediments and Sedimentary Rocks

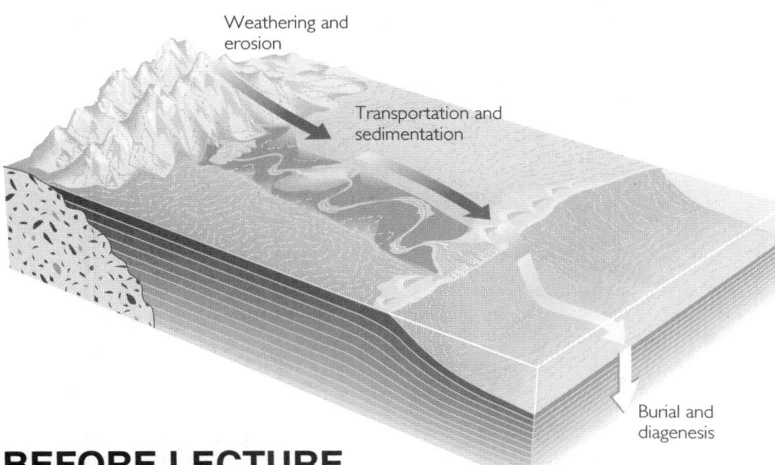

Figure 7.1
The sedimentary stages of the rock cycle embrace several overlapping processes: physical and chemical weathering, erosion, transportation and sedimentation, and burial. Burial converts the sediment into rock through diagenesis.

BEFORE LECTURE

Pre-Lecture Chapter Preview

Before you attend the lecture on Sedimentary Rocks be sure to spend some time previewing Chapter 7. Use the following questions as a guide. They constitute the basic framework for this chapter. Working with the preview questions before lecture and committing them to memory afterwards should result in your understanding the lecture much better and getting a better set of notes. As always, we start by connecting the new material on sedimentary rocks to the rock cycle.

CHAPTER PREVIEW

- **Where do sediments come from?**
  Hint: See Chapter 3, Rock Cycle.

- **How are sediments produced?**
  Hint: See Chapter 6, Weathering and Erosion.

- **What are the major processes in the formation of sedimentary rocks?**
  Brief answer: Burial, lithification and diagenesis. See Figure 7.1 (above).

- **What are the major types of sediments and sedimentary rocks?**
  Brief answer: Clastic (sandstone) vs. chemical/biochemical (evaporites and limestone).

- **What is the relationship between sediments and environments on the Earth's surface?**
  Hint: See Figures 7.5 and Tables 7.2 and 7.3.

Vital Information from other Chapters

- The Rock Cycle: Chapter 3
- Weathering and Erosion: Chapter 6

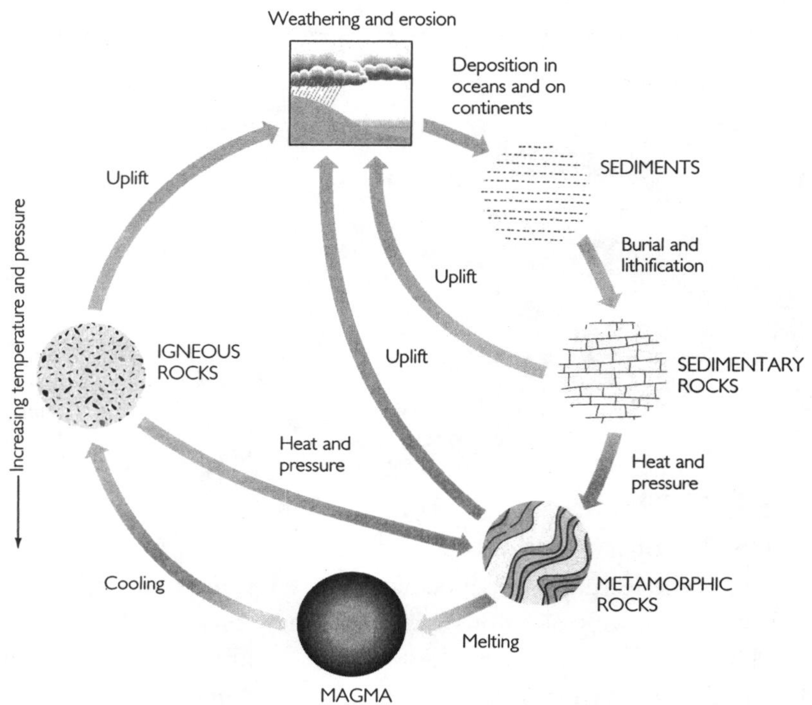

Figure 3.10 / The Rock Cycle

Website and CD Preview

http://www.whfreeman.com/presssiever

- **Common Sedimentary Environments Exercise:** At the Website do the Interactive Exercises called *Common Sedimentary Environments* and *Identifying Sedimentary Environments* for Chapter 7.

DURING LECTURE

One goal for lecture should be to leave class with a good set of answers to the preview questions. To avoid getting lost in details keep the big picture in mind: Chapter 7 tells the story of how sedimentary rocks are created by the rock cycle. Focus on Figure 7.1. It draws together the entire chapter by providing an overview of the sedimentary stages of the rock cycle. It may be helpful to have Figure 7.1 with you during class so that you can refer to it during lecture.

Your lecturer may show slides of various sedimentary environments. He or she may also bring samples of various sedimentary rocks to class. You will get more out of these if you sit close to the front row where you can see the sample rocks as your instructor discusses them. Remember to focus carefully on clues provided for recognizing a particular sample in the field. As you listen, try to formulate at least one good question you can ask during the lecture.

AFTER LECTURE

Review Notes

The perfect time to review your notes is right after lecture, while the material is still fresh in your mind. Review to be sure you got all the key points and wrote them down in a form that will be readable later.

NOTE REVIEW CHECKLIST

- ✔ All notes legible? (Rewrite so they read easily.)
- ✔ Important points clearly identified? You should now have headers in your notes that tie to each of the questions in the *BEFORE LECTURE/Chapter Preview.*
- ✔ Holes (missing material) filled in from memory?
- ✔ Areas where you don't remember what was said marked for a follow-up session with your instructor, tutor, or study partner?
- ✔ Possible test questions indicated in the margin (TQ)?
- ✔ Additional visual material. *Suggestions for Chapter 7: Be sure to sketch in Figure 7.1. It is the key figure for this chapter. Visual and kinesthetic learners may find it helpful to add sketches of some of the key sedimentary environments such as bedded sandstone (see Figure 7.6).*
- ✔ Reworked notes into a form that is efficient for your learning style?
- ✔ Created a brief "big picture" overview of this lecture (using a sketch or written outline)? *Suggestions for Chapter 7: Figure 7.1 provides a pretty good overview sketch for this chapter. What could you add to the sketch to make it even more useful as a learning device?*

Intensive Study Session

You learn geology much as you would build a house. The first step is to construct the frame. Without a frame you would have nothing to attach the siding to. In geology, each chapter is the frame to which the next chapter will be attached. You must thoroughly master the concepts of each chapter because the next chapter will be based on that framework.

Schedule at least one hour after each lecture to master the key concepts. Mastery is not gained by just reading your text. Mastery occurs as the result of asking yourself questions (and answering them). The following Website and CD activities are designed to help you reach mastery level quickly.

Website and CD Activities and Tools

http://www.whfreeman.com/presssiever

- **Website Q & A Practice Multiple Choice Questions:**
  At the Website complete the *Q & A*. Pay particular attention to the explanations for answers. Flashcards at the Website will help you learn new terms.

PRACTICE EXERCISES AND STUDY QUESTIONS

(Answers and explanations are at the end of this chapter.)

Exercise 1: Common Sedimentary Environments

Hint: Practice the Website Interactive Exercise: *Common Sedimentary Environments*.

http://www.whfreeman.com/presssiever

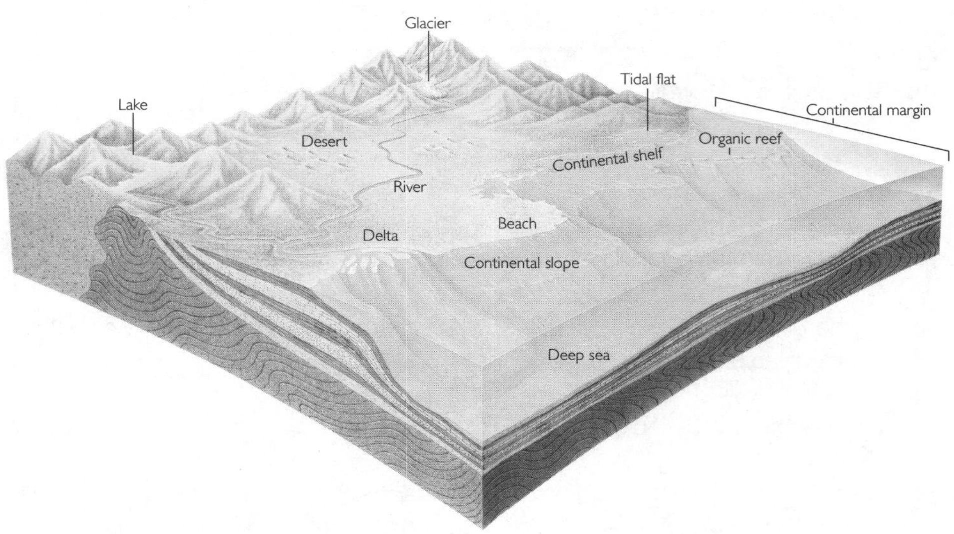

Figure 7.5 / Common sedimentary environments.

Using the above figure as a guide, fill in the blanks in the table below with the clastic or chemical sediment (e.g., sand, silt, mud, salts, carbonate, peat) that best matches the environment of deposition.

Environment of Deposition	Name(s) of sediment(s) typically deposited in each environment:
alpine or glacial river channel	*silt, sand and gravel*
dunes in a desert	
flood plain along a broad river bend	
river delta along a marine shoreline	
continental shelf	
deep sea adjacent to a continental shelf	
shoreline beach dunes	
tidal flats	
organic reef	

Exercise 2: Grain-Sizes in Sedimentary Rocks

Using the terms for *Sediments and Rocks* listed in Table 7.4 fill in the blanks with the appropriate name of the sediment and rock type that matches the typical particle size of the *Common Object* listed in the table below.

Grain Size	Common Object	Sediment	Rock Type
Coarse ↑	football to bus	*boulder*	
	plum or lime		*conglomerate*
	pea or bean		
	course ground pepper or salt		*sandstone*
	fine ground pepper or salt		
↓ Fine	talcum powder or baby powder		

Exercise 3: Clastic and Chemical Sediments and Sedimentary Rocks

Given the description in the *Statement* column, fill in the blanks with the *Sediment Type* and *Sedimentary Rock* in the table below. Use the following terms for *Sediment Type* and *Sedimentary Rock*:

Sediment Type
biochemical
chemical
clastic

Sedimentary Rock
chert
conglomerate
dolostone
evaporite
limestone
peat
phosphorite
sandstone
shale
siltstone

Statement	Sediment Type	Sedimentary Rock
composed largely of rock fragments		
precipitated in the environment of deposition		
important source of coal		
often formed by diagenesis	*chemical*	*dolostone or phosphorite*
formed from abundant skeleton fragments of marine or lake organisms, such as coral, sea shells, foraminifers		
produced by physical weathering		

EXAM PREP

If you have used regular intensive study sessions to master each chapter, now you get a pay back! Review will proceed smoothly and take far less time.

The following *Chapter Summary* and *Practice Exam* will simplify review. Read the *Chapter Summary* to begin your review session. It provides a helpful overview or "big picture" for sedimentary rocks that should get you back into the material. Next, try the *Practice Exam Questions*. Take it just like you would the midterm and then score it to determine how you stand in regard to mastery of this chapter. Most importantly, review any items you missed. Identify and correct any misconceptions you discover. For efficient review of all chapters covered on the midterm use the *Eight-Day Study Plan* (see Appendix).

CHAPTER 7 SUMMARY

- Weathering and erosion produce the clastic particles and dissolved ions that compose sediment. Water, wind and ice transport the sediment to where it is deposited. Burial, lithification and diagenesis harden sediments into sedimentary rocks.
- The two major types of sediments are clastic and chemical/biochemical. Clastic sediments are formed from rock particles and mineral fragments. Chemical and biochemical sediments originate from the ions dissolved in water. Chemical and biochemical reactions precipitate these dissolved ions from solution.
- The names of clastic sediments and sedimentary rocks are based primarily on the size of the grains within the rock. The names of chemical and biochemical sediments and sedimentary rocks are based primarily on their composition.

Website and CD Activities and Tools

http://www.whfreeman.com/presssiever

More detailed characteristics of metamorphic rocks are summarized at the *Sedimentary rock properties data bank*, an Interactive Geology Tool at the website. The *Photo Gallery* for Chapter 7 at the website and on your CD contains a selection of images of various metamorphic rocks.

PRACTICE EXAM QUESTIONS FOR CHAPTER 7

(Answers and explanations are at the end of this chapter.)

1. Which of the following rock groups includes only clastic sedimentary rocks?
   A. dolomite, gypsum, and limestone
   B. cherts, sandstone, and shale
   C. dolomite, coal, and limestone
   D. shale, sandstone, and conglomerate

2. Which sequence of rock names has the correct arrangement in order of decreasing particle diameters?
   A. conglomerate, shale, sandstone
   B. shale, siltstone, sandstone
   C. sedimentary breccia, shale, sandstone, claystone
   D. conglomerate, sandstone, claystone

3. The grains in a sandstone may include
   A. rock fragments
   B. quartz
   C. feldspar
   D. mica
   E. all of the above

4. Sedimentary rocks are produced through the following sequence of events
   A. erosion, weathering, transportation, deposition, burial and diagenesis
   B. weathering, erosion, transportation, burial, diagenesis, and deposition
   C. erosion, weathering, deposition, transportation, and cementation
   D. weathering, erosion, transportation, deposition, burial and diagenesis

TEST TAKING TIP

When answering multiple choice questions like #4, treat each alternative as a true/false question. Rule out any alternative that is false. For example, A and C are false because erosion must follow weathering. B is false because deposition must precede burial and diagenesis. Therefore, D must be the correct answer.

5. Dolomite is the primary mineral found in dolostone. It is formed by
   A. foraminifera extracting minerals from seawater
   B. diagenetic alteration of calcite
   C. direct precipitation in lake water
   D. coral reef exposure to direct sunlight

6. What are the two most important diagenetic processes that transform loose sediments into hard sedimentary rocks?
   A. compaction, cementation
   B. transportation, burial
   C. erosion, transportation
   D. uplift, erosion

7. What are two chemical or biochemical sedimentary rocks formed by diagenetic process?
   A. dolostone, phosphorite
   B. limestone, chert
   C. sandstone, limestone
   D. siltstone, graywacke

8. Where do most sediments get deposited?
   A. on the continental shelf and adjacent ocean floor
   B. in lakes
   C. along streams
   D. in deserts
   E. in mountains

9. In which of the following would you least expect to find cross-bedding?
   A. tidal flat sediments with abundant ripple marks
   B. sand dunes
   C. river sand in a river channel
   D. evaporites

10. The most widespread environment of deposition for carbonates in the world today is
    A. the deep sea, e.g. the Arctic Ocean
    B. the tidal flat environment, e.g. the Mississippi Delta
    C. in caves and caverns
    D. warm, shallow-water marine environment, e.g. Florida Keys

11. A sandstone made of pure quartz grains is likely to be derived primarily from a source area with significant outcrops of
    A. granite
    B. limestone
    C. sandstone
    D. basalt

12. Chemical weathering is most dominant in
    A. warm, dry climates
    B. warm, wet climates
    C. cool, dry climates
    D. cool, wet climates

STUDY TIP
Review
Chapter 6

13. The coast of Alaska is known for its high mountainous relief and active volcanoes and glaciers that reach to the sea. What types of sediments would you expect to be commonly deposited off shore from this landscape?
    A. arkose
    B. carbonate sand
    C. quartz sand
    D. graywacke

14. When a granite is subject to intense chemical weathering, it most likely will result in a sediment composed of
    A. feldspar and clay
    B. quartz and calcium carbonate
    C. quartz and clay
    D. quartz and feldspar

HINT
See
Table 7.1

15. Given that feldspar is the most abundant silicate mineral in the crust of the Earth, the most common sedimentary rock is
    A. sandstone
    B. conglomerate
    C. limestone
    D. shale

ANSWERS AND EXPLANATIONS FOR EXERCISES AND QUESTIONS

Exercise 1: Common Sedimentary Environments

alpine or glacial river channel / *silt, sand and gravel*

dunes in a desert / sand

flood plain along a broad river bend / silt and clay

river delta / mud

continental shelf / sand, silt, mud, carbonate

deep sea adjacent to a continental shelf / fine sand, silt, mud

shoreline beach dunes / sand

tidal flats / sand, silt, mud, carbonate

organic reef / carbonate

Exercise 2: Grain-Sizes in Sedimentary Rocks

boulder / conglomerate

cobble or gravel / *conglomerate*

pebble / conglomerate

sand / *sandstone*

silt / siltstone

clay or mud / claystone, mudstone or shale

Exercise 3: Clastic and Chemical Sediments and Sedimentary Rocks

clastic / conglomerate

chemical or biochemical / evaporite or limestone

biochemical / peat

chemical / dolostone or phosphorite

biochemical / limestone and chert

clastic / conglomerate, sandstone, siltstone or shale

1. D. Shale, sandstone and conglomerate are clastic rocks. Dolomite, chert, limestone, coal and gypsum are chemical/biochemical.
2. D. (see Table 7.4)
3. E. Sand is a term that refers solely to a particular range of grain sizes. It does not imply a specific composition. Quartz and feldspar are common constituents of sand grains because they are common minerals in the Earth's continental crust. Carbonates like limestone, dolostone, and fragments of seashells can also make up sand grains.
4. D. (see Figure 7.1)
5. B. There is no known marine life that precipitates dolomite. Dolomite is precipitated inorganically in seawater, not fresh water.

6. A.
7. A.
8. A.
9. E. Cross bedding is a feature produced as sediment is deposited by currents of air or water. Since evaporites are formed by the precipitation of sediment, dissolved in water, cross-bedding does not form in evaporites.
10. D. Carbonate sediments are deposited in warm water. Carbonates dissolve in cold water because their solubility is linked to the amount of carbon dioxide dissolved in the water and the solubility of carbon dioxide increases as the temperature of the water decreases. Therefore, carbonate sediments are not deposited in the cold ocean waters of the Arctic and Antarctic oceans. Nor are carbonate sediments deposited on the deep ocean floor. Some carbonate sediments may be deposited in deltaic environments depending on water temperatures. However, clastic sediments are a more characteristic type of sediment deposited at the mouths of rivers. The beautiful formations in caves are made from carbonate minerals, but the amount of carbonate precipitated as cave formations is insignificant compared to carbonate deposition in warm, shallow marine environments, like the Florida Keys.
11. C. One might be inclined to answer 'A', based on Table 7.1. Granitic and gneissic rocks are the ultimate source of quartz grains. However, pure quartz sands usually require more than one cycle of weathering/erosion/deposition/lithification to form, so that all less stable minerals and rock fragments are winnowed away.
12. B. (see Table 6.1)
13. D. (see Figure 7.16)
14. C. (see Table 7.1)
15. D. (see Table 7.1) The weathering of silicate minerals, except quartz, typically produces clay. Review *From Feldspar to Clay* in Chapter 6.

Chapter 8
Metamorphic Rocks

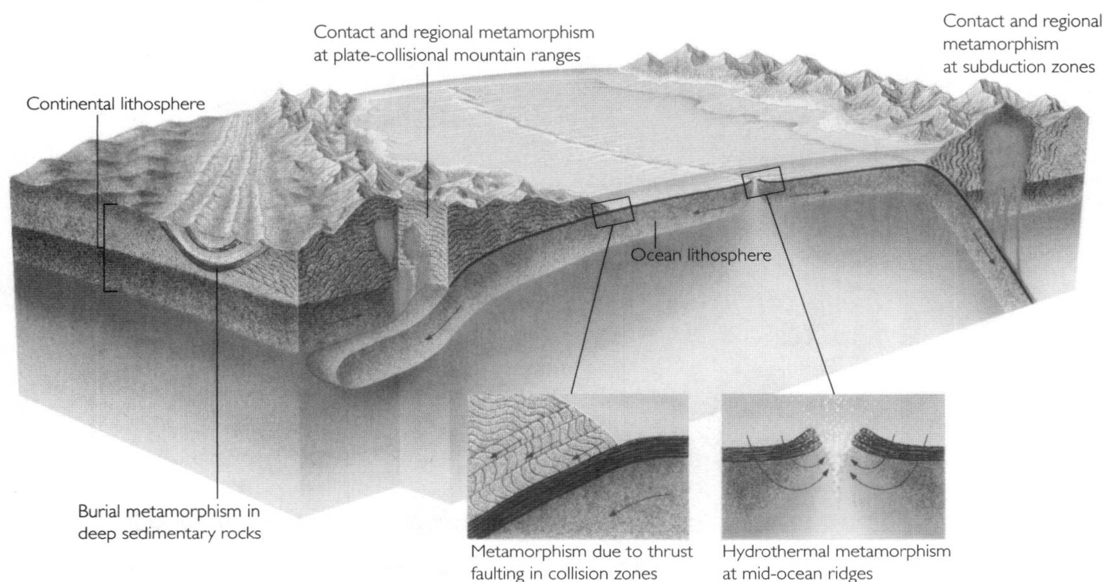

Figure 8.3 / Metamorphic rocks are formed in four main plate-tectonic settings: subduction zones, continental collisions, mid-oceanic ridges, and deeply subsiding regions on continents. Contact metamorphism associated with more localized magmatic and volcanic processes is not emphasized in this figure.

BEFORE LECTURE

Before you attend lecture be sure to spend some time previewing the chapter. For an efficient preview use the questions below.

CHAPTER PREVIEW

- **What causes metamorphism?**
  Brief answer: Metamorphism — alteration of preexisting rocks in the solid state — is caused by increases in pressure and temperature and by reaction with chemical components introduced by migrating fluids.

- **What are the various kinds of metamorphism?**
  Brief answer: Regional and contact metamorphism are the most common.

- **What are the chief types of metamorphic rocks?**
  Brief answer: Metamorphic rocks fall into two major textural classes called foliated (minerals oriented in some preferred direction, like the grain in wood) and nonfoliated (no preferred mineral orientation).

- **How is metamorphism linked to plate tectonics?**
  Hint: Refer to Figure 8.3 above and in the textbook.

Vital Information from other Chapters

- Review the following sections in Chapter 3: *Metamorphic Rocks* and *The Rock Cycle as a Subsystem of the Earth System*. These sections are very short and well worth your review.

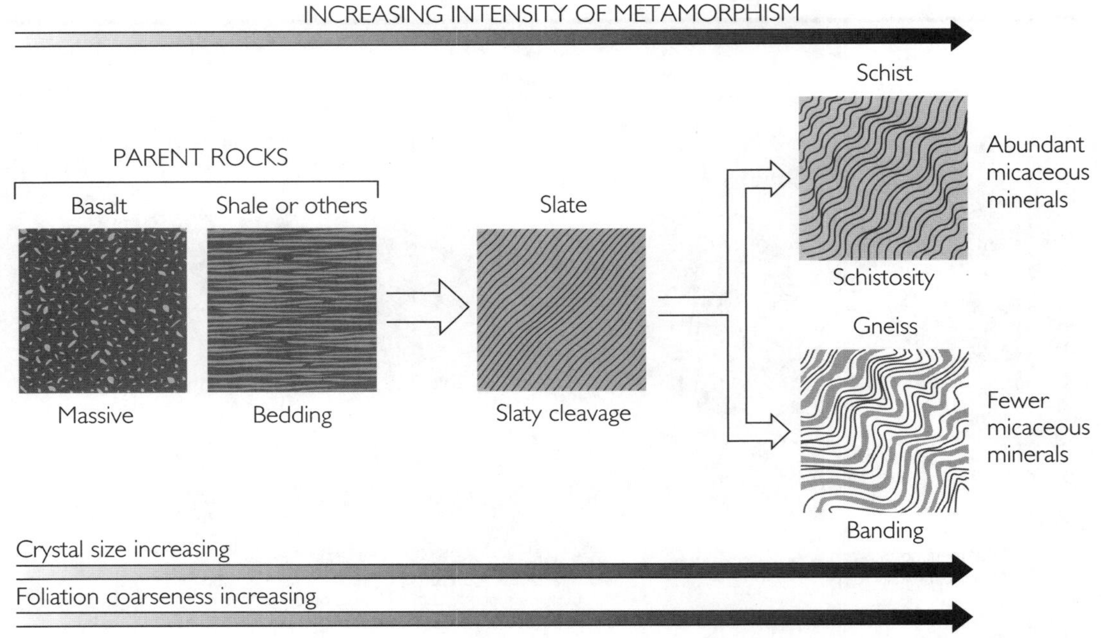

Figure 8.7 / Classification of foliated rocks. Cleavage, schistosity, and banding do not generally correspond to the original bedding direction of sedimentary rocks.

DURING LECTURE

Keeping up with a fast-speaking lecturer can be quite a challenge.

- Take as many notes as you can.
- Don't stop writing if you want to ponder a concept (you can do that later) or when you get confused.
- If you miss something, leave a space so you can fill it in.

Abbreviations save time, allowing you to get more of the important material into your notes. Here are some abbreviations that you could use for this chapter:

- *meta rk* or *met rx* (metamorphic rock)
- *sedi rk* or *sed rx* (sedimentary rock)
- *ign rk* or *ign rx* (igneous rock)
- *fol* (foliated rock)
- *ufol* (unfoliated rock)
- *gran* (granular)
- *sl* (slate)
- *shl* (shale)
- *shst* (shist)

Try these as you take notes. Feel free to develop abbreviations of your own. The important thing is to develop a short-hand that is meaningful to you, while being rapid and easy to use and remember.

AFTER LECTURE

Review Notes Checklist

Review your notes right after lecture while the material is fresh in your mind.

NOTE REVIEW CHECKLIST

✔ All notes legible? (Rewrite so they read easily.)

✔ Important points clearly identified? You should now have headers in your notes that tie to each of the questions in the *BEFORE LECTURE/Chapter Preview.*

✔ Holes (missing material) filled in from memory?

✔ Areas where you don't remember what was said marked for a follow-up session with your instructor, tutor, or study partner?

✔ Possible test questions indicated in the margin (TQ)?

✔ Additional visual material. *Suggestion for Chapter 8: Metamorphic rock is classified in part by texture. Develop and draw simple sketches of textures (slaty cleavage, phyllite, schist, gneiss) to help you remember the grades of metamorphism. Close study of Figures 8.7 and 8.8 will help you see how to do this.*

✔ Reworked notes into a form that is efficient for your learning style? Are you a kinesthetic learner? If so you may find it helpful to rewrite your notes as questions that you will answer as you review.

✔ Created a brief "big picture" overview of this lecture (using a sketch or written outline)? *Suggestion for Chapter 8: Figure 8.3 is key to understanding the tectonic settings that drive metamorphism. Sketch a simplified version that clearly shows the four tectonic settings. Write a caption for this figure in your own words.*

Intensive Study Session

As always, schedule an hour or two after the lecture on *Metamorphic Rocks* for intensive study and mastery of the key ideas. The *Practice Exercises* and *Study Questions* are designed to help you get to mastery level quickly.

Website and CD Activities and Tools

http://www.whfreeman.com/presssiever

- **Website Q & A Practice Multiple Choice Questions:**
  At the Website complete the *Q & A.* Pay particular attention to the explanations for answers. Flashcards at the Website will help you learn new terms.

The important thing is not to stop questioning.
—Albert Einstein

PRACTICE EXERCISES AND STUDY QUESTIONS

(Answers and explanations are at the end of this chapter.)

Exercise 1: Classification of Metamorphic Rocks Based on Texture

Complete the table below by filling in the blank spaces.

Hint: Refer to *Metamorphic Textures* section of your textbook and Table 8.1.

PARENT ROCK	METAMORPHIC ROCK	TEXTURE (foliated / nonfoliated)
Shale		foliated
Quartz-rich sandstone		
Granite		
Limestone		
	greenstones or amphibolite	

STUDY TIP: PUTTING IT ALL TOGETHER

Now you have been inroduced to all three major rock types: igneous, sedimentary and metamorphic. This is a good time to assemble what you have learned into a comparison chart. Doing so is an excellent way to insure you remember details about these rock types. *Exercise 2* will help you do it.

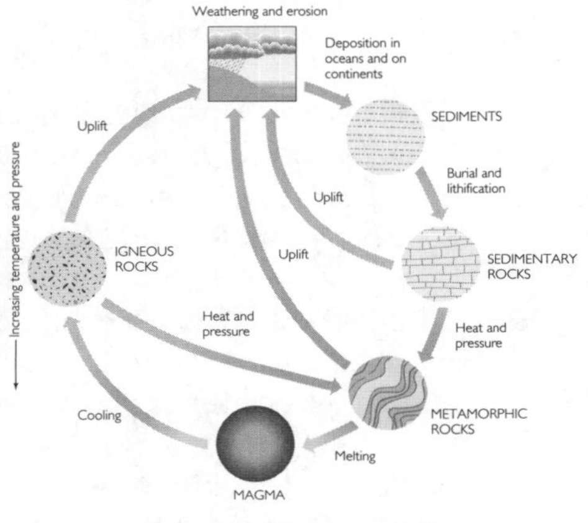

Figure 3.10 / The Rock Cycle

Exercise 2: Comparing Igneous, Sedimentary, and Metamorphic Rocks

Complete the table below by filling in the blank spaces.

Major Mineral Composition	Texture	Rock Type igneous,sedimentary, metamorphic	Rock Name e.g., granite, sandstone, marble
calcium carbonate	nonfoliated		
quartz, K and Na feldspar, mica, and amphibole	phaneritic		
clay	fine-grained clastic		
pyroxene, calcium feldspar, and olivine			basalt
quartz	nonfoliated		
pebbles and cobbles of a variety of rock types			
fragments of seashells and fine mud		sedimentary	
quartz, muscovite, chlorite, garnet			schist

1. When existing rocks undergo metamorphism, they become changed by
   A. the weathering process at or near the surface
   B. color and hardness
   C. melting and crystallization from that melt
   D. the application of heat, pressure and shearing

2. Mechanical deformation and ______________________ are the principal mechanisms responsible for changes that take place during metamorphism.
   A. chemical deprivation
   B. recrystallization
   C. crushing and grinding
   D. solidification from a melt

3. Generally there are two types of metamorphism:
   A. deep and shallow metamorphism
   B. foliated and non-foliated metamorphism
   C. regional and contact metamorphism
   D. mechanical and chemical metamorphism

4. Foliation typically does not occur in contact metamorphism because
   A. contact metamorphism does not produce minerals, such as the micas, which are needed for foliation to occur
   B. the rock is usually only heated, not squeezed and deformed, while it is being metamorphosed
   C. the parent rock is usually not the correct type to produce foliation
   D. the pressure is usually too high to allow foliation to occur

5. Foliation occurs due to
   A. confining pressure
   B. directed pressure
   C. high water content
   D. high temperature

EXAM PREP

Use the materials in this section to help you prepare for your midterm or final. First, the *Eight-Day Study Plan* tells you how to conduct a systematic and thorough midterm review. *The Chapter Summary* and *Practice Exam Questions* for Chapter 8 are review tools that will make the process more efficient.

Eight-Day Study Plan

Adapted with permission from the University Learning Center, University of Arizona.
(Make several copies of this Study Plan and use one for every exam you take.)

Here is a guide you can use to prepare for your midterm. Everyone develops their own approach to preparing for exams, so feel free to adapt these ideas to your particular needs and situation. The basic idea is to conduct your preparation in a systematic fashion with focus on the most important material. Our plan accomplishes this by dividing the material equally and suggesting how to incorporate the *Exam Prep* materials provided in this study guide for each text chapter. You begin the plan eight days prior to the exam.

Day 8: Get organized!
Step 1: Clarify the task. Determine what sort of midterm you will take by briefly answering the following questions:

1. This exam will cover (list each chapter to be covered):

 ____

 ____

2. Material and kinds of skills to be particularly emphasized (list chapters/ideas/skills your instructor said would be particularly important):

 ____

 ____

3. The test-question format(s) will be (check all that apply):

 ___ Multiple Choice
 ___ True / False
 ___ Essay
 ___ Thought Problems
 ___ Other (specify) ____________

3. Review session is scheduled for (date) ____________ . Be sure to attend.

Step 2 Divide the material you must review into four equal parts: A, B, C, D.

Day 7 Begin your review. Review all material in Part A.

DO THE FOLLOWING FOR EACH CHAPTER IN PART A:

1. **Chapter Summary:** To get yourself started, read the *Chapter Summary* (*Exam Prep* section of this guide) for the chapter you want to review.
2. **Practice Exam Questions:** Answer *Practice Exam Questions* (*Exam Prep* section of this guide) to see where you are with the material. Force yourself to answer all questions for the chapter without referring to the answer key. Correct only after you have tried all items. Be sure to review carefully any items you missed. Correct the misconception that resulted in error.
3. **Class Notes:** Review your class notes and annotations you made in the text margin by asking yourself questions.
4. **Focus on Visual Materials and Key Figures:** This may also be a good time to re-do some of the *Practice Exercises* in this study guide. Many *Practice Exercises* are designed to help you master the visual concepts of geology. Review visual material in your notes. Test yourself by seeing if you can reconstruct key figures from memory.
5. **Self Test:** Spend as high a proportion of your study time as possible asking yourself questions.

Day 6: Review Part B. Repeat instructions for Day 7, this time reviewing Part B. If you have problems with material see your instructor at the next open office hour.

Day 5: Part C. Repeat instructions for Day 7, this time reviewing Part C. If you have problems with material see your instructor at the next open office hour.

Day 4: Part D. Repeat instructions for Day 7, this time reviewing Part D. If you have problems with material see your instructor at the next open office hour.

Day 3: Review all parts – A, B, C, D – fully. Prioritize your time. Focus on important material that will be covered. Work hardest where you are least sure of your self. If you have problems with material see your instructor at the next open office hour.

Day 2: Review all parts – A, B, C, D – fully. Prioritize your time. If you have problems with material see your instructor at the next open office hour.

Night Before: Be sure you get the amount of sleep you need to be alert and perform at your best. You don't need to cram. Just stay focused.

Zero Hour: You have prepared well. Allow yourself be confident. Stay focused and confident during the exam. Use your best test-taking strategies.

CHAPTER 8 SUMMARY

- Metamorphism is the alteration in the solid state of preexisting rocks, including older metamorphic rocks. Increases in temperature and pressure and reactions with chemical-bearing fluids cause metamorphism. Metamorphism typically involves a rearrangement (recrystallization) of the chemical components within the parent rock. Rearrangement of components within minerals is facilitated by: higher temperatures which increase ion mobility within the solid state; higher confining pressure compacts the rock; directed pressure associated with tectonic activity can cause the rock to shear (smear) which orients mineral grains and generates a foliation; and chemical reactions with migrating fluids may remove or add materials and induce the growth of new minerals.
- The two major types of metamorphism are regional metamorphism, associated with orogenic processes that build mountains, and contact metamorphism, caused by the heat from an intruding body of magma. Other less common kinds of metamorphism are: cataclastic metamorphism which typically occurs along faults where rock is crushed and altered; hydrothermal metamorphism associated with the circulation of hot fluids like the hot water beneath Yellowstone; and burial metamorphism, associated with subsiding regions on continents (refer to Figure 8.3).
- Metamorphic rocks fall into two major textural classes: the foliated (displaying a preferred orientation of minerals, analogous to the grain within wood) and the nonfoliated. The composition of the parent rock and the grade of metamorphism are the most important factors controlling the mineralogy of the metamorphic rock. Metamorphism usually causes little to no change in the bulk composition of the rock. The kinds of minerals and their orientation do change. Mineral assemblages within metamorphic rocks are used by geoscientists as a guide to the original composition of the parent rock and the conditions during metamorphism.
- Metamorphic rocks are formed in four main plate-tectonic settings: subduction; continental collisions; mid-oceanic ridges; and deeply subsiding regions on the continents. Figure 8.16 summarizes the metamorphic conditions for each of these plate-tectonic settings.

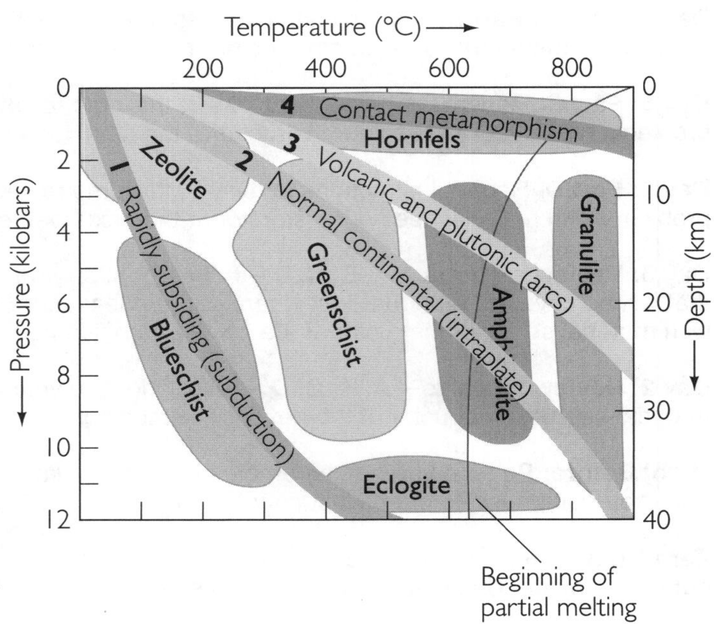

Figure 8.16
The routes taken by increases in pressure and temperature in various tectonic settings superimposed on metamorphic facies. Curve 1 shows rapidly increasing pressure with a slow rise in temperature, characteristic of the forearcs of subduction zones. Curve 2 is the path followed in continental regions away from plate boundaries, places where burial metamorphism occurs. Curve 3 is the higher-temperature route followed in regions of volcanic arcs, typical of regional metamorphism. Curve 4 is the high-temperature and low-pressure path of contact metamorphism at shallow and moderate depths.

Website and CD Activities and Tools
http://www.whfreeman.com/presssiever
The Interactive Exercise *Metamorphism and Plate Tectonics* at the Website reinforces the link between metamorphism and plate tectonics. More detailed characteristics of metamorphic rocks are summarized at the *Metamorphic Rock Properties Data Bank*, an Interactive Geology Tool at the Website. The *Photo Gallery* for Chapter 8 at the Website and on your CD contains a selection of images of various metamorphic rocks.

PRACTICE EXAM QUESTIONS FOR CHAPTER 8

(Answers and explanations are at the end of this chapter.)

1. Metamorphic rocks are distinguishable from igneous and sedimentary rocks because their constituent grains often
   A. interlock, forming a continuous mosaic
   B. have quite different chemical compositions
   C. tend to be lined up in a particular direction or plane
   D. are rounded and cemented together

2. Chemical compositions of metamorphic rocks are determined by the
   A. pressures to which they have been subjected
   B. effects of both temperature and pressure
   C. temperature to which they have been raised
   D. composition of the original igneous or sedimentary rocks

3. Metamorphism affects
   A. only older igneous rock
   B. any younger igneous and metamorphic rock
   C. only older sedimentary rock
   D. any older igneous, sedimentary, or metamorphic rocks

4. Which metamorphic sequence correctly shows increasing grain size?
   A. schist - gneiss - phyllite - slate
   B. slate - phyllite - schist - gneiss
   C. gneiss - phyllite - slate - schist
   D. phyllite - slate - gneiss - schist

5. Starting with the lowest temperature zone, which series of index minerals show the correct order of metamorphic zones?
   A. chlorite, biotite, garnet, sillimanite
   B. garnet, chlorite, biotite, kyanite
   C. biotite, garnet, chlorite, sillimanite
   D. chlorite, biotite, sillimanite, kyanite, garnet

6. On the Moon, the major cause of metamorphism is
   A. thermal metamorphism
   B. contact metamorphism
   C. meteor impacts
   D. cold temperatures

7. Of the metamorphic rocks listed below, which one is formed at the highest temperature?
   A. slate
   B. marble
   C. schist
   D. eclogite

8. The difference between a gneiss and a granite is that the gneiss mainly
   A. has a different bulk chemical composition
   B. shows a distinct foliation
   C. has a different mineral composition
   D. is generally less coarse-grained

9. Schist and slate are distinguishable in that
   A. schist is fine grained, whereas slate is coarse grained
   B. schist is foliated, whereas slate is not foliated
   C. slate is fine grained, whereas schist is coarse grained
   D. slate is foliated, whereas schist is not foliated

10. What kind of rock is mylonite?
   A. an intensely sheared metamorphic rock
   B. a hydrothermal metamorphic rock
   C. a metamorphosed limestone
   D. a contact-metamorphosed shale

11. If gneiss or another metamorphic rock is heated to a degree that it begins to melt,
    A. the quartz, K-feldspar, and Na-rich plagioclase would start to melt first
    B. Na-Ca plagioclase, biotite, and minerals like garnet would melt first, leaving a residue rich in felsic minerals
    C. all the minerals in the rock would start to melt at essentially the same temperature to form a magma of the same composition of the gneiss
    D. the ferromagnesian minerals would start to melt first

Hint: Refer to the Bowen's Reaction Series in Figure 4.11.

12. As magma intrudes into a host or country rock (that pre-existing rock in contact with the intrusion) it forms a new rock. What do we call this process?
    A. regional metamorphism
    B. contact (thermal) metamorphism
    C. recrystallization
    D. schistosity formation

13. High pressures and low temperatures result in regional metamorphism, forming a rock called
    A. marble
    B. kyanite
    C. blue-schist
    D. sillimanite

Hint: Refer to the photomicrographs of schists in the *Photo Gallery* on your CD.

14. Regional metamorphism is found in association with
    A. lava flows
    B. hot springs
    C. very low pressures
    D. subduction zones and cores of mountain ranges

15. Confining pressures that are extremely high will form rocks with
    A. marble
    B. foliation
    C. veins
    D. denser minerals

ANSWERS AND EXPLANATIONS FOR EXERCISES AND QUESTIONS

After Lecture

Exercise 1: Classification of Metamorphic Rocks Based on Texture

PARENT ROCK	METAMORPHIC ROCK	TEXTURE (foliated / nonfoliated)
Shale	Slate or argillate	foliated
Quartz-rich sandstone	Quartzite	nonfoliated
Granite	Gneiss	foliated
Limestone	Marble	nonfoliated
Basalt (mafic lavas and ash)	greenstones or amphibolite	nonfoliated

Exercise 2: Comparing Igneous, Sedimentary, and Metamorphic Rocks

Major Mineral Composition	**Texture**	**Rock Type** igneous,sedimentary, metamorphic	**Rock Name** e.g., granite, sandstone, marble
calcium carbonate	nonfoliated	metamorphic	marble
quartz, K and Na feldspar, mica, and amphibole	phaneritic	igneous (plutonic)	granite
clay	fine-grained clastic	sedimentary	mudstone, shale
pyroxene, calcium feldspar, and olivine	aphanitic porphyritic	igneous (volcanic)	basalt
quartz	nonfoliated	metamorphic	quartzite
pebbles and cobbles of a variety of rock types	clastic	sedimentary	conglomerate
fragments of seashells and fine mud	bioclastic biochemical	sedimentary	limestone
quartz, muscovite, chlorite, garnet	foliated	metamorphic	schist

1. D. If the rock is melted, it is an igneous rock. Some metamorphic rocks get hot enough to "sweat" quartz like migamtite.
2. B. Mechanical deformation, like along a fault zone, and recrystallization due to higher temperatures are typical causes of metamorphism.
3. C. See the textbook section entitled, *Kinds of Metamorphism.*
4. B. Contact metamorphism occurs in the adjacent country rock that is in direct contact with the intruding hot magma body.
5. B. The preferred orientation of minerals in foliated metamorphic rocks is due to shearing caused by directed pressure—typically associated with tectonic forces.

Exam Prep

1. C. Not all metamorphic rocks exhibit an alignment of mineral grains, but foliation is a distinctive characteristic of metamorphic rocks.
2. D. Although mineralogy may be altered by metamorphism, typically there is little to no change in bulk composition of the rock.
3. D. Even preexisting metamorphic rocks can be metamorphosed.
4. B. Slate - phyllite - schist - gneiss is the correct sequence from fine to progressively coarser-grained metamorphic rocks. All rocks listed in this sequence are foliated.
5. A. Refer to Figure 8.11 in your textbook.
6. C. Meteorite impacts of the lunar surface are an example of dynamic metamorphism.

7. D. Refer to Figure 8.16.
8. B. Gneiss has the same mineralogical composition as a granite. Its distinctive characteristic is foliation. Granite is a common parent rock for gneiss.
9. C. Both slate and gneiss are foliated metamorphic rocks. Being lower in grade, slate is much finer grained.
10. A. Mylonite is an intensely sheared (smeared) rock. Even though it became soft enough to flow (shear), it did not melt.
11. A. Refer to Bowen's Reaction Series, Figure 4.11 in Chapter 4.
12. B. The intrusion of a hot magma body causes contact metamorphism.
13. C. Refer to Figure 8.16.
14. D. Refer to Figure 8.3.
15. D. Under higher confining pressures the crystal structure of minerals collapses into a denser form which results in a new mineral. A classic example of this is the transformation of graphite to diamond.

"Study is not reading. Study is asking yourself questions".
—a good friend who earned lots of A's in college

Chapter 9

The Rock Record and the Geologic Time Scale

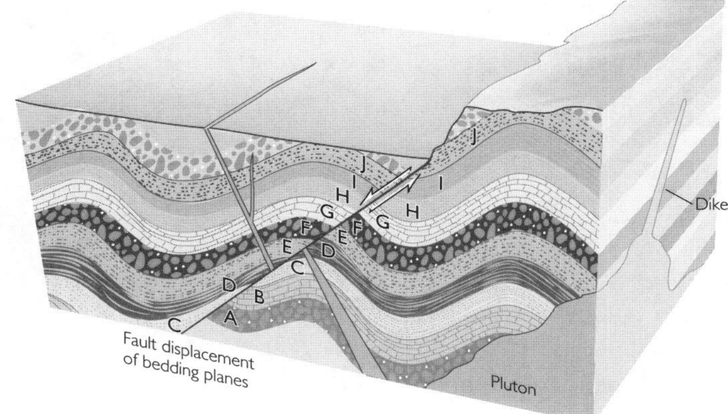

Figure 9.9
Superposition and cross-cutting relationships provide a basis for ordering the geologic events represented by the rocks and geologic features in this cross-section.

BEFORE LECTURE

Before you attend lecture, be sure to spend some time previewing the chapter. For an efficient preview use the questions below. *Chapter Preview* questions constitute the basic framework for understanding the chapter. Preview works best if you do it just before lecture. With the main points in mind you will understand the lecture better. This in turn will result in a better and more-complete set of notes. How much time should you devote to preview? Obviously, more time is better than less. But even a brief (five or ten minute) preview session just before lecture begins will produce a result you will notice. For a refresher on why previewing is so important see the Appendix: *How to Study Science.*

CHAPTER PREVIEW

- **How can the relative ages of rocks be determined for an outcrop?**
  Brief answer: The principles of superposition and cross-cutting relationships provide a basis for establishing the relative age of a sequence of rocks at an outcrop.

- **How can the relative ages for rock outcrops at two or more locations be determined?**
  Brief answer: The principle of superposition, fossils (faunal and floral succession), and radiometric dates of rock units provide a basis for establishing how rock outcrops at different localities may be related to each other, even if they are hundreds or thousands of miles apart.

- **What is the Geologic Time Scale and how is it calibrated?**
  Brief answer: The Geologic Time Scale is the internationally accepted reference for the sequence of events represented by Earth's rock record.

- **How is the Geologic Time Scale and geochronological methods like radiometric dating applied to geologic problems?**
  Brief answer: We understand Earth's history to the degree to which we can place the record of geologic events in time. The Geologic Time Scale is the accepted standard for how geologic time is subdivided.

Vital Information from other Chapters

- *Where We See Rocks* section of Chapter 3.
- *Atomic Structure of Atoms* in Chapter 2 before reading about radiometric dating methods.

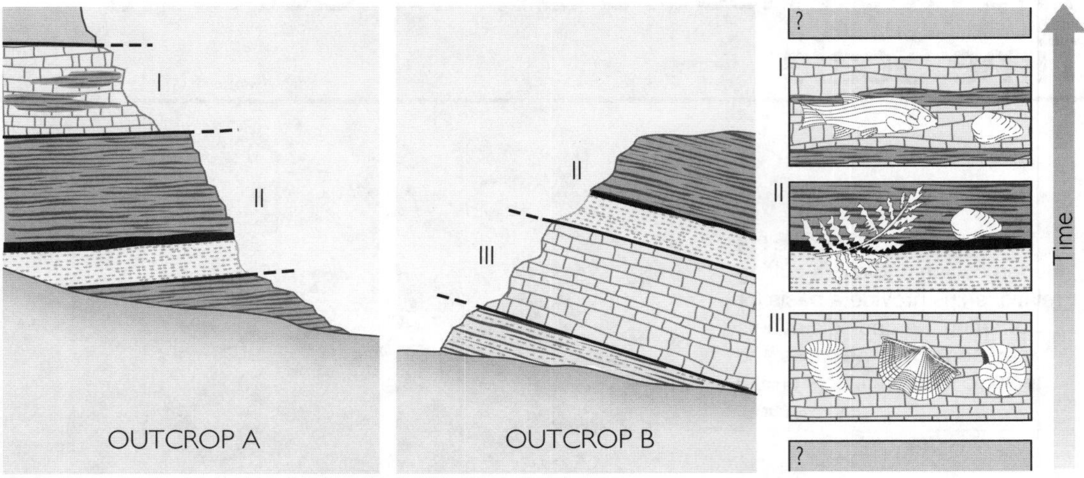

Outcrops may be separated by a long distance

Figure 9.5 / These two outcrops of sedimentary rock may be separated by a long distance. Using the sequence (superposition) of the rock units and the fossils (faunal succession) within them, one can draw a composite of the two that shows Formations I and II both overlying Formation III. Based on the principle of superposition, Formation I is the youngest — it's on top — and Formation III is the oldest of the three layers — it's on the bottom.

Strata, no matter how different they may appear,
are of the same age if they contain the same assemblage of fossils.
—William "Strata" Smith (1769-1839)

DURING LECTURE

Here are some tips that should help you get a good set of notes for this lecture.

Big Picture

The big picture for this lecture is the entire geological time scale! Geological time is wondrously huge! So huge that, at first, it seems incomprehensible. Your lecturer will describe the 4.5 **Big** billion year expanse of geological time and may use examples and exercises designed to help you grasp geological time.

New Terms

It is hard to talk about the geological time scale without referring to its time intervals, so you will be barraged with new and unfamiliar terms: epochs, periods, eras, eons, Holocene, Pleistocene, Pliocene, etc. To avoid getting lost, keep Figure 9.13 (The Geologic Time Scale) close at hand. Refer to Figure 9.13 to check terms as you need to. Better yet, look over this chart before lecture. Note how different sized chunks of time are used: huge chunks (eons) are sufficient for events early in Earth's history — smaller chunks (epochs) for intervals of time closer to the present.

The Succession of Geologic Events

You will learn how the relative ages of rocks are determined. By the end of lecture you will be able to use two basic principles (superposition and cross-cutting relationships) to determine the relative age of a sequence of rocks such as that shown in Figures 9.5 and 9.9.

Dating Methods

The third chunk of material in this lecture deals with absolute dating of rocks. Carbon 14 has a short half life. It works well for dating younger samples of tissue attached to bone, charcoal, and wood because these materials all contain carbon. Other isotopes (Uranium-238, Potassium-40, Rubidium-87) have much longer half lives and are used to date rocks that are much older. Table 9.1 explains how half-life is related to the effective dating range of each method.

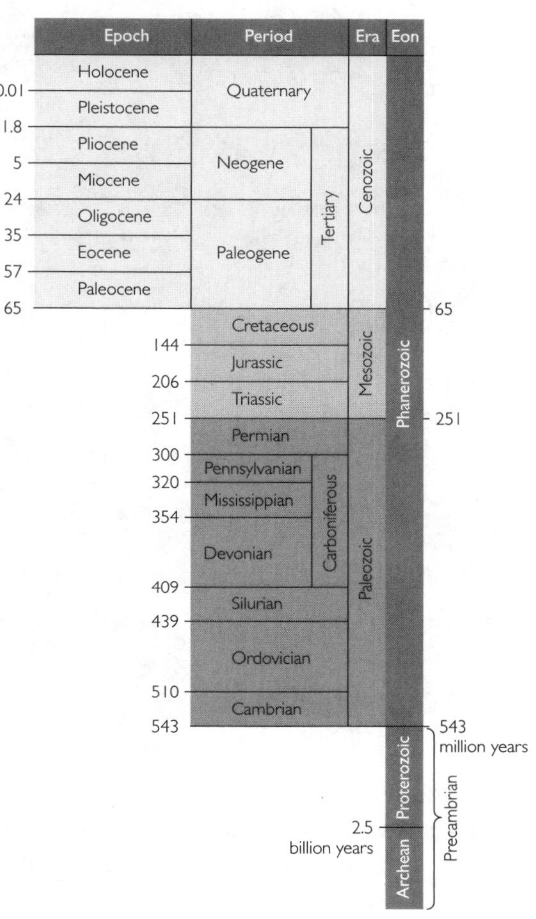

Figure 9.13 / The Geologic Time Scale. Numbers at the sides of the columns are in millions of years before present. [*from S. M .Stanley, Earth System History, New York: W. H. Freeman 1999*]

AFTER LECTURE

The perfect time to review your notes is right after lecture. The checklist on the next page contains both general review tips and specific suggestions for this chapter.

Intensive Study Session

After each lecture you need to thoroughly master the concepts covered. You need to master the geological time scale during this study session! Does that sound like a challenge? Unless you happen to have spent your summer working on a paleontology dig (tossing around terms like Paleocene, Eocene, Oligocene, etc.), the epochs of the Cenozoic may look like a steep memory curve. Here are some clues, tips and strategies to help you learn the Geological Time Scale

Markers

Marker events are simply interesting things that happened: animals or plants that evolved, creatures that dominated the earth, large extinction events. Look at Figure 9.18. Peg some marker events you already know about. *Example*: Can you guess one of the periods during which dinosaurs were dominant? The movie Jurassic Park has made this an easy question. When did complex life begin? Find some other marker events of particular interest to you. Maybe you are surprised at how early some events occurred: e.g. the age of the oldest fossil shells. Marker events will help you remember the Geologic Time Scale.

Study the Groupings

If you are a global learner it may be helpful to start with the big picture. Study the groupings in Figure 9.13 before getting down to the details. Eons are the biggest time chunks. Only the most recent (Phanerozoic eon) is broken down further. There are three eras: old life, middle life and new life, with names whose Greek stems reflect these meanings. All eras are further divided into periods. But only for the most recent era (Cenozoic, or new life) are periods divided into epochs. All epochs end in "cene".

NOTE REVIEW CHECKLIST

- ✔ All notes legible? (Rewrite so they read easily.)
- ✔ Important points clearly identified? You should now have headers in your notes that tie to each of the questions in the *BEFORE LECTURE/Chapter Preview.*
- ✔ Holes (missing material) filled in from memory?
- ✔ Areas where you don't remember what was said marked for a follow-up session with your instructor, tutor, or study partner?
- ✔ Possible test questions indicated in the margin (TQ)?
- ✔ Additional visual material. *Suggestions for Chapter 9: Sketching concepts is a good way to be sure you understand them. For Chapter 9 see if you can represent each the following concepts in a simple sketch: 1) superposition, 2) cross-cutting, 3) unconformity, 4) angular unconformity. Insert each sketch on the sketch page opposite to the notes you took for that concept.*
- ✔ Reworked notes into a form that is efficient for your learning style?
- ✔ Created a brief "big picture" overview of this lecture (using a sketch or written outline)?

Word Stems

Word stems are clues to meaning. Greek and Latin stems are used a great deal by scientists. You can look them up in any good dictionary. A few helpful stems for the geological time scale include:

Paleo <Greek: "old"
Ceno < Greek: "new," plus *zoic* (all epochs of the Cenozoic epoch end in "cene")
zoic < Greek: "life"

Catch Phrases

When there are long lists of unfamiliar terms to learn (like the epochs of the Cenozoic) many learners find it helpful to make use of catch phrases to help them remember. To try this out: do *Exercise 1* in *Practice Exercises and Study Questions.*

Website and CD Activities and Tools

http://www.whfreeman.com/presssiever

- **Website Q & A Practice Multiple Choice Questions:**

  At the Website complete the *Q & A*. Pay particular attention to the explanations for answers. Flashcards at the Website will help you learn new terms. Complete the *Field Relationships for Relative Time Dating* Interactive Exercise at the Website. Also at the Website is the *Geologic Time Scale* Interactive Tool which gives a brief summary of the life and geologic history for each division of geologic time.

PRACTICE EXERCISES AND STUDY QUESTIONS

(Answers and explanations are at the end of this chapter.)

Exercise 1: Geologic Time Scale Mnemonic

Construct a mnemonic device for remembering the geologic time scale names. The first letter of each word must match the first letter of the corresponding period or epoch in the proper order. You may use your native language — English, Spanish, German, French, Chinese — but do not mix words from different languages.

Examples (refer to Figure 9.13 for the Geologic Time Scale):

Here's a good mnemonic by a student for the periods of the Geologic Time Scale: *Chronically Overworked Student Decks Monotonous Physics Professor To Justify Contradictory Test Questions.*

Here's a good mnemonic by a student for the epochs of the Cenozoic: *Please Eat Our Mushroom Pot Pie Hot.*

Now it is your turn to invent your own memorial mnemonic to help you remember the Geologic Time Scale.

____

____

1. The principle of superposition holds that for any unfolded series of sedimentary layers,
   A. overlying strata extend over a broader area than any layers beneath them
   B. the layer at the top of the pile is always younger than those beneath it
   C. sediments generally accumulate in the vertical sandstone-shale-limestone sequence
   D. the layer at the top of the pile is always older than those beneath it

2. From youngest to oldest, the correct sequence of eras dividing the Phanerozoic eon is:
   A. Paleozoic, Cenozoic, Mesozoic
   B. Mesozoic, Cenozoic, Paleozoic
   C. Mesozoic, Paleozoic, Cenozoic
   D. Cenozoic, Mesozoic, Paleozoic

3. The epochs of the Tertiary period progress from oldest to youngest in which sequence?
   A. Eocene, Oligocene, Paleocene, Miocene, Pliocene
   B. Paleocene, Eocene, Oligocene, Miocene, Pliocene
   C. Paleocene, Eocene, Miocene, Pliocene, Oligocene
   D. Paleocene, Eocene, Oligocene, Miocene, Pliocene, Pleistocene

4. From oldest to youngest, the correct order of periods for the Paleozoic era is:
   A. Cambrian, Ordovician, Devonian, Silurian, Mississippian, Permian, Pennsylvanian
   B. Cambrian, Ordovician, Silurian, Devonian, Mississippian, Pennsylvanian, Permian
   C. Cambrian, Silurian, Ordovician, Pennsylvanian, Mississippian, Devonian, Permian
   D. none of the above

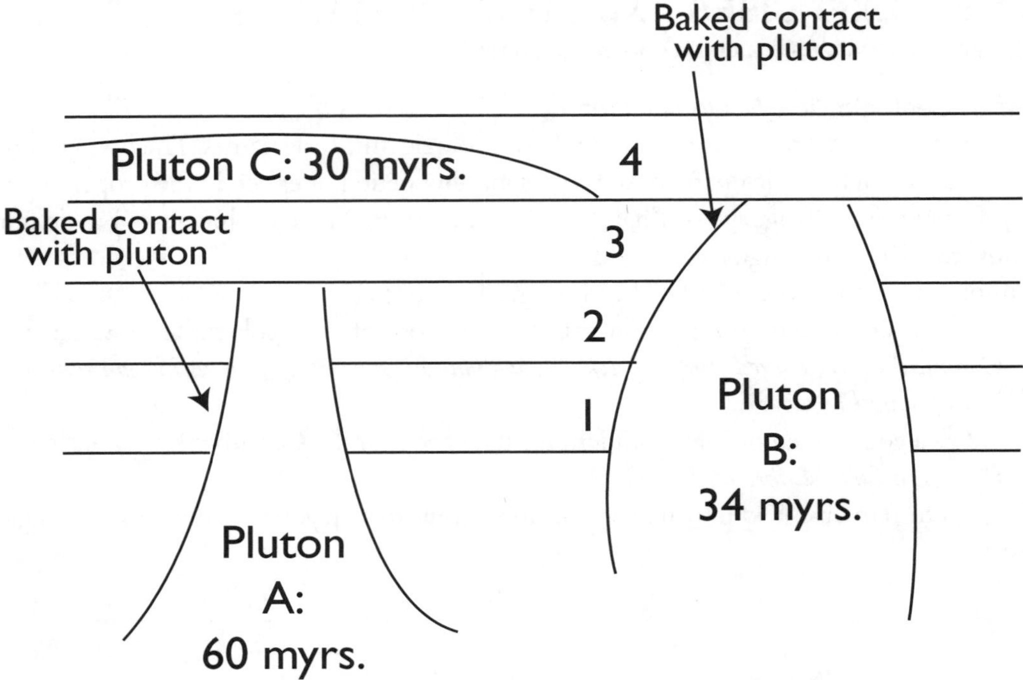

5. From the diagram above, one can infer age limits for rock layer 3 of
   A. between 34 and 60 million years
   B. between 30 and 60 million years
   C. less than 20 million years
   D. more than 60 million years

6. Sandstone and shale rock layers immediately below and above an angular unconformity imply a history of
   A. erosion, deposition, deformation, erosion
   B. erosion, deformation, deposition, erosion
   C. deformation, erosion, deposition, deformation
   D. deposition, deformation, erosion, deposition

Hint: Refer to Figure 9.8

7. What makes the isotopes of a given element different from each other?
   A. their atomic numbers
   B. their number of electrons
   C. their number of neutrons
   D. their number of protons

8. Naturally occurring ________________________ that decay(s) into other materials at known rates can be used to estimate the actual age of a rock.
   A. organic matter
   B. minerals
   C. radioactive elements
   D. silicon

9. The Phanerozoic time is divided into three eras: (1) the interval of old life, (2) the interval of middle life, and (3) the interval of modern life. These intervals correspond to (from oldest to youngest)
   A. Archean, Mesozoic, Paleozoic
   B. Paleozoic, Precambrian, Proterozoic
   C. Precambrian, Cambrian, Neocambrian
   D. Paleozoic, Mesozoic, Cenozoic

10. Which radiometric dating method would be most effective to determine the age of charcoal at an archaeological site?
   A. rubidium-strontium
   B. radiocarbon
   C. uranium-lead
   D. potassium-argon

Hint: Refer to Table 9.1 in your textbook.

EXAM PREP

Materials in this section are most useful during your preparation for midterm and final exams. For optimal performance, midterm preparation should begin about eight days before the exam (See Appendix: *How to Study Geology* for details). The basic idea is a systematic review of material divided into short study sessions.

If you have used regular intensive study sessions to master the material, now you get a pay back! Review for your exam will proceed smoothly and take far less time.

The following *Chapter Summary* and *Practice Exam* should simplify review still further. Read the *Chapter Summary* to begin your session. It provides a helpful overview that should get you back into the material. Next, try the *Practice Exam*. Take it just as you would a midterm to see how you stand in regard to mastery of this chapter. After you answer the questions, score them. Finally, and most important of all: review any question that you missed. Identify and correct the misconception that resulted in missing the item.

CHAPTER 9 SUMMARY

- The principles of superposition and cross-cutting relationships provide a basis for establishing the relative age of a sequence of rocks at an outcrop. Using these two principles, geologists can order (what happened first, second, third, so on) the geologic events represented by the rocks and geologic features in one outcrop. The principle of original horizontality for sedimentary layers provides a basis for identifying sequences of sedimentary rocks affected by tectonic forces after they were deposited.
- To reconstruct the geologic history of the Earth, geologists also need to correlate the geologic events represented by rocks at one locality with the geologic events represented by rocks at other localities. The principle of superposition, fossils (faunal and floral succession), and radiometric dates of rock units provide a basis for establishing how rock outcrops at different localities may be related to each other, even if they are hundreds or thousands of miles apart.

- The Geologic Time Scale is the internationally accepted reference for the sequence of events represented by Earth's rock record. It was constructed over about the last 200 years by geologists using mainly fossils, superposition and cross cutting relationships to establish the relative ages for thousands of rock outcrops around the world. In the last 50 years, the Geologic Time Scale has been calibrated using radiometric methods.
- We understand Earth's history to the degree to which we can place the record of geologic events in time. The Geologic Time Scale is the accepted standard for how geologic time is subdivided.

Time is simply nature's way of keeping everything from happening at once.
— *graffiti*
Department of Geosciences
University of Arizona, Tucson

Website and CD Activities and Tools
http://www.whfreeman.com/presssiever
Complete the *Geologic Time Scale* Interactive Exercise at the Website. Also at the Website is the *Geologic Time Scale* Interactive Tool which gives a brief summary of the life and geologic history for each division of geologic time. The Geology Essay entitled *Historical Geology* provides some examples of how geologists tackle interesting subjects. The *Photo Gallery* on the CD and at the Website contains images and information that will help you visualize and explore various concepts presented in your textbook.

PRACTICE EXAM QUESTIONS FOR CHAPTER 9

(Answers and explanations are at the end of this chapter.)

Exercise 1: Determining the Succession of Geologic Events

The block diagram below illustrates the geology of an area in Argentina. Answer the following questions regarding the geological history of this area.

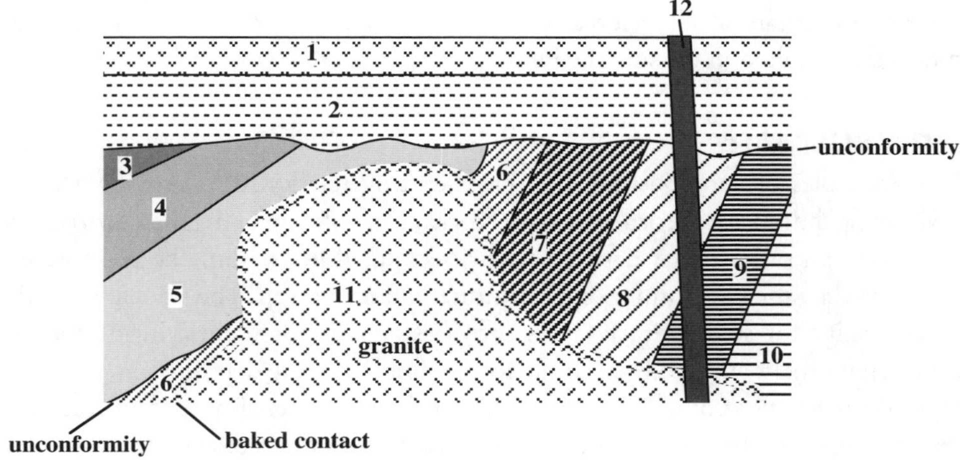

Notes:
Unit 2 contains clasts of units 3 through 10
Unit 5 contains clasts of units 6 through 10
Units 5 through 10 are baked along their contacts with unit 11
Units 1,2, 8, 9, and 11 are baked along their contacts with dike 12

A. Which unit is the youngest rock in this area? 1 2 3 4 5 6 7 8 9 10 11 12

B. Which unit is the oldest rock in this area? 1 2 3 4 5 6 7 8 9 10 11 12

C. Part 1. Is unit 3 older than unit 12? ___Yes ___No ___Not possible to know
Part 2. Explain the logic you used to answer the above question (C.1.)

___

___

D. Part 1. Is unit 1 younger than unit 11? ___Yes ___No ___Not possible to know
Part 2. Explain the logic you used to answer the above question (D.1.)

___

___

E. In an attempt to further work out the geologic age relationships in this area, samples of units 1, 11, and 12 were collected for radiometric dating to determine ages of these igneous rocks. The resulting counts of radioactive parent atoms and daughter atoms are listed in the table below.

ROCK UNIT	# PARENT ATOMS	# DAUGHTER ATOMS
1	500	500
11	250	750
12	750	250

Using this information, unit 1 is

1. older than unit 12, but younger than unit 11
2. younger than unit 12, but older than unit 11
3. older than unit 11, but younger than unit 12

Exercise 2:

In the illustration below, a geologic outcrop reveals three layers of sedimentary rocks, one fault and a single igneous dike intrusion. Using this illustration, order the sequence of geologic events from youngest to oldest.

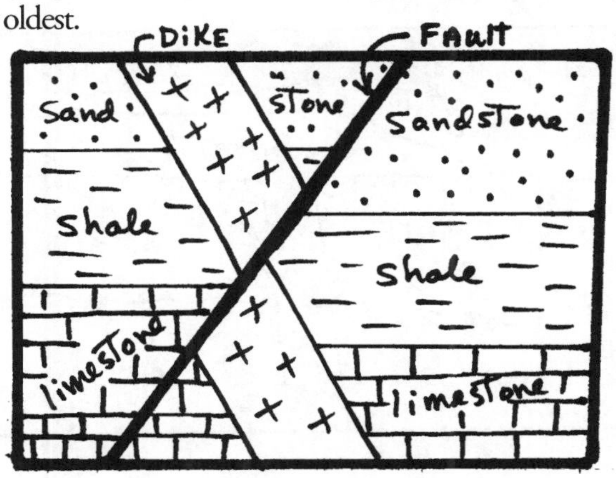

Youngest ___

___

___

___

Oldest ___

1. Only geologically young materials can be dated using radioactive C-14 isotopes because
   A. they have a very low disintegration constant
   B. they have a very short half-life
   C. after a very short time all the C-14 atoms have decayed
   D. they are a very rare isotope

2. If the half-life of some radioactive element is one billion years, and a mass of rock originally contained 1000 atoms of the radioactive element, how many atoms of the radioactive element would be left after three billion years had passed?
   A. 500 atoms
   B. 250 atoms
   C. 125 atoms
   D. no radioactive atoms

3. Small pieces of charcoal from an ancient ruin yield a carbon-14 date of 3000 years. This age best represents the approximate interval of time that has elapsed since
   A. a fire burned the wood
   B. humans inhabited the ruin
   C. humans cut the wood
   D. the wood died

4. Radiometric dates have been attached to the Geologic Time Scale by the determination of
   A. radiometric ages of igneous rocks younger and older than sedimentary formations
   B. radiometric ages of shales
   C. radiometric ages of fossil skeletons
   D. radiometric ages of metamorphosed sediments

5. For the most part, radiometric dates for rocks only represent the last time the rock
   A. crystallized from a magma or was metamorphosed
   B. was eroded
   C. became cemented
   D. was deposited

6. Which sample of basalt in the diagram is likely to yield the most accurate K/Ar radiometric date?
   A. A
   B. B
   C. C

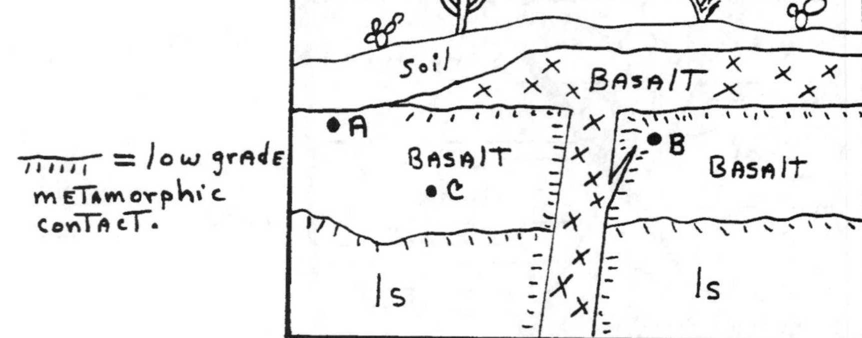

7. A layer of conglomerate contains cobbles of an igneous rock. One of the cobbles was dated radiometrically at 35 million years old. From this radiometric date, the conglomerate layer can be inferred to be
   A. more than 35 million years old
   B. less than 35 million years old
   C. 35 million years old
   D. none of the above

8. One method that geologists use to study buried sediments and unconformities is
   A. seismic stratigraphy
   B. radiometric stratigraphy
   C. depositional stratigraphy
   D. metamorphic stratigraphy

Hint: Refer to Figure 9.10 in your textbook.

9. If a rock is heated by metamorphism and the daughter atoms generated by the decay of the radioactive parent atoms migrate out of a mineral that is subsequently radiometrically dated, the date will be ____________________ the actual age.
   A. younger than
   B. older than
   C. the same as
   D. none of the above

10. In the illustration below, what is the most recent geological event depicted?
   A. Eruption of the lava
   B. Faulting
   C. Intrusion of the pluton
   D. Deposition of shales, sandstones and limestones

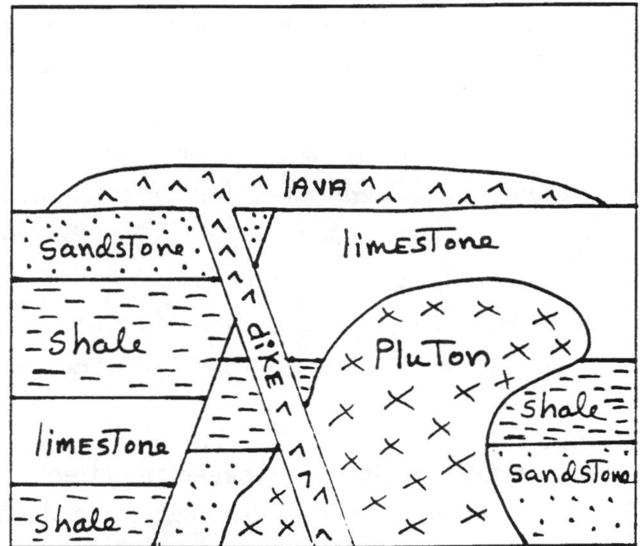

11. Pieces of charcoal were found in a paleosoil layer covering an ancient fire pit with stone tools including arrowheads and axes. The charcoal was radiometrically dated using carbon-14 and yielded an age of approximately 10,500 years before present. What can you infer about the age of the archaeological site?
    A. The archaeological site is about 10,500 years old.
    B. The archaeological site is younger than 10,500 years.
    C. The archaeological site is older than 10,500 years.
    D. The age of the archaeological site is unresolved, but may be older than 10,500 years.

ANSWERS AND EXPLANATIONS FOR EXERCISES AND QUESTIONS

After Lecture

1. B. The rock layer at the top of an undeformed sequence of rock layers is the youngest in the sequence (refer to *Interpreting the Grand Canyon Sequence).*
2. D. Cenozoic / Mesozoic / Paleozoic (refer to Figure 9.13).
3. B. Paleocene / Eocene / Oligocene / Pliocene are the epochs of the Tertiary Period in the Geologic Time Scale. Remember that the Pleistocene and Holocene are epochs within the Quaternary Period (refer to Figure 9.13).
4. B. Cambrian/Ordovician/Silurian/Devonian/Mississippian/Pennsylvanian/Permian (refer to Figure 9.13).
5. A. Rock layer 3 is deposited on top of layer 2 and the 60 million year old pluton A. Based on superposition, layer 3 is younger than 60 million years. Layer 3 is also cut by the 34 million year old pluton B. Therefore based on this cross-cutting relationship, layer 3 has to be older than 34 million years.
6. D. Deposition / deformation / erosion / deposition, (refer to Figure 9.8).
7. C. Refer to *Radioactive Atoms: The Clocks in Rocks* section of Chapter 9.
8. C. Refer to *Radioactive Atoms: The Clocks in Rocks* section of Chapter 9.
9. D. Refer to Figure 9.13 in your textbook.
10. B. Radiocarbon can be used to radiometrically date carbon containing materials like charcoal if the sample is younger than about 70,000 years old, (refer to Table 9.1).

Exam Prep

Exercise 1: Determining the Succession of Geologic Events

A. 12. Even though the black dike does not cut all rock units in the diagram, it does cut the granite and layer 1 which is above all other layered rock units. Therefore, the dike is younger than the granite and layer 1. If the dike was not shown cutting the granite, one would not be able to resolved the relative ages of the dike and the granite. Radiometric dates on the two igneous rock bodies would resolve their ages.

B. 10. Layer 10 is the oldest in the outcrop (diagram). It is at the bottom of a tilted sequence of layered rocks that go from layer 10 (the oldest) to layer 3 (the youngest of the sequence), assuming that this whole sequence has not been completely overturned.

C. Yes. Layer 3 is older than the dike (unit 12). Even though the dike does not cut layer 3, it does cut the unconformity which cuts layer 3 and it also cuts layers 2 and 1 which are above layer 3.

D. Not possible to know. Since the black dike cuts both the granite (unit 11) and layer 1, we can say that layer 1 and the granite are older than the dike. However, the granite may have intruded into the rock layers before or after the deposition of layer 1. There is no way of telling whether the granite is older or younger than layers 4, 3, 2, or 1. A radiometric date on the granite and layer 1, which is a lava flow, would help to resolve this question.

E. 1. Unit 1 is older than unit 12 but younger than unit 11. You do not even have to calculate radiometric ages to answer this question. Just look at the trend in the abundance of the radioactive parent atoms or accumulation of the daughter atoms. The rock with the greatest amount of parent atoms (and least amount of daughter) is the youngest. The rock with the lowest amount of radioactive parent atoms and the largest accumulation of daughter atoms is the oldest. Since the abundance of radioactive parent atoms in unit 1 falls between units 11 and 12, the radiometric age of unit 1 would be some number of years between unit 11 and 12. Unit 11 has the smallest amount of radioactive parent atoms remaining, therefore, it is the oldest.

Exercise 2

This is a clear illustration of the principle of superposition and cross-cutting relationships at work. Using the principle of superposition, we see that the limestone must have been deposited before the shale, and the shale before the sandstone. Likewise, these beds must exist before being first cut by the dike and second by the fault.

Youngest	faulting
	dike intrudes
	deposition of sandstone
	deposition of shale
Oldest	deposition of limestone

1. B. Because radiocarbon has a short half-life (5730 years), the level of radiocarbon decays below the detectable limit in samples older than about 70,000 years. Refer to Table 9.1 in your textbook.
2. C. Given a half-life of one billion years, for every billion years that elapses the number of radioactive atoms decreases by half. Therefore, after one billion years there are 500 left from the original 1000 atoms; after two billion years there are 250 out of the 500 left; and after three billion years there are 125 out of the 250 left.
3. D. To correctly interpret this question it is important to distinguish between observation (data) and inference. The radiocarbon age of 3000 years is data derived from laboratory

analysis of the charcoal sample. The result of the analysis represents the approximate time that has elapsed since the organism died (in this case wood) and was no longer exchanging carbon dioxide with the environment. It is reasonable to infer that the charcoal formed when inhabitants at the site burned wood for cooking and heating. It is also reasonable to assume that the wood was not dead for a long time before it was harvested for fuel. If these assumptions are correct, then the radiocarbon date represents the approximate time for the occupation of the site. However, the charcoal could have formed from a forest fire decades or centuries earlier than human occupation. Or, the wood harvested for fuel may have been dead and laying on the ground for hundreds or even thousands of years. In arid regions where decay is very slow, dead wood can lie on the ground for hundreds to thousands of years. If "old" wood happened to be used to fuel the fires at this site, then the radiocarbon age could be significantly older than the time of occupation.

4. A. The Geologic Time Scale was constructed over the last 200 years by geologists using mainly fossils, superposition and cross cutting relationships to establish the relative ages for thousands of rock outcrops around the world. In about the last 50 years, the Geologic Time Scale has been calibrated using radiometric methods to date mostly igneous and metamorphic rocks.
5. A. Review *Radioactive Atoms: The Clocks in Rocks* section of Chapter 9.
6. C. Weathering and alteration of rock samples by metamorphism can cause a redistribution of radioactive parent atoms and the daughter atoms produced by decay of the parent. This redistribution usually causes uncertainty in the radiometric date. Therefore, geologists typically radiometrically date the freshest rock samples that have been the least affected by subsequent geologic events. Sample C in the middle of the lava flow is the least likely to be affected by weathering like sample A, or contact metamorphism like sample B.
7. B. Since the radiometrically dated cobble is included in the conglomerate layer, the layer has to be younger than the cobble — the cobble had to form before it could become a part of the conglomerate. The radiometric date represents a maximum age for the conglomerate which has to be younger than any included component. If geologists really wanted to get a better estimate for the age of the conglomerate, they would need to do radiometric dates for as many cobbles of different igneous rocks as possible to find the date of the youngest cobble included in the layer. In this way, one could focus in on the approximate age of the sedimentary layer using the radiometric dates on igneous inclusions and the principle of included fragments.
8. A. Refer to Figure 9.10.
9. A. Loss of daughter atoms, such as lead, generated from the decay of radioactive parent atoms, such as uranium, would result in a date that was younger than the actual time that has elapsed since the rock solidified. One important assumption that is made when interpreting radiometric dates is that the mineral or rock has remained a closed system — no elements have been added to or removed from the mineral except by radioactive decay. If this assumption holds, then the radiometric date typically represents the time that has elapsed since the rock solidified from a magma.

10. A. Cross-cutting relationships tell us that the dike and lava flow must be the last event to have occurred. Using the principle of superposition we can see the first event must be the deposition of the lower shale, followed by the limestone-shale-sandstone sequence, so these cannot be the most recent event. Cross-cutting relationships again inform us that the fault and pluton are the next events, though which is more recent than the other is inconclusive, as neither cuts the other. Finally, the dike cuts all the preceding rock units, extruding the lava flow at the surface—the final rock unit. Refer to the sections on *The Stratigraphic Record* in your text for a discussion on these concepts.
11. D. D is the best answer although one might be inclined to answer C based on superposition. Archeologists will commonly use radiocarbon ages to bracket the age of an archaeological site if no datable material is found within the site itself. Again, it is important to distinguish between observation (the data) and inferences (the interpretation). The radiocarbon date represents the time that has elapsed since the wood died. Perhaps the wood died during a forest fire and the charcoal was washed away from the burned area and deposited in a layer that covered the fire pit. If this is the case, then the radiocarbon date would be younger than the fire pit. Although charcoal is soft, it is chemically very stable and can remain well preserved in sediments. It is also possible that the charcoal weathered out of some sedimentary layers that are actually much older than the fire pit and ended up incorporated in the layer that covers that pit. In this case, the radiocarbon date would be misleading because the charcoal was actually significantly older than the fire pit but ended up in a soil layer above the fire pit because it was recycled from eroding older sediments.

No vestige of a beginning,
no prospect of an end.
—James Hutton (1726 – 1797)

Chapter 10

Folds, Faults, and Other Records of Rock Deformation

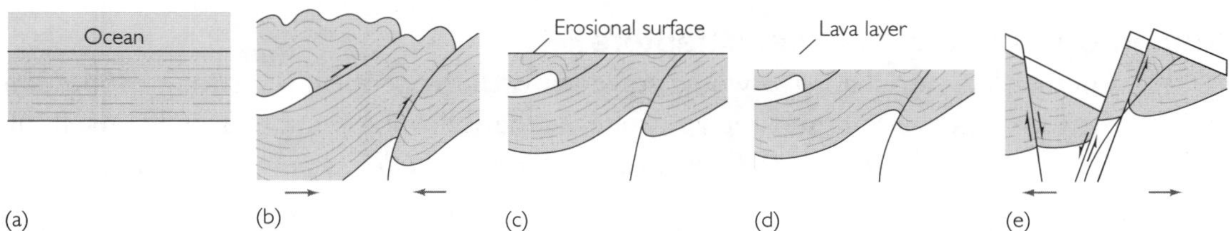

Figure 10.27 / Stages in the development of the Basin and Range provinces of Nevada, Utah, Arizona and California. Geologists see only the last stage (e) and to attempt unravel episodes of tensional and compressional deformation in order to reconstruct the earlier stages (a – d) in the geologic history of the region.

BEFORE LECTURE

Before you attend lecture be sure to spend some time previewing the chapter. For an efficient preview use the questions below. *Chapter Preview* questions constitute the basic framework for understanding the chapter. Preview works best if you do it just before lecture. With the main points in mind you will understand the lecture better. This in turn will result in a better and more-complete set of notes.

How much time should you devote to preview? Obviously, more time is better than less. But even a brief (five or ten minute) preview session just before lecture begins will produce a result you will notice. For a refresher on why previewing is so important see the Appendix: *How to Study Geology*.

CHAPTER PREVIEW

- **How do rocks deform (bend and break)?**
  Brief answer: Rocks typically break (fault) when temperature is low, burial is shallow, and the force is applied quickly. Rocks typically bend (fold) when temperature is higher, burial is deeper, and the force is applied over a long period.

- **What geologic features are produced when rocks deform?**
  Brief answer: Folds, faults and joints are common geologic structures produced when rocks are deformed.

- **What do geologic structures produced by rock deformation tell geologists about the history of a region?**
  Brief answer: The types of folds, faults and joints and their spatial orientation provides geologists with clues for deciphering the kinds of forces affecting a region over time.

STUDY TIP

Rock Deformation is a particularly visual lecture. Slide material on folds and faults can be confusing if you have never seen these geologic features before, so be sure to preview the figures before attending this lecture. For an overview, start with Figures 10.6 and 10.27. Then study Figure 10.22 until you are clear about the difference between a normal and reverse fault.

Vital Information from Other Chapters

- Another look at Chapter 3 would serve you well. Since there is a strong connection between deformation and metamorphism, the following sections in Chapter 8 are recommended for review: *Deformational Textures* and *Plate Tectonics and Metamorphism.*

Website and CD Preview

http://www.whfreeman.com/presssiever

- At the Website, the Interactive Exercise *Tectonic Forces in Rock Deformation,* based on Figure 10.6, introduces you to the most basic terminology and concepts within this chapter.

DURING LECTURE

One goal for lecture should be to leave class with a good set of answers to the *Preview Questions.*

- To avoid getting lost in details keep the "big picture" in mind: Chapter 10 tells the story of three kinds of forces (compressive, tensional and shearing) and how geologists find evidence of these forces in rock structures (folds and faults).
- This is a particularly visual lecture. Slide material on folds and faults can be confusing if you have never seen these geologic features before, so be sure to preview the figures before attending this lecture. For an overview, start with Figures 10.6 and 10.27. Then study Figure 10.22 until you are clear about the difference between a normal and reverse fault.
- Bring to class Figures 10.6, 10.27 and 10.22 either as photocopies or book-marked in your textbook for quick reference. Refer to them during lecture.

NOTE-TAKING TIP

Figures of faults and folding can be drawn very simply . . . once you understand them. But until you do, make it easy on yourself. Photocopy Figures 10.6, 10.27, and 10.22. Three-hole punch them for easy insertion into your 3-ring notebook. Having them already in your notebook means you won't be distracted by drawing them during lecture.

AFTER LECTURE

Education is a voyage in self-discovery.
—Laurence M. Gould

The perfect time to review your notes is right after lecture. The following *Note Review Checklist* contains both general review tips and specific suggestions for this chapter.

NOTE REVIEW CHECKLIST

- ✔ All notes legible? (Rewrite so they read easily.)
- ✔ Important points clearly identified? You should now have headers in your notes that tie to each of the questions in the *BEFORE LECTURE/Chapter Preview.*
- ✔ Holes (missing material) filled in from memory?
- ✔ Areas where you don't remember what was said marked for a follow-up session with your instructor, tutor, or study partner?
- ✔ Possible test questions indicated in the margin (TQ)?
- ✔ Additional visual material. *Suggestions for Chapter 10: Test your understanding by adding simple sketches of each of the following: syncline, anticline, normal fault, reverse faults, strike fault. Insert these sketches into your notes. You may want to add captions to help you keep things straight. (For example, I associate the word "sink" with the "s" in syncline. "Sink" is what a syncline resembles.)*
- ✔ Reworked notes into a form that is efficient for your learning style?
- ✔ Created a brief "big picture" overview of this lecture (using a sketch or written outline)? *Hint: The material is visual, so the summary should be visual.*

Intensive Study Session

After each lecture you need to thoroughly master the concepts covered. You need to do this before you attend the subsequent lecture. The ideas of geology are like a stack of boxes. Each new idea rests on all ideas (boxes) stacked beneath it.

> *You can't stack something on something that isn't there.*
> —*from "The Skyscraper Construction Manual"*

Schedule at least one hour after lecture for intensive study. Use this time to master key concepts. By now you know well that mastery is not gained by just reading your text. To master geology you must ask yourself questions and answer them. Use the Website and CD Activities and Tools suggested below along with the *Practice Exercises* and *Study Questions* to insure you master this chapter. Do as many of these as you have time for during your scheduled study session. Pay particular attention to exercises recommended by your instructor during lecture.

Website and CD Activities and Tools

http://www.whfreeman.com/presssiever

- **Website Q & A Practice Multiple Choice Questions:**
  At the Website complete the *Q & A.* Pay particular attention to the explanations for answers. Flashcards at the Website will help you learn new terms. At the Website, do the

Interactive Exercise on *Types of Folds* based on Figure 10.12. Images of many different geologic structures are available in the *Photo Gallery* on your CD. Explore these images of geologic structures and their captions to become familiar with what folds, faults and joints can look like on both a large and small scale in the field.

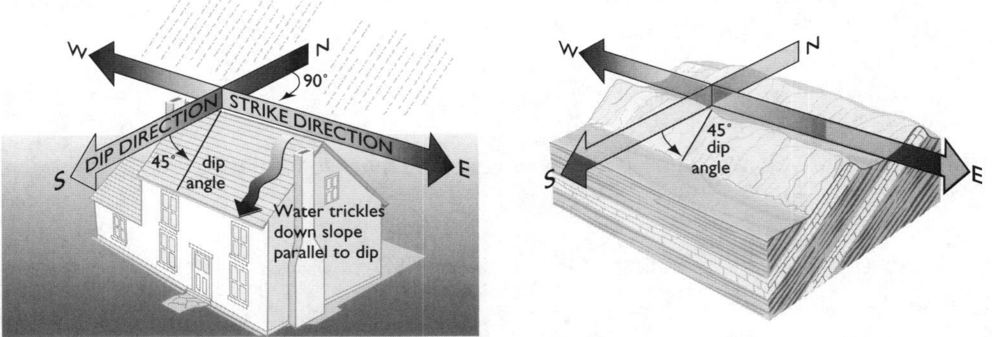

Figure 10.4 / Geologists use the strike and dip of a rock formation to describe its orientation in their field notes and on geologic maps.

PRACTICE EXERCISES AND STUDY QUESTIONS

(Answers and explanations are at the end of this chapter.)

Exercise 1: Silly Putty

Silly Putty is a popular teaching aid (and toy) with geologists because at room temperature it exhibits all three kinds of deformation characteristic of solids. If you pull on the putty quickly, it will snap into two pieces. It is easy to bend and mold the putty into many shapes. Plus, if you throw a ball of it on the floor, the ball will bounce. Compare the properties of silly putty with the behavior of rocks by completing the table below.

Behavior of Silly Putty	**Behavior of Rock**	**Type of Force**	**Geologic Structure produced by this style of deformation**
snaps into pieces		tensional	
bends	ductile		
bounces	ELASTIC Rocks do exhibit elastic behavior. (More on this when we study earthquakes.)	COMPRESSIONAL The ball of putty is compressed by the impact with the floor.	NOTE: Earthquakes are attributed to the elastic properties of rocks.

1. A rock that breaks suddenly in response to the application of force is
   A. elastic
   B. plastic
   C. brittle
   D. ductile

2. The two measurements that define the orientation of an exposed rock layer are
   A. strike and dip
   B. strike and slip
   C. slip and dip
   D. fold axis and tilt

3. The sense of motion along the San Andreas fault in California is
   A. left-lateral strike slip
   B. right-lateral strike slip
   C. dip slip
   D. thrust

4. As tensional forces are applied to a continental region, the resulting geologic feature will be a/an
   A. anticline
   B. rift valley
   C. thrust fault
   D. dome

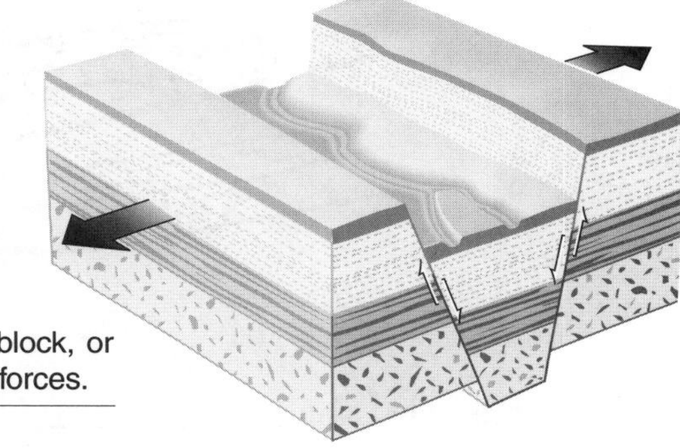

Figure 10.25 / A down-faulted block, or rift valley, results from tensional forces.

5. When no up and down or sideways motion can be detected in the fracture of a rock, it is called a
   A. stress plane
   B. joint
   C. fault
   D. rupture

6. This fold is called a/an ____________________________.

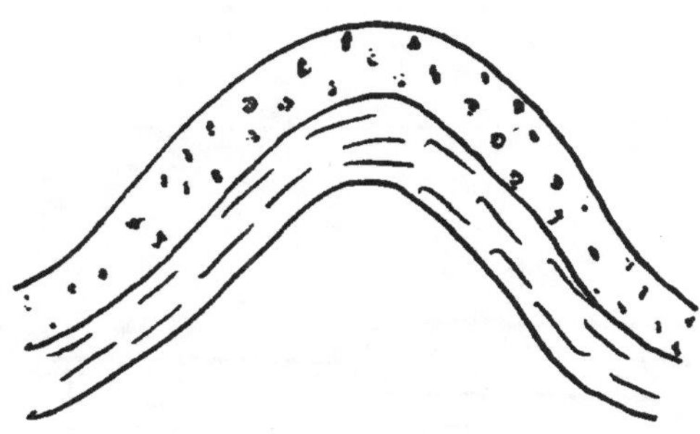

7. This fold is called a/an ___________________________.

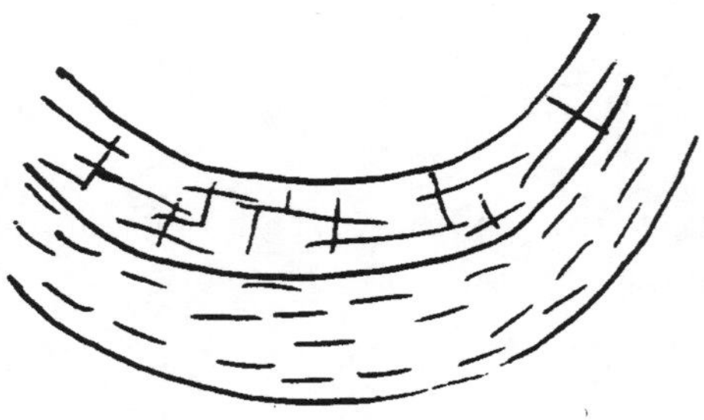

8. This fault is called a/an ___________________________.

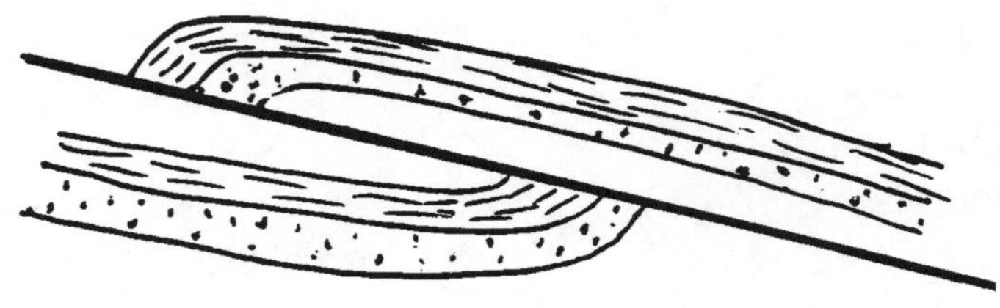

9. This fault is called a/an ___________________________.

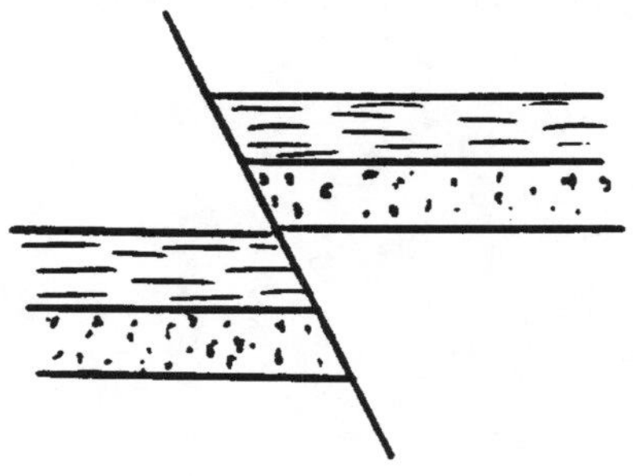

10. This fault is called a/an ___________________________.

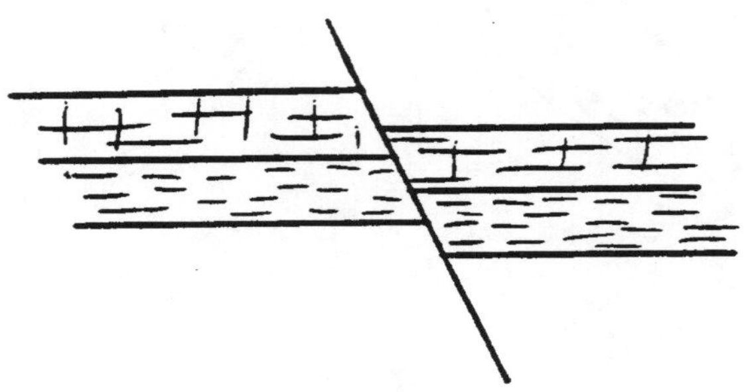

EXAM PREP

Materials in this section are most useful during your preparation for midterm and final exams. For optimal performance, midterm preparation should begin about eight days before the exam (see Appendix: *How to Study Geology* for details). The basic idea is a systematic review of material divided into short study sessions.

If you have used regular intensive study sessions to master the material, now you get a pay back! Review for your exam will proceed smoothly and take far less time.

The following *Chapter Summary* and *Practice Exam* should simplify review still further. Read the *Chapter Summary* to begin your session. It provides a helpful overview that should get you back into the material. Next, try the *Practice Exam*. Take it just as you would a midterm: to see how you stand in regard to mastery of this chapter. After you answer the questions, score them. Finally, and most important of all: review any question that you missed. Identify and correct the misconception that resulted in missing the item.

CHAPTER 10 SUMMARY

- All rocks may bend (behave ductilely) and break (behave brittlely) in response to the application of forces. Laboratory experiments have revealed that whether a rock exhibits ductile or brittle behavior depends on its composition, temperature, depth of burial (confining pressure) and rate with which tectonic processes apply force.
- Ductile behavior is more likely when a rock is exposed to higher temperatures, deeper burial, slower application of tectonic forces, and is a sedimentary rock. Brittle behavior is favored when rocks are cooler, closer to the Earth's surface, exposed to more rapid application of tectonic forces, and is an igneous or high grade metamorphic rock.
- Folding is a result of ductile deformation. From the type of fold and its orientation, geologists can interpret the orientation of the tectonic forces and characteristics of the rock layers during deformation.
- Faulting and jointing are a result of brittle deformation. Jointing occurs when a rock fractures but there is little movement along the fracture planes. Faults are fractures along which there is appreciable movement (offset). The type and orientation of faults and joints provides valuable information about the tectonic forces and the characteristics of the rock layers at the time of deformation.
- The type of fold or fault provides a basis for geologists to interpret the type of tectonic force acting on the rock during deformation. Tectonic forces can be of three types: compressive, tensional and shearing. Compressive forces dominate at convergent boundaries where plates collide or subduct. Tensional forces dominate at divergent boundaries where plates are pulled apart. Shearing forces dominate at transform faults where plates slide horizontally past each other.
- Geologic structures such as folds, faults and joints occur on all scales from microscopic to the size of a mountain-side. Geologists deduce the geologic history of a region in part by unraveling the history of deformation, thereby reconstructing what the rock units looked like before deformation. Regional deformational fabrics can help geologists decipher the plate tectonic history for the region.

Website and CD Activities and Tools
http://www.whfreeman.com/presssiever

Images of many different geologic structures are available in the *Photo Gallery* on your CD and at the Website. Explore these images of geologic structures to become familiar with what folds, faults and joints can look like on both a large and small scale in the field.

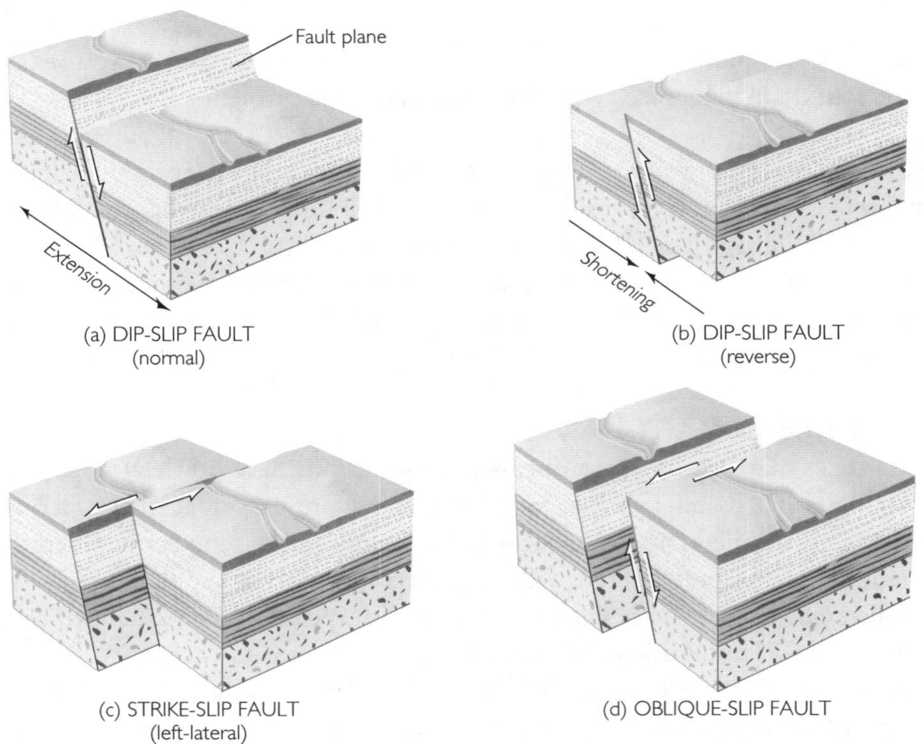

Figure 10.22 / Types of faults. (a) Normal faults, caused by tensional forces, resulting in extension of the crust. (b) Reverse faults, caused by compressive forces, resulting in shortening of the crust. (c) Strike-slip faults are associated with shearing forces. (d) Oblique slip suggests a combination of shear and compression or tension.

PRACTICE EXAM QUESTIONS FOR CHAPTER 10

(Answers and explanations are at the end of this chapter.)

Exercise 1: Geologic Structures

For each of the following five illustrations of deformed rocks, name the (A) geologic structure, e.g., normal fault, syncline; (B) type of force, e.g., compressional, tensional, shearing force, responsible for producing each geologic structure; and (C) the plate tectonic boundary, e.g., convergent, divergent or shear, with which the geologic structure is commonly associated.

VERTICAL SECTION

A. Geo structure ______________________

B. Type of force ______________________

C. Commonly associated plate tectonic boundary

____

D. Geo structure ______________________

E. Type of force ______________________

F. Commonly associated plate tectonic boundary

____

G. Geo structure ______________________

H. Type of force ______________________

I. Commonly associated plate tectonic boundary

____

J. Geo structure ______________________

K. Type of force ______________________

L. Commonly associated plate tectonic boundary

____

M. Geo structure ______________________

N. Type of force ______________________

O. Commonly associated plate tectonic boundary

____

Exercise 2: Anticline vs. Syncline

A. Briefly describe the diagnostic differences between an anticline and a syncline.
B. Draw a picture of a typical outcrop pattern for a plunging syncline exposed at the surface.
Hint: Figures 10.5 and 10.14 will be very helpful references.

A. ____

____

____

B. Drawing of an outcrop pattern for a plunging syncline:

1. Of the following conditions, which one promotes ductile deformation of rocks?
   A. old igneous rocks within the interior of a continent
   B. rocks deforming at a relatively low temperature
   C. stress building up rapidly to a very high level
   D. rocks subjected to large confining pressures and high temperatures

2. When deformed, which of the following rocks is more likely to fracture brittlely, instead of flow ductilely?
   A. basalt
   B. shale
   C. pure limestone
   D. muddy sandstone

3. As molten rock cools near or at the surface, it can develop shrinkage fractures that extend vertically down into the rock body. These crisscrossing, regularly patterned fractures create long thin rods of rock we call
   A. shrinkage palisades
   B. tension faults
   C. columnar joints
   D. elongate joints

Hint: Refer to the photo at the beginning of Chapter 4.

4. Which of the following geologic structures is caused by tensional forces?
   A. thrust fault
   B. reverse fault
   C. anticline
   D. normal fault

5. Thrust faults commonly form
   A. around hot spots
   B. where continents are colliding
   C. where continents are pulling apart
   D. along a transform fault

6. Of the following rock beds, which will have a strike direction but no dip direction?
   A. bed inclined 45 degrees from horizontal
   B. vertical bed
   C. horizontal bed
   D. none of the above

7. Which of the following defines the direction of dip?
   A. a line at right angles to the strike line
   B. a line north of the strike line
   C. a line parallel to the strike line
   D. a line parallel to plunge

Hint: Refer to Figure 10.4 in your textbook.

8. From an airplane you notice that the outcrops of tilted layers of rock make a distinct zigzag pattern across a plain. You reasonably conclude that the
   A. area has been cut by numerous normal faults
   B. layers are folded into a series of plunging folds
   C. layers are folded into a series of non-plunging folds
   D. layers have been tilted so that all the layers dip in the same direction

Hint: Refer to Figure 10.15 in your textbook.

ANSWERS AND EXPLANATIONS FOR EXERCISES AND QUESTIONS

After Lecture

Exercise 1: Silly Putty

Behavior of Silly Putty	Behavior of Rock	Type of Force	Geologic Structure produced by this style of deformation
snaps into pieces	brittle	tensional	fault or joints
bends	ductile	compressional	folds
bounces	ELASTIC Rocks do exhibit elastic behavior. (More on this when we study earthquakes.)	COMPRESSIONAL The ball of putty is compressed by the impact with the floor.	NOTE: Earthquakes are attributed to the elastic properties of rocks.

1. C. Solids that break are called brittle. Whether a solid behaves brittlely depends on its composition, temperature, confining pressure, and the rate of application of directional forces.
2. A. Refer to Figure 10.4 in your textbook.
3. B. The San Andreas is a strike-slip fault. It also represents a transform plate boundary. Refer to the section in Chapter 10 entitled *Faults* and Figure 10.21.
4. B. Refer to Figure 10.25.
5. B. Joints are fractures along which there is no appreciable movement. Faults are fractures along which there is appreciable offset.
6. anticline
7. syncline
8. thrust fault
9. reverse fault
10. normal fault

Exam Prep

Exercise 1: Geologic Structures

A. reverse fault
B. compressional
C. convergent

D. anticline
E. compressional
F. convergent

G. normal fault
H. tensional
I. divergent

J. strike-slip fault
K. shearing
L. transform

M. syncline
N. compressional
O. convergent

Exercise 2: Anticline vs. Syncline

A. In an anticline the youngest rock layer is on the outside of the fold, where as in a syncline the youngest rock layer is in the middle of the fold.
B. Refer to Figure 10.14

1. D. All other conditions mentioned would favor brittle behavior.
2. A. Basalt is an igneous rock with interlocking fine-grained silicate minerals which have a high melting point. Therefore, basalt tends to behave brittlely. Rocks with high clay or calcium carbonate content tend to behave ductilely.
3. C. Columnar jointing is fracturing caused by shrinkage during cooling of an igneous rock.
4. D. Normal faults are a result of brittle behavior in response to tension.
5. B. Thrust faults are caused by compressional forces typical of what is generated at convergent boundaries like continental collisions.
6. D. Remember, although a horizontal plane has no dip-direction, it will not have a strike direction either. Vertical planes dip vertically, and plunge is the direction that a fold axis will "dip," not the direction that planes dip. Refer to Figure 10.4.
7. A. Refer to Figure 10.4.
8. B. Refer to Figure 10.15.

Chapter 11
Mass Wasting

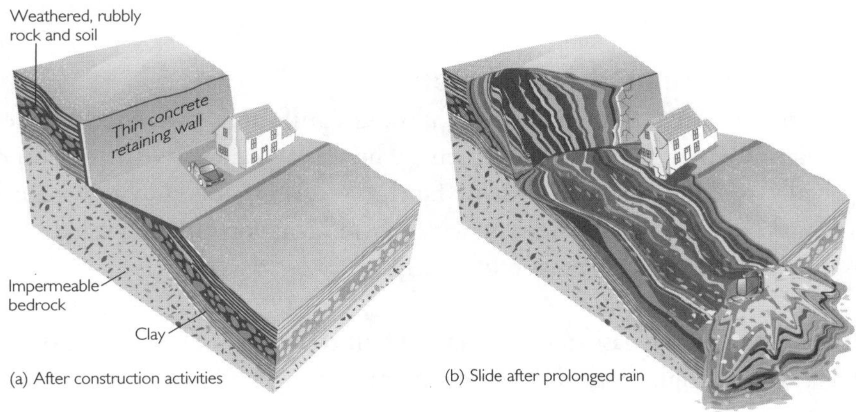

Earth Policy Box 11.1 / This slope is unstable because it parallels the dip of the underlying rock layers and rests on a clay layer that would act as a lubricant if it became waterlogged. The slope in the back of the house has been oversteepened, and the concrete retaining wall is too thin to hold it.

BEFORE LECTURE

Before you attend lecture be sure to spend some time previewing the chapter. For an efficient preview use the questions below. *Chapter Preview* questions constitute the basic framework for understanding the chapter. Preview works best if you do it just before lecture. With the main points in mind you will understand the lecture better. This in turn will result in a better and more-complete set of notes.

How much time should you devote to preview? Obviously, more time is better than less. But even a brief (five or ten minute) preview session just before lecture begins will produce a result you will notice. For a refresher on why previewing is so important see the Appendix: *How to Study Science.*

CHAPTER PREVIEW

- **What is mass wasting?**
  Brief answer: Mass wasting, also called mass movement, is the downslope movement of rock material.

- **Why do mass movements occur?**
  Brief answer: The three most important factors enhancing the potential for mass movements are: the steepness of the slope, the nature of the rock making up the slope, and the water content.

- **How can damage from mass movements be minimized?**
  Brief answer: Careful engineering and restricting land use can minimize the hazards associated with mass movements.

PREVIEWING TIP FOR CHAPTER 11

It will be very helpful to work on Exercise 1 (in *Practice Exercises and Review*) before going to lecture. Complete the exercise and take it to class with you. Your lecturer will probably show slides to help you understand the different kinds of mass wasting. You will understand these differences better if you have worked on them before class.

Vital Information from Other Chapters

- The composition and internal fabric of rocks significantly influence the rock's strength and the potential for mass movement. Therefore, a review of the composition and especially of the different kinds of fabrics (textures) for igneous, sedimentary and metamorphic rocks will provide you with vital information for understanding the different circumstances that cause mass movements.

- While reviewing the basic rock textures described in Chapters 4, 5, 7 and 8, ask yourself what textures might contribute to a weaker rock and an enhanced potential for mass movement.

DURING LECTURE

One goal for lecture should be to leave class with a good set of answers to the *Preview Questions*.

- To avoid getting lost in details keep the "big picture" in mind: Chapter 11 tells the story of what causes mass wasting: the steepness of the slope, the kind of rock in the slope being moved, and the water content.

- You may not be familiar with the kinds of mass wasting (rock avalanche, creep, earthflows, etc.). To help you develop this familiarity your lecturer may show slides of various kinds of mass wasting. Some of these may be very dramatic. Enjoy the drama and excitement!

But remember, this is about classification. Your job is to understand the differences between kinds of wasting, so do take notes, particularly on comments your instructor may make about how each kind of mass wasting differs from the others. Hint: In general these differences will be about the steepness of the slope, kind of rock in the slope being moved, and the water content.

- If you took our previewing tip and completed *Exercise 1* before coming to class, you can refer to that as your lecturer talks about the different kinds of mass wasting. If you did not complete the exercise, refer to Table 11.2, *Classification of Mass Movements,* during the lecture.

AFTER LECTURE

The perfect time to review your notes is right after lecture. The checklist below contains both general review tips and specific suggestions for this chapter.

NOTE REVIEW CHECKLIST

- ✔ All notes legible? (Rewrite so they read easily.)
- ✔ Important points clearly identified? You should now have headers in your notes that tie to each of the questions in the *BEFORE LECTURE/Chapter Preview.*
- ✔ Holes (missing material) filled in from memory?
- ✔ Areas where you don't remember what was said marked for a follow-up session with your instructor, tutor, or study partner?
- ✔ Possible test questions indicated in the margin (TQ)?
- ✔ Additional visual material. *Suggestions for Chapter 11: Make simple sketches that will help you remember the key aspect of each kind of mass wasting. Hint: Your sketch need not be artistic to be useful. Sketch only the features you need to remember. Example: for a rock avalanche you could draw a steep slope (one line at a 45 degree angle) with a pile of large blocks at the bottom to designate large masses of broken rock (see Figure 11.10).*
- ✔ Reworked notes into a form that is efficient for your learning style?
- ✔ Created a brief "big picture" overview of this lecture (using a sketch or written outline)? *Suggestion for Chapter 11: Write a brief summary of the most important points you have learned from this chapter that might influence your choice of future home sites.*

Intensive Study Session

After each lecture you need to thoroughly master the concepts covered. You need to do this before you attend the subsequent lecture. The ideas of geology are like a stack of boxes. Each new idea rests on all ideas (boxes) stacked beneath it.

Schedule at least one hour after lecture for intensive study. Use this time to master key concepts. By now you know well that mastery is not gained by just reading your text. To master geology you must ask yourself questions and answer them. Use the Website and CD Activities and Tools suggested below, along with the *Practice Exercises* and *Study Questions,* to insure that you master this chapter. Do as many of these as you have time for during your scheduled study session. Pay particular attention to exercises recommended by your instructor during lecture.

Website and CD Activities and Tools

http://www.whfreeman.com/presssiever

- **Website Q & A Practice Multiple Choice Questions:**

  At the Website complete the *Q & A*. Pay particular attention to the explanations for answers. Flashcards at the Website will help you learn new terms. *Rock Mass Movement* and *Unconsolidated Mass Movement* are Interactive Exercises at the Website to help you review the names for different kinds of mass movement. After completing these short exercises, be sure to compare the illustrations with photos of different kinds of mass movement that appear in your textbook and in the *Photo Gallery* on your CD and at the Website. Tables 11.1 and 11.2 will also help you sort out the distinguishing features of the different kinds of mass movement.

PRACTICE EXERCISES AND STUDY QUESTIONS

(Answers and explanations are at the end of this chapter.)

Exercise 1: Inventory of the Different Kinds of Mass Wasting

Press and Siever discuss eight different kinds of mass wasting. As an aid to learning the circumstances that favor each of these types of mass movement, use your textbook to fill in the blanks in the table below. Textbook figures, figure captions and photographs will help you complete the table. Hint: probably you haven't seen many of these features, so be sure to examine the pictures of each form of mass wasting in the text.

KIND OF MASS WASTING	COMPOSITION OF SLOPE	CHARACTERTISTICS
rock avalanche		Speed: fast Slope angle: steep Triggering event(s): earthquakes Notes: Occur in mountainous regions where rock is weakened by weathering, structural deformation, weak bedding, or cleavage planes.
creep		Speed: Slope angle: any angle Triggering event(s): none Notes:
earthflows		Speed: Slope angle: any angle Triggering event(s): intense rainfall Notes: fluid-like movement
debris flow		Speed: Slope angle: any angle Triggering event(s): Notes:

	mostly finer rock materials with some coarser rock debris with large amounts of water	Speed: Slope angle: Triggering event(s):intense rainfall or catastrophic melting of ice and snow by a volcanic eruption Notes: contains large amounts of water
debris avalanche	water-saturated soil and rock	Speed: Slope angle: Triggering event(s): Notes:
slump		Speed: slow Slope angle: any slope Triggering event(s): rainfall Notes:
	surface layers of soil	Speed: slow Slope angle: any angle Triggering event(s): Notes: Occurs only in cold regions when water in the surface layers of the soil alternately freezes and thaws. Water cannot seep into the ground because deeper layers are frozen.

Exercise 2: Assessing Potential Hazards from Mass Wasting

Given the circumstances illustrated in the diagram on the right, assess the slope stability for each housing site. Do so by simply putting one word—POOR, BETTER, or BEST—above each house in diagrams A, B and D. Diagram C has been completed for you as an example.

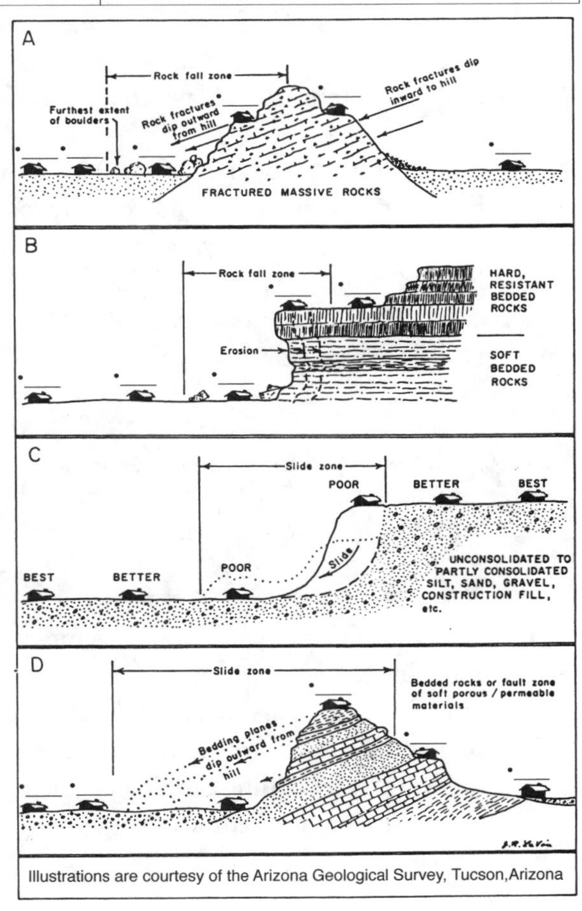

Illustrations are courtesy of the Arizona Geological Survey, Tucson,Arizona

1. What force drives mass wasting?
   A. heat
   B. gravity
   C. friction
   D. convection

2. Mass wasting tends to occur when
   A. a slope becomes steeper due to undercutting by a river or ocean waves
   B. the mass on a slope decreases by draining water from the ground
   C. friction is increased by draining water from the ground
   D. friction is decreased by taking water out of the ground

3. Talus consists largely of
   A. clay and other very fine rock particles
   B. a mixture of powdered rock and ice
   C. coarse, angular rock fragments
   D. alternate layers of sand, silt and clay

Hint: See *Talus Slope Near Hofn, Iceland* in the *Photo Gallery* on your CD.

4. The angle of repose is
   A. that angle at which rock material is most stable
   B. that angle at which lava flows will solidify without spreading out
   C. the angle of a slope that will no longer support large boulders and rock pillars
   D. the maximum slope at which a slope of loose material will lie without cascading down

5. An important factor in mass wasting is the orientation of rock layers, foliation, or jointing. For layered sedimentary and volcanic rocks, which condition is the least stable ?
   A. Rock layers are at right angles to the slope.
   B. Rock layers are parallel to the slope.
   C. Rock layers are horizontal to the slope.
   D. Rock layers stand vertical.

TEST-TAKING TIP

It is sometimes helpful to sketch the alternatives to a test question. For example, if you aren't sure about item 5 you could sketch "rock layers at right angles to the slope," "rock layers parallel to the slope," etc. Sketching can be particularly useful to kinesthetic learners. The movement of writing may unlock memory.

6. Which of the following would be most subject to mass movements (assume the slope and climate is the same in each case)?
   A. high-grade gneiss, with highly contorted foliation
   B. quartz-cemented sandstone, with layering perpendicular to the slope
   C. shale, with bedding dipping parallel to the slope
   D. massive granite bedrock

7. Solifluction usually occurs in
   A. cold regions
   B. very cold regions like Antarctica
   C. any area where there is lots of sunshine
   D. tropical regions

EXAM PREP

Materials in this section are most useful during your preparation for midterm and final exams. For optimal performance, midterm preparation should begin about eight days before the exam (see Appendix: *How to Study Geology* for details). The basic idea is a systematic review of material divided into short study sessions.

If you have used regular intensive study sessions to master the material, now you get a pay back! Review for your exam will proceed smoothly and take far less time.

The following *Chapter Summary* and *Practice Exam* should simplify review still further. Read the *Chapter Summary* to begin your session. It provides a helpful overview that should get you back into the material. Next, try the *Practice Exam*. Take it just as you would a midterm: to see how you stand in regard to mastery of this chapter. After you answer the questions, score them. Finally, and most important of all: review any question that you missed. Identify and correct the misconception that resulted in missing the item.

CHAPTER 11 SUMMARY

- Mass movements are slides, flows, or falls of large masses of rock material down slopes when the pull of gravity exceeds the strength of the slope materials. Such movements can be triggered by earthquakes, absorption of large quantities of water during torrential rain, undercutting by flooding rivers, human activities or other geologic processes.

- The three most important factors enhancing the potential for mass movements are: the steepness of the slope, the nature of the rock making up the slope, and the water content. Although steep slopes are prone to mass movements, slopes of only a few degrees can also fail catastrophically because of these other factors.

- Slopes become unstable when they become steeper than the angle of repose, the maximum slope angle that unconsolidated material will assume. Slopes in consolidated material may also become unstable when they are oversteepened or denuded of vegetation. Erosion by rivers and glaciers and human activities can oversteepen slopes and, thereby, enhance the potential for mass movement.

- The composition, texture and geologic structure of the slope material is another important factor influencing the potential for slope failure. For example, rocks with high clay content tend to be weak and may liquefy. Tilted layers of sedimentary or volcanic rocks are more likely to fail along bedding planes when the bedding parallels the slope. Failure of foliated metamorphic rocks is more likely to occur parallel to the direction of foliation.

- Water absorbed by the slope material contributes to instability in two ways: (1) by lowering internal friction (and thus resistance to flow) and, (2) by lubricating planes of weakness in the slope.

- The hazards and damage associated with mass movements can be minimized by careful geological assessment, engineering and land use policies that restrict development on unstable slopes.

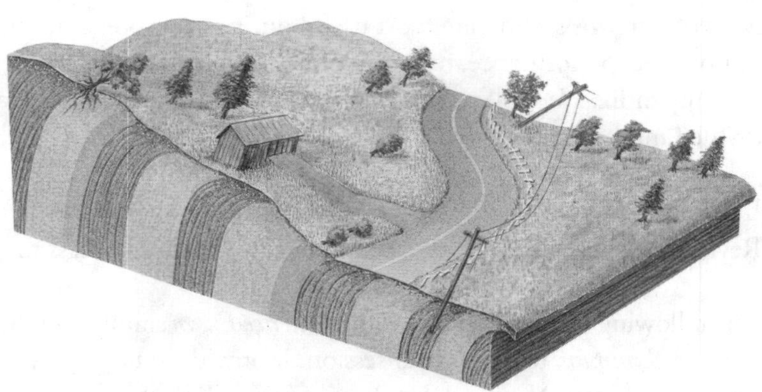

Figure 11.11 Mass movement is not restricted only to steep slopes. Soil creep (shown here), solifluction, earthflows, mudflows, and the liquefaction of clay layers can occur in nearly horizontal layers (also refer to Figure 11.7).

Website and CD Activities and Tools
http://www.whfreeman.com/presssiever
Study the photos of different kinds of mass movement provided in the *Photo Gallery* on your CD, at the Website, and in your textbook. As you study the images, assess what factors were most important in causing slope instability. Remember that the three most important factors enhancing the potential for mass movements are the steepness of the slope, the nature of the rock material in the slope, and the water content.

PRACTICE EXAM QUESTIONS FOR CHAPTER 11

(Answers and explanations are at the end of this chapter.)

Exercise 1: Water's Role in Mass Wasting

Water enhances the potential for mass wasting in many ways. Using your textbook as a guide, briefly describe five different ways water enhances the potential for mass movements.

1. ______________________________

2. ______________________________

3. ______________________________

4. ______________________________

5. ______________________________

Exercise 2: Evaluation of Slope Stability

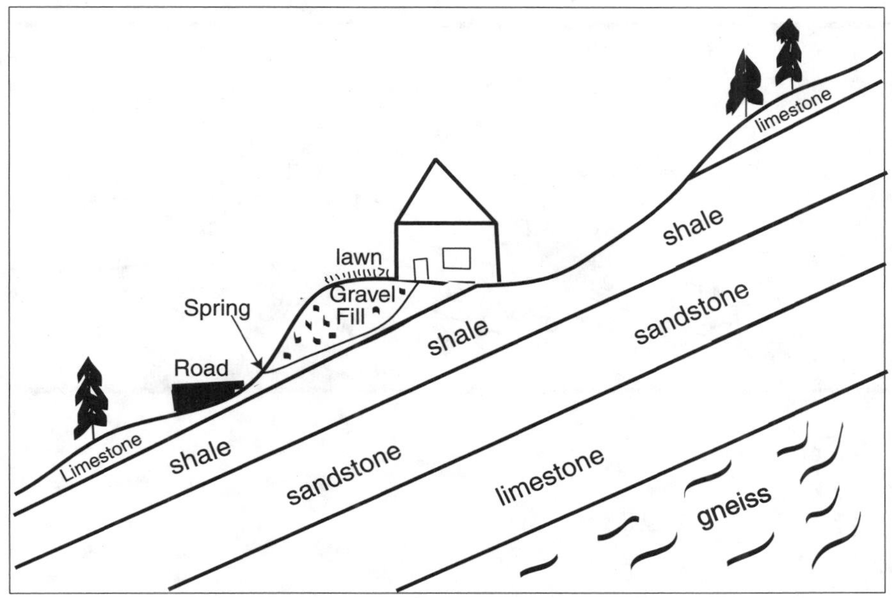

A. Discuss three factors that enhance the potential for mass movement at the home site shown above.

1. ______________________________

____

2. ______________________________

____

3. ______________________________

____

B. Given that the home is already built on this site, briefly discuss two possible ways of reducing the risk of damage to the house due to slope failure.

1. ______________________________

____

2. ______________________________

____

1. Which hillside home site is the best long-term investment?

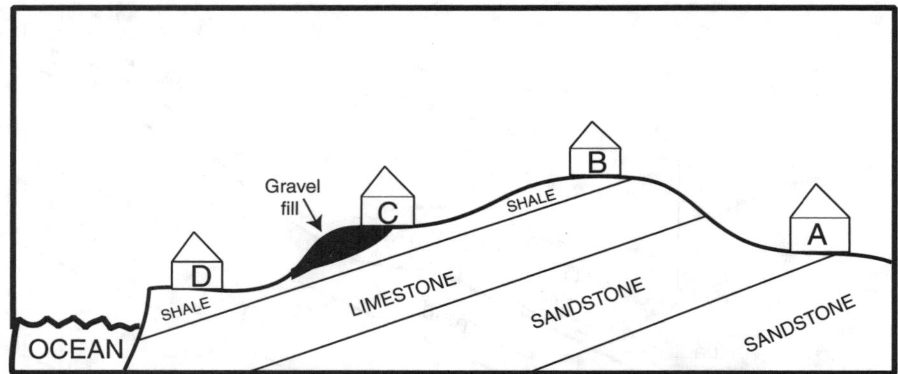

A. Site A
B. Site B
C. Site C
D. Site D

2. What is the most effective way to stabilize an active landslide, given the options below?
A. Piling additional rock and soil material on the landslide near its top.
B. Saturating the landslide itself with water.
C. Draining the water away from and out of the landslide area.
D. Cutting away the toe (base) of the landslide.

3. Roads through mountainous regions tend to be unstable and require more maintenance if they are built on
A. rock layers that dip perpendicular to the hill slope
B. rock layers that dip parallel to the hill slope
C. bedrock such as granite
D. horizontal lava flows

4. Your beautifully landscaped house, built on an idyllic Georgia hillside setting of small, irregularly undulating knolls and depressions, with trees tilted at interesting angles, has developed a bad case of broken and shifting foundation. The probable cause for the foundation problem is
A. melting permafrost
B. the house is built on an active earthflow
C. mud-flow from an active nearby volcano
D. root wedging from the trees

5. Homeowners in California that survived recent wildfires are not "out of the woods." With the approaching rainy season their next problem will be
A. increased potential for mud flows and debris flows
B. accelerated soil erosion
C. flash floods
D. all of the above

ANSWERS AND EXPLANATIONS FOR EXERCISES AND QUESTIONS

After Lecture

Exercise 1: Inventory of the Different Kinds of Mass Wasting

KIND OF MASS WASTING	COMPOSITION OF SLOPE	CHARACTERTISTICS
rock avalanche	large masses of rocky materials	Speed: fast Slope angle: steep Triggering event(s): earthquakes Notes: Occur in mountainous regions where rock is weakened by weathering, structural deformation, weak bedding, or cleavage planes.
creep	soil	Speed: slow Slope angle: any angle Triggering event(s): none Notes: Influenced by the kind of soil, climate, steepness of slope and density of vegetation.
earthflows	soils and fine-grained rock materials, like shales and clay-rich rocks	Speed: moderate Slope angle: any angle Triggering event(s): intense rainfall Notes: fluid-like movement
debris flow	rock fragments supported by a muddy matrix	Speed: fast Slope angle: any angle Triggering event(s): intense rainfall Notes: Contains coarser rock materials compared to earthflows.
mudflow	mostly finer rock materials with some coarser rock debris with large amounts of water	Speed: fast Slope angle: any angle Triggering event(s):intense rainfall or catastrophic melting of ice and snow by a volcanic eruption Notes: contains large amounts of water
debris avalanche	water-saturated soil and rock	Speed: fast Slope angle: steep Triggering event(s): earthquakes Notes: Occur in humid, mountainous regions.
slump	unconsolidated rock material	Speed: slow Slope angle: any slope Triggering event(s): rainfall Notes: Debris slide moves faster than a slump.
solifluction	surface layers of soil	Speed: slow Slope angle: any angle Triggering event(s): freeze / thaw Notes: Occurs only in cold regions when water in the surface layers of the soil alternately freezes and thaws. Water cannot seep into the ground because deeper layers are frozen.

Exercise 2: Assessing Potential Hazards from Mass Wasting

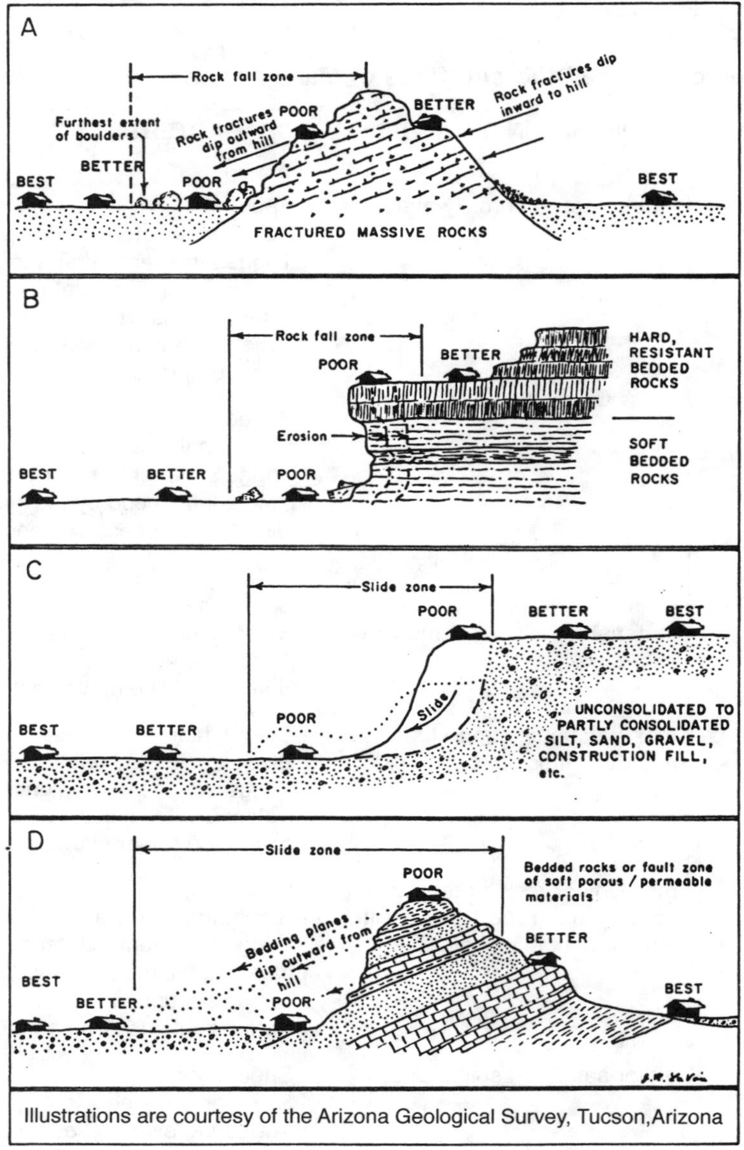

Illustrations are courtesy of the Arizona Geological Survey, Tucson,Arizona

1. B. Mass movements occur when the force of gravity exceeds the strength of the slope materials.
2. A. Undercutting by a river or waves will oversteepen a hillslope and enhance the potential for slope failure. Since water can act as a lubricant and also adds weight to the slope materials, draining it will reduce the weight of the slope material and increase friction, thereby reducing the potential for slope failure.
3. C. Talus refers to the blocks of rock that collect at the base of a steep slope or cliff.
4. D. The angle of repose for most loose sands is about 35°. The angle of repose varies significantly with a number of factors, one of which is the size of the particles (refer to Figure 11.1).
5. B. Bedding, joint planes or a foliation fabric all are potential planes of weakness within rock.

The orientation of any of these fabrics parallel to the hillslope compromises slope stability (refer to Figure 11.20).

6. C. Refer to Figure 11.20.
7. A. Solifluction is a result of repeated freezing and thawing.

Exam Prep

Exercise 1: Water's Role in Mass Wasting

Water enhances and triggers the potential for mass wasting in many ways. Using your textbook as a guide, briefly list different five ways water enhances the potential for mass movements. Below are five reasonable answers.

1. Water lubricates, especially if the ground is saturated (all pore spaces are filled with water), by reducing the internal friction between rock particles. In unconsolidated rock material a small amount of water increases surface tension which actually helps to weakly "glue" the damp, loose material together. Too much water keeps the particles apart and allows them to move freely over one another. Therefore, saturated sand, in which all pore space is occupied by water, runs like a fluid and collapses to a flat pancake shape (refer to Figure 11.3).
2. Water (hydrostatic) pressure may become great enough to separate the grains or promote the slippage of beds past one another. Refer to the section in your textbook entitled *Water Content*.
3. Water is a major agent in physical and chemical weathering which promotes mass wasting. Weathering promotes mass wasting by chemically and physically weakening rock.
4. Freezing and thawing is a specific role water plays in causing solifluction, a kind of mass movement. Refer to the discussion in your textbook on *Unconsolidated Mass Movements* and Figure 11.18.
5. Undercutting and oversteepening of hillslopes by erosion is another way water enhances mass wasting. Flowing water in rivers and frozen water in glaciers are powerful agents of erosion. Rivers typically erode on the outside of bends in the river. Erosion can undercut and oversteepen the river bank and adjacent hillslope.
6. Water-saturated rock materials rich in clays or loose sand may be transformed into fluid slurries by a process called liquefaction. Refer to the section in your textbook entitled *Triggers for Mass Movements*.

Exercise 2: Evaluation of Slope Stability

A. Discuss three factors that enhance the potential for mass movement at the home site shown.

- The house is built on a cut and fill foundation. A cut and fill foundation is particularly susceptible to slope failure because it is very loose material that has been bulldozed into place.
- The house and associated possessions add weight to the slope.
- Watering the lawn will enhance the potential for slope failure.
- The presence of a spring indicates that the slope beneath the house is saturated with water. Water adds weight to the slope and acts as a lubricant.

- Traffic on the road below the house adds weight and creates vibrations in the ground which may compromise slope stability.
- The slope consists of sedimentary rocks that have a dip parallel to the slope. Bedding planes are zones of weakness within these layers; therefore, this rock layer orientation enhances the potential for slope failure.
- The orientation of the rock fabric within the slope enhances the potential for slope failure. Shale is an especially soft and weak sedimentary rock. Slope failure is likely to occur along the shale layer which dips parallel to the slope. Even the foliation within the gneissic bedrock parallels the hillslope. Planes of weakness within a rock typically occur parallel to the rock's textural fabric.
- If the slope was undercut by the road builders, this will also compromise the slope stability.

B. Given that the home is already built on this site, briefly discuss two possible ways of reducing the risk of damage to the house due to slope failure.

This is an inherently unstable slope. A slope ordinance probably should have restricted building on this slope. However, given that the house is already there, what might be done to decrease the risk for slope failure? Reasonable approaches include:

- Don't water the lawn.
- Drain water from the slope above and off the roof away from the gravel fill beneath the house.
- At some expense, rock bolts could be installed to help stabilize rock layers.
- Maintain a good cover of vegetation on the slope above and below the house.
- Put in a retaining wall along the road below the house and be sure that wall does not restrict the drainage of water out of the slope.

1. A. Site D may have the best view of the shoreline and Site B the best view, but Site A is on the most stable ground.
2. C. Draining the water from the landslide area would help to reduce the weight of the slope materials and reduce the potential lubricating effects of water.
3. B. Bedding planes, joint planes or textural fabrics like foliation are zones of weakness within rock. If they parallel the hillslope, the potential for slope failure is enhanced.
4. B. Refer to *Unconsolidated Mass Movements* in your textbook.
5. D. The barren slopes left by the wild fires will enhance the potential for all of the hazards listed.

Chapter 12
The Hydrologic Cycle and Groundwater

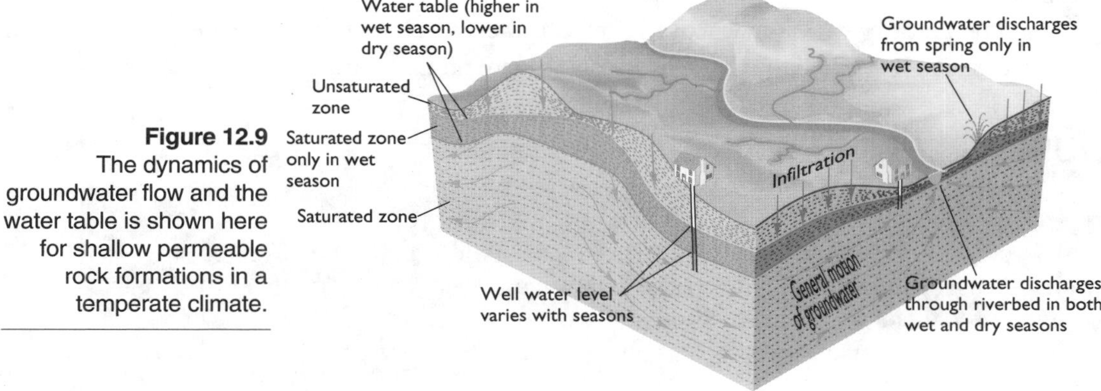

Figure 12.9 The dynamics of groundwater flow and the water table is shown here for shallow permeable rock formations in a temperate climate.

BEFORE LECTURE

Before you attend lecture be sure to spend some time previewing the chapter. For an efficient preview use the questions below. *Chapter Preview* questions constitute the basic framework for understanding the chapter. Preview works best if you do it just before lecture. With the main points in mind you will understand the lecture better. This in turn will result in a better and more-complete set of notes. How much time should you devote to preview? Obviously, more time is better than less. But even a brief (five or ten minute) preview session just before lecture begins will produce a result you will notice. For a refresher on why previewing is so important see the Appendix: *How to Study Science.*

CHAPTER PREVIEW

- **How does water move around and in the Earth?**
  Brief answer: The hydrologic cycle is a model for the movement of water on Earth (refer to Figure 12.2).

- **How does water move below the ground surface?**
  Brief answer: Porosity and permeability are the principle factors that control the infiltration and flow of ground water.

- **What factors govern our use of groundwater resources?**
  Brief answer: Water, like soil and air, is one of our most basic natural resources. Some of the most important factors governing our ability to use groundwater are: the depth to the ground water table, springs and artesian systems, the balance between recharge and discharge, Darcy's Law, and water quality.

- **What geologic processes and features are associated with groundwater?**
  Brief answer: Groundwater is the geologic agent responsible for caves, karst topography and the formations that decorate caves. Groundwater is a vital component of geothermal systems.

Vital Information from Other Chapters

- Chapters 12 and 13 are a package. A quick look through both chapters will provide you with a more complete and effective preview before you start lectures on these topics. Hint: For a rapid, efficient review of Chapter 13 use the *Chapter Preview* questions.

Website and CD Preview

http://www.whfreeman.com/presssiever

- The *Hydrologic Cycle* Interactive Exercise at the Website is a short exercise that provides an excellent preview of the whole hydrologic system on Earth. Refer to Figure 12.2 as a guide.

DURING LECTURE

WARM-UP ACTIVITY

This is a very interesting chapter! Give it a chance to speak to you by spending 5 to 10 minutes just before lecture browsing for topics and pictures that interest you. What figure grabs your attention? What issue interests you the most? After browsing, ask yourself what you would most like to learn from this chapter/lecture.

One goal for lecture should be to leave class with a good set of answers to the *Preview Questions*.

- To avoid getting lost in details, keep the "big picture" in mind: Chapter 12 is a survey of water in and around the earth. It tells the story of how water moves in around the Earth in the hydrologic cycle in a manner that insures balance.
- Focus on understanding Figure 12.2, The Hydrologic System.

AFTER LECTURE

The perfect time to review your notes is right after lecture. The checklist below contains both general review tips and specific suggestions for this chapter.

NOTE REVIEW CHECKLIST

✔ All notes legible? (Rewrite so they read easily.)

✔ Important points clearly identified? You should now have headers in your notes that tie to each of the questions in the *BEFORE LECTURE/Chapter Preview.*

✔ Holes (missing material) filled in from memory?

✔ Areas where you don't remember what was said marked for a follow-up session with your instructor, tutor, or study partner?

✔ Possible test questions indicated in the margin (TQ)?

✔ Additional visual material. *Suggestions for Chapter 12: Sketch a simple version of Figure 12.7, Porosity of Rock Materials. Photocopy a copy of Exercise 1 after you have completed the chart. This will be a great aid for exam review: it summarizes all you will need to know about porosity.*

✔ Reworked notes into a form that is efficient for your learning style?

✔ Created a brief "big picture" overview of this lecture (using a sketch or written outline)? *Hint: Your own sketched version of the hydrologic cycle (Figure 12.2) would provide a good visual summary of the chapter. You might also try writing a paragraph answer in your own words to each of the Preview Questions. Want feedback? Compare your answers to the Chapter Summary.*

Intensive Study Session

Schedule at least one hour after lecture for intensive study. Use the Website and CD Activities and Tools suggested below along with the *Practice Exercises* and *Study Questions* to insure you master this chapter. Do as many of these as you have time for during your scheduled study session. Pay particular attention to exercises recommended by your instructor during lecture.

Website and CD Activities and Tools

http://www.whfreeman.com/presssiever

- **Website Q & A Practice Multiple Choice Questions:**

  At the Website complete the *Q & A*. Pay particular attention to the explanations for answers. Flashcards at the Website will help you learn new terms. The *Photo Gallery* on your CD and at the Website contains images illustrating a variety of the topics in this chapter.

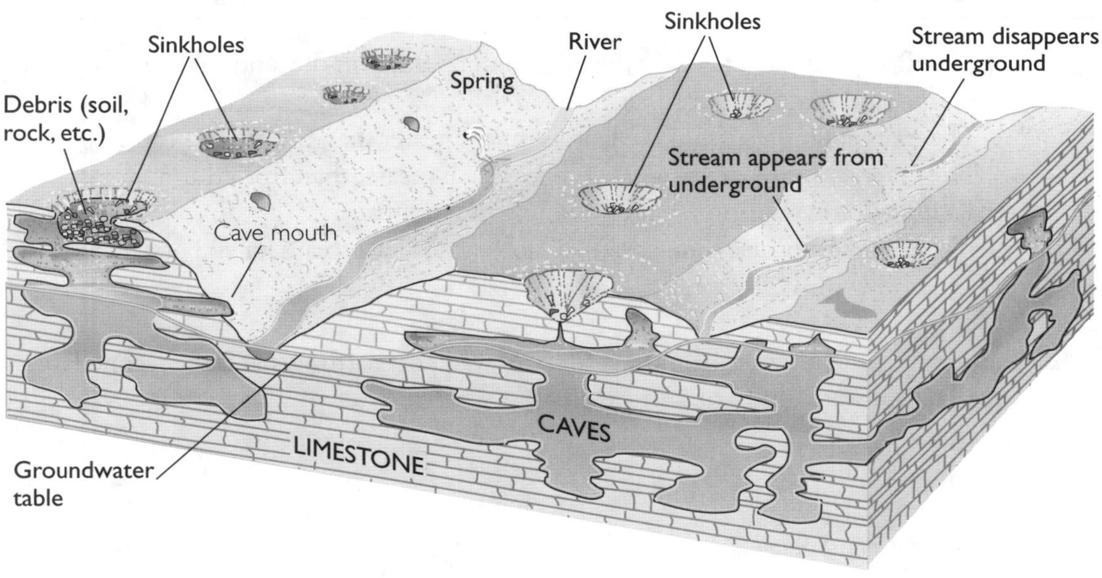

Figure 12.20 / Some major features of karst topography.

PRACTICE EXERCISES AND STUDY QUESTIONS

(Answers and explanations are at the end of this chapter.)

Exercise 1: Evaluating Rock Materials as Potential Aquifers

You have recently purchased a rustic country cabin and need to drill a new well for a dependable water supply for the cabin. The geology around your cabin is complex due to ancient mountain building events. Because the rocks are folded and faulted, it is difficult to predict what rock might be encountered as the water well is drilled. Which of the following rock materials has the potential of yielding ground water to your well?

Hint: Keep in mind that generally, permeability increases as porosity increases, but not always. Permeability also depends on the sizes of the pores, how well they are connected, and how tortuous a path the water must travel to pass through the material (refer to Figure 12.7).

ROCK MATERIAL	POROSITY (high, medium, low)	POTENTIAL AS AN AQUIFER (good, moderate, poor)
loose, well-sorted, coarse sand		
silt and clay	low	
granite and gneiss		poor
highly fractured granite		
sandstone	medium	
shale		
highly jointed limestone		moderate to good

1. In the hydrologic cycle, how does the evaporation rate from the land surface compare to the evaporation rate off the oceans?
   A. The evaporation rate from the land is much greater than from the oceans.
   B. The evaporation rate from the oceans is much greater than from the land.
   C. Evaporation rates from the land and oceans are equal.
   D. There is no reasonable comparison. It is too wet over the oceans for evaporation to occur.

Hint: Refer to Figure 12.2.

2. The oceans contain by far the most amount of water on Earth. What is the second largest reservoir for water on Earth?
   A. lakes
   B. groundwater
   C. rivers
   D. polar ice and glaciers

3. The oceans contain about how much of the water in the hydrosphere?
   A. 96%
   B. 80%
   C. 50%
   D. 35%

4. What happens to the porosity as the grain size gets smaller?
   A. It increases.
   B. It remains unchanged.
   C. It decreases.
   D. None of the above.

5. The water table is
   A. the top of the unsaturated zone
   B. the top of the saturated zone
   C. generally present only in moist climates
   D. the contact between an aquifer and underlying, impermeable layer of rock

6. The ability of a solid, such as rock, to allow fluids to pass through it is
   A. discharge
   B. capillary fringe
   C. permeability
   D. porosity

7. An icicle-like deposit hanging from the ceiling of a cave is a
   A. stalactite
   B. karst formation
   C. stalagmite
   D. quartzite

8. Of the following rock types, which is the most susceptible to ground water solution, therefore making it a formation most likely to have caves?
   A. granite
   B. sandstone
   C. limestone
   D. shale

9. A rock or soil layer that is water-bearing is
   A. a perched water table
   B. a zone of aeration
   C. a stratum
   D. an aquifer

10. The potential for geothermal energy is highest in a region that has numerous
   A. surface lakes
   B. caves
   C. hot springs
   D. sinkholes

11. An aquiclude is
    A. a confined aquifer
    B. always located at the top of the water table
    C. a rock layer that provides a good flow of water into a well
    D. an impermeable rock layer that does not allow water to flow through it

12. Which of the following would make the best aquifer?

	Porosity	Permeability
A. Rock A	5%	high
B. Rock B	10%	medium
C. Rock C	30%	low
D. Rock D	35%	medium

Americans now drink more soda pop than water from the kitchen tap — 47 gallons of soda pop to only 37 gallons of water per person soda each year.
—*World Watch, 1990*

EXAM PREP

Materials in this section are most useful during your preparation for midterm and final exams. For optimal performance, midterm preparation should begin about eight days before the exam (see Appendix: *How to Study Geology for details*). The basic idea is a systematic review of material divided into short study sessions.

If you have used regular intensive study sessions to master the material, now you get a pay back! Review for your exam will proceed smoothly and take far less time.

The following *Chapter Summary* and *Practice Exam* should simplify review still further. Read the *Chapter Summary* to begin your session. It provides a helpful overview that should get you back into the material. Next, try the *Practice Exam*. Take it just as you would a midterm: to see how you stand in regard to mastery of this chapter. After you answer the questions, score them. Finally, and most important of all: review any question that you missed. Identify and correct the misconception that resulted in missing the item.

CHAPTER 12 SUMMARY

- The hydrologic cycle is a flow chart or model for the distribution and movements of water on and below the surface of the Earth. The major reservoirs for the hydrologic cycle are oceans, glaciers, groundwater, lakes and rivers, the atmosphere, and biosphere in decreasing volumes. Water moves in and out of these reservoirs by various pathways and at varying rates. Over the short term, a balance is maintained among the major reservoirs at and near the Earth's surface. However, climate change, longer-term tectonic processes such as mountain building and human activity can alter the rate of water movement between reservoirs and impact the size of the reservoirs.

- Groundwater is estimated to be the third largest reservoir of water on Earth. The infiltration of water into the ground and groundwater flow are largely controlled by the porosity and permeability of the rock materials and topography. A groundwater aquifer is in dynamic balance between recharge (the amount of water that infiltrates into the aquifer) and discharge which can occur from springs or wells.

- Darcy's Law describes the groundwater flow rate in relation to the slope of the water table and the permeability of the aquifer.

- Human demand for groundwater has increased to a level where pumping discharges from many aquifers exceeds the natural rates of recharge. As a result, aquifers are being depleted and groundwater tables are lowering to a point where dependable, quality groundwater is becoming more and more of a challenge to supply.

- Water quality may be compromised by both natural and human sources of contamination. Various factors like recharge rate and aquifer size influence the amount of effort and effectiveness of attempts to clean up contamination.

- Caves, sinkholes and associated karst topography are sa result of the dissolution of carbonate rocks (limestone) by groundwater. Karst topography is most well-developed in regions of high-rainfall, abundant vegetation, an underlying extensively fractured limestone, and an appreciable hydrologic gradient to enhance groundwater flow rates. Environmental problems associated with karst regions include surface subsidence from collapse of underground space and catastrophic cave-ins and sinkhole formation.

- All rocks below the groundwater table are saturated with water. With increasing depth, porosity and permeability typically decreases as confining pressure increases. Water temperature increases progressively with increasing depth and, as a result, the water dissolves more solids. Hot springs and geysers are surface expressions of the circulation of hydrothermal waters over a magma body or along a deep-seated fault.

Per capita domestic water use in the United States is two to four times as great as that of Western Europe, where users pay as much as 350 percent more for their water.
—Press and Siever

Website and CD Activities and Tools
http://www.whfreeman.com/presssiever
The *Photo Gallery* on your CD and at the Website contains images illustrating a variety of the topics in this chapter.

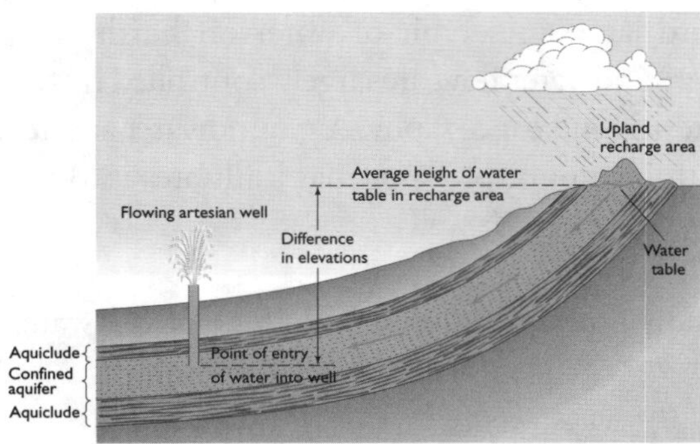

Figure 12.11 / An artesian system requires an aquifer that is confined by two aquicludes (beds of low permeability). The artesian well flows in response to the difference in natural pressure between the height of the water table in the recharge area and the bottom of the well. An artesian spring occurs if the confined aquifer is cut by a fracture along which water can flow to the surface or cut by the Earth's surface itself.

Figure 12.12 / A perched water table can form due to many different geologic situations — in this case, by a shale aquiclude located above the main water table in a sandstone aquifer. A spring occurs where the land surface intersects the perched water table.

PRACTICE EXAM QUESTIONS FOR CHAPTER 12

(Answers and explanations are at the end of this chapter.)

Exercise 1: Evaluating Groundwater Wells

Fill in the blanks below with either "high," "low" or "none" for your evaluation of the potential characteristic of each well.

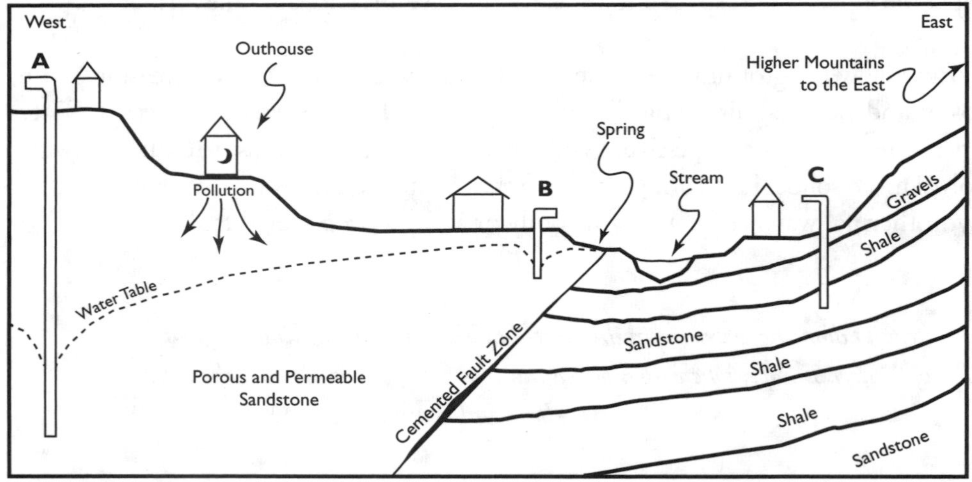

Well A — potential for 1. pollution ________________

2. artesian flow ________________

3. discharge ________________

Well B — potential for 4. pollution ____________________

5. artesian flow ____________________

6. discharge ____________________

7. long-term supply ____________________

Well C — potential for 8. pollution ____________________

9. artesian flow ____________________

10. discharge ____________________

1. If all other conditions are equal, ground water will move faster
   A. where sand grains are very well cemented
   B. through loose sand than through clay
   C. where permeability of the aquifer is lower
   D. through clay than through sand

Hint: Refer to Figure 12.7.

2. A perched water table will most likely develop on top of
   A. shale
   B. highly fractured granite
   C. gravel
   D. sandstone

3. Rivers and streams that flow all year long, even during long periods without rain, are likely to be fed by
   A. sinkholes
   B. springs
   C. wells
   D. karst conditions

Hint: Refer to Figure 12.10.

4. Stalactites, stalagmites and other cave formations are formed as
   A. ground water evaporates upon entering a cave
   B. ground water dissolves limestone
   C. temperature change induces the precipitation of the calcium carbonate
   D. loss of dissolved CO2 gas from the ground water causes precipitation of calcium

5. An artesian well will flow if the
   A. top of the well is lower than the water table in the recharge area
   B. top of the well higher than the water table in the recharge area
   C. bottom of the well is lower than the land surface in the recharge area
   D. bottom of the well is lower than the water table in the recharge area

Hint: Refer to Figure 12.11.

6. Which well will exhibit artesian flow?

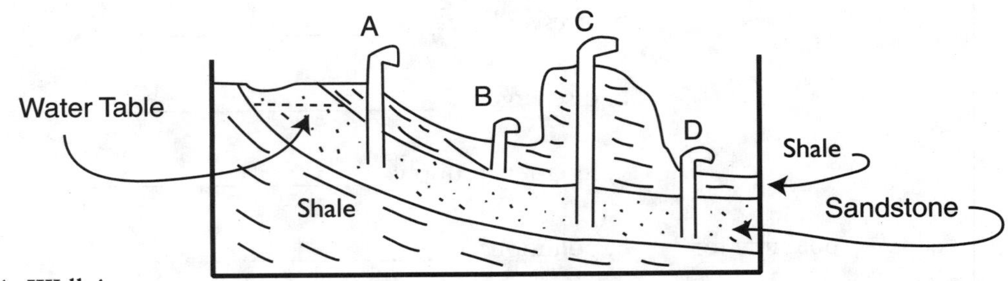

A. Well A
B. Well B
C. Well C
D. Well D

Hint: Refer to Figure 12.11.

7. Which materials would make the best aquifer:
A. clay and silt
B. gravel and sand
C. unfractured granite
D. highly cemented sandstone

8. At a shallow depth, a well will most likely encounter a good water supply in which of the following locations?
A. in granite on a ridge top
B. in sandstone on a ridge top
C. in a shale in a valley bottom
D. in sandstone in a valley bottom

Hint: Make a sketch illustrating each situation.

9. Which well is most likely pumping polluted water?

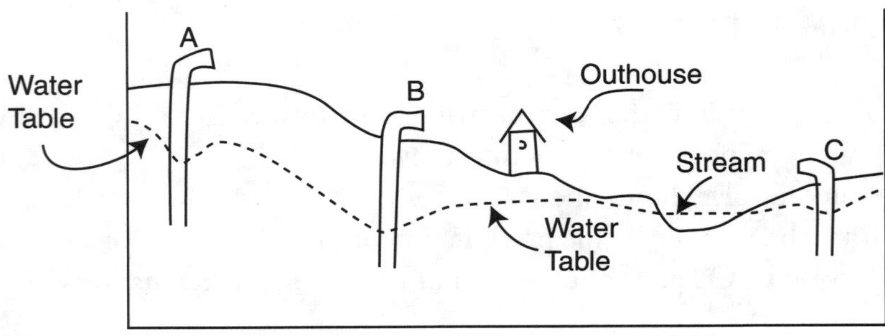

A. Well A
B. Well B
C. Well C
D. None

10. Of the wells illustrated in the diagram below, which one will produce the greatest water over the longest time?

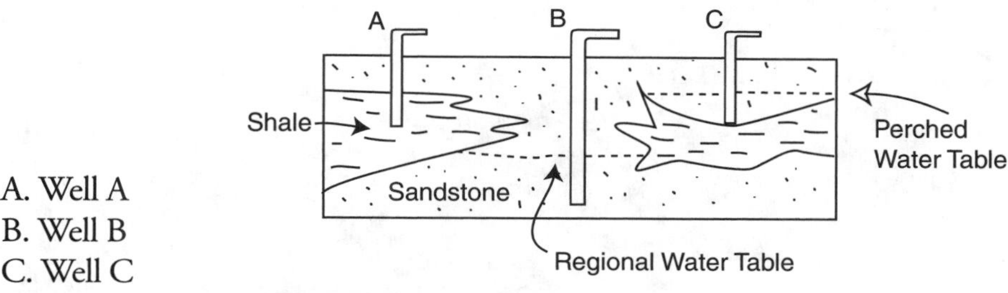

A. Well A
B. Well B
C. Well C
D. All of the wells will have high productivity.

11. If water is pumped from well K faster than natural recharge replenishes it, the result will most likely be

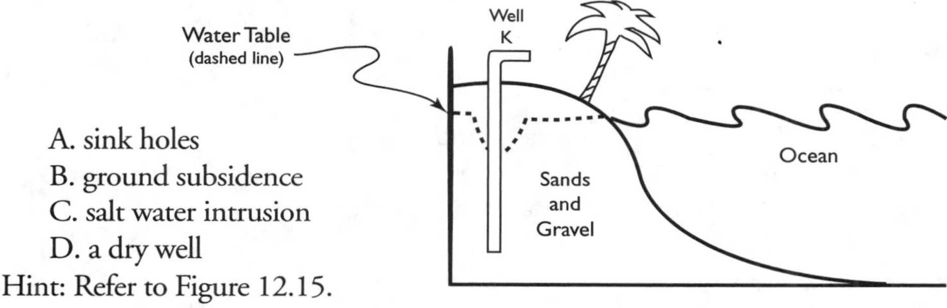

A. sink holes
B. ground subsidence
C. salt water intrusion
D. a dry well

Hint: Refer to Figure 12.15.

12. Which water source illustrated below is LEAST likely to be polluted?

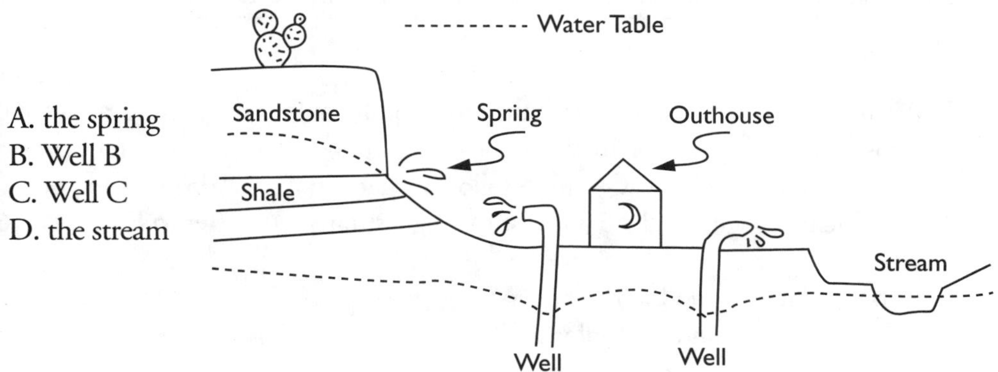

A. the spring
B. Well B
C. Well C
D. the stream

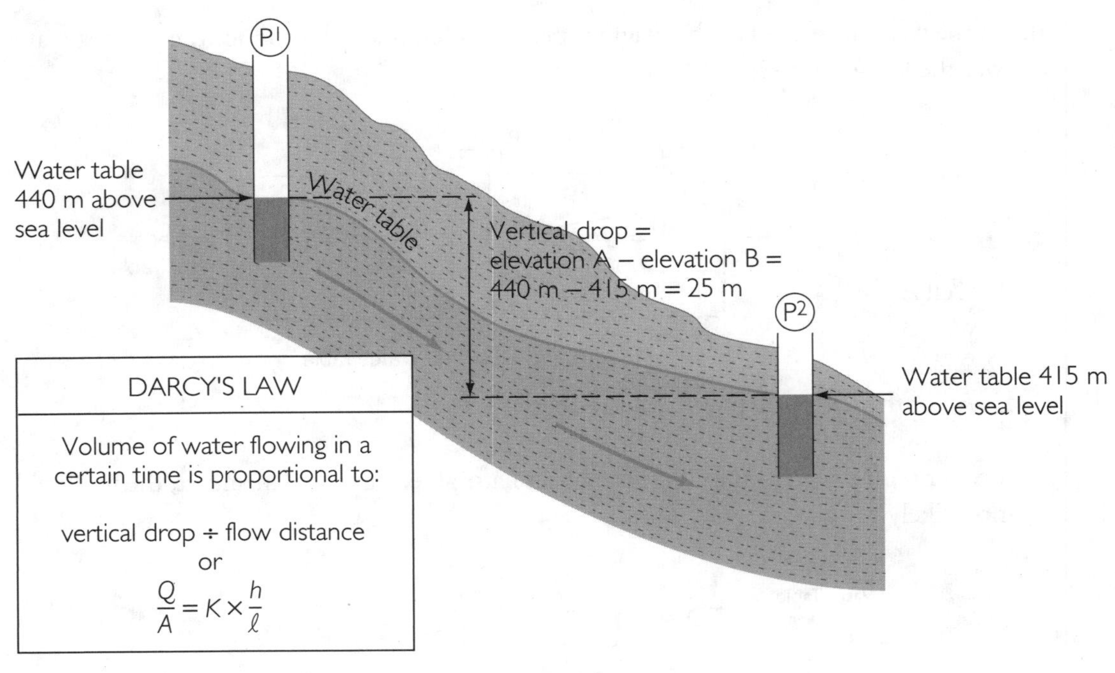

Q: Volume of water flowing in a given time

A: Cross-sectional area through which water flows

K: Hydraulic conductivity (a measure of permeability)

h: Vertical drop between two points

ℓ: Distance the flow travels

Figure 12.16 / Darcy's Law describes the rate of groundwater flow down a slope between two points, P^1 and P^2. The rate of flow is proportional to the difference in height between the high and low points of the slope divided by the flow distance between them and K, a constant proportional to the permeability of the aquifer.

13. Assume that you are dealing with the same aquifer at four different localities and that A (cross-sectional area through which the water flows) and K (the hydraulic conductivity) are the same for each site. Given the following data on vertical drop (h) and flow distance (l), which well will produce water at the highest rate (Q)? Refer to Figure 12.16.

    A. h = 20 meters and l = 500 meters
    B. h = 30 meters and l = 1 kilometer
    C. h = 300 meters and l = 6 kilometers
    D. h = 600 meters and l = 100 kilometers

ANSWERS AND EXPLANATIONS FOR EXERCISES AND QUESTIONS

After Lecture

Exercise 1: Evaluating Rock Materials as Potential Aquifers

ROCK MATERIAL	POROSITY (high, medium, low)	POTENTIAL AS AN AQUIFER (good, moderate, poor)
loose, well-sorted, coarse sand	high	good
silt and clay	low	poor
granite and gneiss	low - Interlocking grains of silicate minerals provides for little pore space.	poor
highly fractured granite	medium - Fracturing can significantly increase pore space and improve permeability.	moderate
sandstone	medium - The cement that holds the sand grains together reduces pore space. Nevertheless, sandstones are typically good aquifers.	moderate to good
shale	low - Fracturing will increase pore space, but permeability may still remain low. Shales are typically aquicludes.	poor
highly jointed limestone	medium - Fracturing and the formation of a cavern system within the limestone can greatly enhance the porosity and permeability of limestones. Caverns serve as an open plumbing system for groundwater.	moderate to good

1. B. Evaporation from the oceans is more than six times more than from the land surface (refer to Figure 12.2).
2. D. Polar ice caps and glaciers contain about 2.97% of all water on Earth. This is the second largest reservoir for water (refer to Figure 12.1).
3. A. Refer to Figure 12.1.
4. C. Rock particles typically can compact more efficiently as grain-size decreases. This reduces porosity and permeability (refer to Figure 12.7).
5. B. Refer to Figure 12.8.

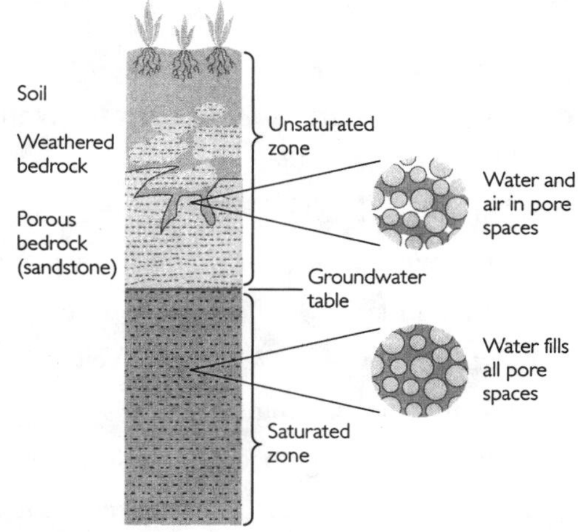

Figure 12.8 / The groundwater table is the boundary between the unsaturated zone and the saturated zone.

6. C. Permeability is the ability of a solid to allow fluids to pass through it. Generally, permeability increases as porosity increases, but not always. Permeability also depends on the sizes of the pores, how well they are connected, and how tortuous a path the water must travel to pass through the material.
7. A. Stalactite holds "tight" to the ceiling.

> **Memory Tip:**
> **You can easily remember the difference between a stalagmite and a stalactite because stalactites hold "tight" to the ceiling and stalagmites "might" reach the ceiling of a cave.**

8. C. Carbonate rocks, like limestone, are susceptible to dissolution. A wet climate favors cave formation.
9. D. An aquifer is a rock that will yield a good flow of groundwater.
10. C. Refer to *Water Deep in the Crust* in Chapter 12.
11. D. Shales are typically aquicludes.
12. D. Porosity and permeability both influence how much water an aquifer can produce. Porosity represents the total amount of pore space in which groundwater can be stored. Permeability is linked to porosity and represents the ability of water to flow through the rock. Some shales and lava flows have moderate to high porosity but very low permeability because the pore spaces are not interconnected. Some rocks like granite have moderate permeability because they are extensively fractured. However, porosity can be low because the fracturing is very tight.

Exam Prep

Exercise 1: Evaluating Groundwater Wells

Well A — potential for:

1. Pollution – High because the water table slopes towards Well A and up slope is the outhouse. The cone of depression produced by pumping Well A enhances the potential for pollution by creating a larger gradient in the water table between Well A and the outhouse.
2. Artesian flow – None because Well A will exhibit no artesian flow because the aquifer is not confined nor sloping.
3. Discharge – High because the aquifer is a porous and permeable sandstone of large volume and Well A is drilled deep into the aquifer.

Well B — potential for

4. Pollution – Assuming the house does not release pollutants, low because the water table slopes away from Well B and the outhouse is down gradient. The cone of depression around Well B is small and an issue.
5. Artesian flow – None because Well B will exhibit no artesian flow because the aquifer is not confined nor sloping.
6. Discharge – As long as the water table does not lower, discharge from Well B will be potentially high.
7. Long-term supply – Low because Well B is shallow. A lowering of the water table, due to a change in climate or due to pumping from Wells A and B at rates that exceed recharge, could compromise the productivity of Well B.

Well C — potential for

8. Pollution – Low because there is not source of pollution shown for the confined sandstone aquifer.
9. Artesian flow – High because the aquifer for Well C is a tilted and confined layer of sandstone and the recharge area for the aquifer appears to be higher in elevation than the top of the well.
10. Discharge – High because with high mountains to the east there should be good recharge of the confined sandstone aquifer.

1. B. Loose sand will typically exhibit high porosity and permeability because the round sand grains cannot pack together very efficiently (refer to Figure 12.7).
2. A. Refer to Figure 12.12.
3. B. Rivers fed by springs from a shallow groundwater table that intersects the stream channel are called effluent (refer to Figure 12.10).
4. D. Loss of carbon dioxide gas from groundwater as it enters the cave atmosphere, or due to agitation during impact or flow, will induce the precipitation of calcium carbonate. Over time, this process forms stalactites and stalagmites.
5. A. Refer to Figure 12.11.
6. D. Well D. Refer to Figure 12.11.
7. B. Sand and gravel will typically exhibit a high porosity and permeability.
8. D. Sandstones are typically good aquifers. Since water flows down hill, it will tend to collect and recharge aquifers located beneath topographically low spots like a valley.
9. B. Well B. The cone of depression around well B has lowered the water table enough that effluent leached from the outhouse will flow towards well B, even though the topography slopes toward the river. Refer to Figures 12.13 and 12.22.
10. B. Well B will be the most productive water well because it taps into a larger sandstone aquifer. Well A is not likely to be productive since it bottoms out in a shale which typically has very low permeability. Well C will yield water since it bottoms out in the sandstone, above a

shale unit that creates a perched water table. The volume of water from Well C will be limited.

11. C. Refer to Figure 12.15.
12. A. The spring is least likely to be contaminated by the outhouse because it is produced from a perched water table well above and independent of the aquifer that is affected by the pollutants from the outhouse.
13. C. h/l for A = 0.04, B = 0.03, C = 0.05, and D = 0.006.

The frog does not drink up the pond in which it lives.
—Native American proverb

Chapter 13

Streams: Transport to the Oceans

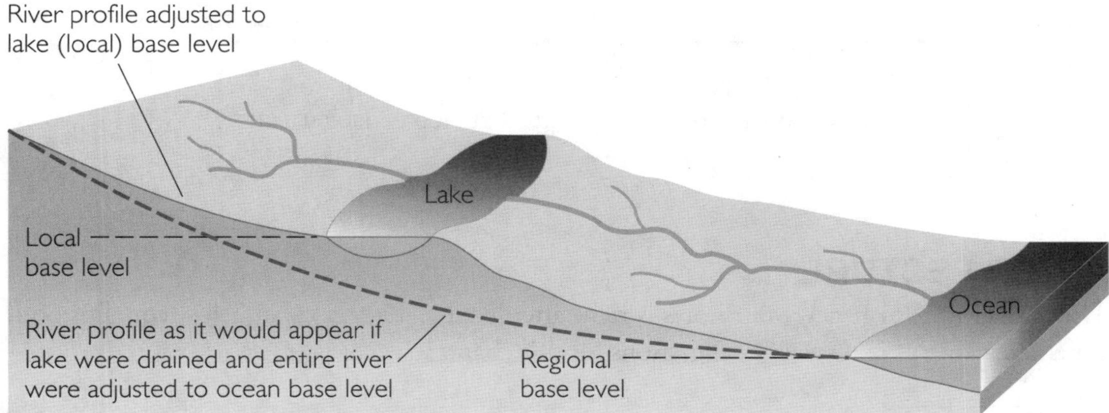

Figure 13.17 / A stream's longitudinal profile (dashed line) is a plot of the stream gradient (slope of the stream channel). The lake represents a local base level and the ocean is the regional base level. In each river segment, the river adjusts to the lowest level to which it can erode, which is determined by the local or regional base levels.

BEFORE LECTURE

Before you attend lecture be sure to spend some time previewing the chapter. For an efficient preview use the questions below

CHAPTER PREVIEW

- **How is flowing water in streams able to erode solid rock and to transport and deposit sediments?**
  Brief answer: Flow water erodes rock by physical and chemical weathering processes. Turbulent flow is responsible for transporting sediments. When a stream flow slows, it loses its competence to carry sediment and deposits it.

- **How do stream valleys and their channels and floodplains evolve?**
  Brief answer: *Stream Dynamics*, an animation on your CD, will provide an overview of the channel evolution and sediment transport down a river system.

- **How is the stream's gradient (slope), velocity (speed), discharge (amount of water), and sediment transport linked?**
  Brief answer: The dynamic equilibrium model can be used to predict the behavior of a stream system as major variables characterizing a stream system change.

- **How do drainage networks work as collection systems and deltas as distribution systems for water and sediment?**
  Brief answer: Drainage patterns depend on topography, rock type and geologic structure. Deltas are major sites of deposition of sediment.

Vital Information from Other Chapters

- Chapter 12 and 13 are a package. Be sure to review material in the first half of Chapter 12.

DURING LECTURE

One goal for lecture should be to leave class with a good set of answers to the preview questions.

> Mark Twain, noting how muddy the Missouri was, proclaimed it "too thick to navigate but too thin to cultivate.

AFTER LECTURE

The perfect time to review your notes is right after lecture. The checklist below contains both general review tips and specific suggestions for this chapter.

NOTE REVIEW CHECKLIST

- ✔ All notes legible? (Rewrite so they read easily.)
- ✔ Important points clearly identified? You should now have headers in your notes that tie to each of the questions in the *BEFORE LECTURE/Chapter Preview.*
- ✔ Holes (missing material) filled in from memory?
- ✔ Areas where you don't remember what was said marked for a follow-up session with your instructor, tutor, or study partner?
- ✔ Possible test questions indicated in the margin (TQ)?
- ✔ Additional visual material. *Suggestions for Chapter 13: Be sure you get the important figures into your notes (see visual note-taking suggestions above). If you noted chapter figure numbers during lecture it will be easy to fill in material you missed.*
- ✔ Reworked notes into a form that is efficient for your learning style?
- ✔ Created a brief "big picture" overview of this lecture (using a sketch or written outline)?

Intensive Study Session

After lecture you need to thoroughly master the processes by which rivers evolve and transport material. Schedule at least one hour after lecture for intensive study. Use this time to master key concepts. By now you know well that mastery is not gained by just reading your text. To master geology, you must ask yourself questions and answer them. Use the Website and CD Activities and Tools suggested below along with the *Practice Exercises* and *Study Questions* to insure you master this chapter. Do as many of these as you have time for during your scheduled study session. Pay particular attention to exercises recommended by your instructor during lecture.

Website and CD Activities and Tools

http://www.whfreeman.com/presssiever

- **Website Q & A Practice Multiple Choice Questions:**
  At the Website complete the *Q & A*. Pay particular attention to the explanations for answers. Flashcards at the Website will help you learn new terms. The Interactive Exercise *Marine Delta* at the Website will help you learn the basic components of a large marine delta. *Stream Dynamics*, an animation on your CD, will provide an overview of the channel characteristics and sediment transport down a river system.

PRACTICE EXERCISES AND STUDY QUESTIONS

(Answers and explanations are at the end of this chapter.)

Exercise 1: Stream Velocity

Stream velocity is a dependent variable that governs stream behavior—whether a stream is dominantly eroding and transporting or depositing sediments along a section of the channel. Various independent variables (factors) influence stream velocity and therefore can affect the behavior of a certain stretch of stream channel. The major independent variables affecting velocity are listed in the table below. Complete the table by describing how changes in each factor affects stream velocity.

VARIABLE AFFECTING STREAM VELOCITY	RELATIONSHIP OF VARIABLE TO STREAM VELOCITY	ANALOGY
Gradient - the slope of the stream channel		Do you tend to walk faster down a steep slope or a more gradual slope? — faster down a steep slope
Discharge - the amount of water in the stream channel		Will you move into a new house slower or faster if you have more people helping you? — faster
Sediment load		Will you travel faster or slower if you are carrying a heavier load? — slower
Channel characteristics:		
• channel roughness	As channel roughness increases, velocity will decrease.	Cross country hiking without a trail tends to slow one down.
• channel shape	The stream has more contact with the channel surface if the channel shape is very wide or very narrow. More contact with the channel will increase drag and decrease velocity.	When you have more contact with the ground surface, like when you crawl you move slower than if the only contact you have is with the bottom of your feet.

Exercise 2: Relationship between Stream Flow and Groundwater

Why do streams in desert regions typically flow intermittently and streams in more temperate regions, like New England, flow year-round? Well-labeled diagrams illustrating each situation with a brief discussion is an excellent way to answer this question. Hint: Refer to Figure 12.10 in Chapter 12.

1. What is the largest river in North America?
   A. Columbia River
   B. Colorado River
   C. St. Lawrence River
   D. Mississippi River

2. Where do rivers obtain their power to erode and transport sediments?
   A. heat
   B. gravity
   C. electricity
   D. friction

3. The volume of water which flows past a given point along the stream channel in a given interval of time is the
   A. velocity
   B. discharge
   C. capacity
   D. gradient

4. If the gradient (slope) of a stream is increased, what happens to the velocity of the water?
   A. velocity increases
   B. velocity decreases
   C. velocity remains unaffected
   D. velocity may increase or decrease

5. Stream erosion and deposition are primarily controlled by a river's
   A. width
   B. velocity
   C. depth
   D. channel shape

6. You are canoeing a river in remote Alaska and your GPS reads an elevation of 2500 feet. After paddling for five days, you calculate that you have traveled 200 miles. Your GPS now reads an elevation of 2300 feet. What is the stream's gradient in feet/mile for the stretch you just canoed?
   A. 1 foot/mile
   B. 2 feet/mile
   C. 5 feet/mile
   D. 10 feet/mile

7. A trellis drainage pattern forms on
   A. horizontal lava flows
   B. tilted sedimentary rock layers of varying resistance
   C. a dome
   D. horizontal sedimentary rocks

Hint: Refer to Figure 13.23.

8. Stream competence is measured by
   A. the largest particle size that the stream can transport in its bed load
   B. the amount of material in the dissolved load
   C. the maximum width of the channel along the floodplain
   D. the total amount of suspended and bed load

9. Particles tend to settle out of the suspended load in the following order
   A. clay, sand, pebbles
   B. pebbles, sand, silt
   C. sand, pebbles, cobbles
   D. clay, pebbles, sand

10. Active erosion in a meander bend takes place
   A. in the center of the stream
   B. along the outer bank of a bend
   C. along the inside bank of a bend
   D. near a stream's headwaters

11. The Kali Gandaki River cuts one of the deepest gorges on Earth right through the Himalayan Mountain Range. Briefly describe two ways how a river can cut through a mountain range. Hint: Refer to Figures 13.24 and 13.25.

    A. ____

    ____

    B. ____

    ____

EXAM PREP

Materials in this section are most useful during your preparation for midterm and final exams. For optimal performance, midterm preparation should begin about eight days before the exam (see Appendix: *How to Study Geology* for details). The basic idea is a systematic review of material divided into short study sessions.

If you have used regular intensive study sessions to master the material, now you get a pay back! Review for your exam will proceed smoothly and take far less time.

> **Tip for Exam Prep:**
> Think of Chapters 12 and 13 as a package.
> Review them as one integrated unit.

The following *Chapter Summary* and *Practice Exam* should simplify review still further. Read the *Chapter Summary* to begin your session. It provides a helpful overview that should get you back into the material. Next, try the *Practice Exam*. Take it just as you would a midterm: to see how you stand in regard to mastery of this chapter. After you answer the questions, score them. Finally, and most important of all: review any question that you missed. Identify and correct the misconception that resulted in missing the item.

CHAPTER 13 SUMMARY

- Streams erode, transport and deposit sediments. The turbulence of streams allows water to transport sediment by suspension, saltation, rolling and sliding. The tendency for particles to be carried in suspension is countered by gravity, pulling them to the bottom and measured by the settling velocity. Deposition of sediments occurs when the velocity of the stream decreases.

- The physical features (drainage pattern, stream channel, floodplain, meander bends in the channel, alluvial fans and deltas) of a stream system evolve over time.

- The longitudinal profile represents the stream gradient. It is a plot of the elevation of the stream channel bottom at different distances along the stream's course. The longitudinal profile is controlled by local (the river or lake into which the stream flows) and regional (the ocean) base levels (refer to Figure 13.17). Streams cannot cut below base level, because base level is the "bottom of the hill".

- Whether a stream is dominantly eroding or depositing its load (sediments) is determined by stream velocity. Stream velocity in turn depends on the stream's gradient (slope), discharge (amount of water in the stream), load (sediment in transport), and channel characteristics. A stream's drainage patterns, the stream channel, and floodplain evolve in response to changes in stream velocity, gradient, sediment load, discharge and the characteristics of the bedrock over which the stream flows. Alluvial fans form at mountain fronts, in response to an abrupt widening of the stream valley and a change in slope.

- Drainage networks exhibit different patterns depending on topography, rock type, and geologic structure in the drainage area. Near its mouth, a river tends to branch downstream into distributary channels forming a delta. Deltas are major sites of sediment deposition. Where waves, tides, and shoreline currents are strong, deltas may be modified or even absent. Tectonics controls delta formation by uplift in the drainage basin and subsidence in the delta region.

Website and CD Activities and Tools
http://www.whfreeman.com/presssiever
The Interactive Exercise *Marine Delta* at the Website will help you learn the basic components of a large marine delta. *Stream Dynamics*, an animation on your CD, will provide an overview of the channel characteristics and sediment transport down a river system. Also look over the many photographs of rivers and floods provided in the *Photo Gallery* on your CD and Website.

PRACTICE EXAM QUESTIONS FOR CHAPTER 13

(Answers and explanations are at the end of this chapter.)

1. Entrenched meanders like those shown for the San Juan River in Figure 13.9 are evidence of
   A. a decrease in stream gradient
   B. a decrease in discharge
   C. a change, such that the river has renewed ability to erode
   D. a decrease in stream velocity

2. If a dam is placed across a stream that has been carrying a large volume of sediment, the stream would probably
   A. deposit downstream and erode upstream from the dam
   B. erode downstream and not change upstream from the dam
   C. erode downstream and gradually deposit upstream from the dam
   D. not change downstream but would deposit upstream from the dam

Hint: Refer to Figure 13.18.

3. If crushed rock and coal waste are regularly dumped into a stream, this leads to
   A. erosion upstream and deposition downstream from the dump
   B. erosion both upstream and downstream from the dump
   C. deposition both upstream and downstream from the dump
   D. deposition upstream and erosion downstream from the dump

4. If flood control engineers straighten out a meandering stream channel, the stream will probably
   A. flow more slowly in the straightened stretch
   B. deposit in the straightened stretch
   C. downcut in the straightened stretch
   D. downcut and flow more rapidly in the straightened stretch

5. If a flood is classified as a 50-year flood, it is meant that
   A. it has been 50 years since the last flood that large occurred
   B. a flood at least that large has occurred every year within the last 50 years
   C. the flooded area is safe from a serious flood for at least 50 years
   D. a flood that large occurs on the average of once every 50 years but also has a chance of occurring during any year

   Hint: Refer to Figure 13.15.

6. Stream velocity generally increases downstream even though there is a decrease in stream gradient because
   A. the river channel typically gets rougher downstream
   B. channels typically meander less downstream
   C. the amount of sediment decreases downstream
   D. stream discharge typically increases downstream as tributaries contribute their water

7. If the regional base level of a stream is lowered
   A. the stream will deposit to raise the base level to its former position
   B. the stream will begin to downcut at its headwaters
   C. the stream will begin to downcut at its downstream end, and downcutting will progress upstream until the stream channel is graded with the new base level.
   D. a change in base level will have no effect on the stream

8. Where streams emerge from narrow mountain canyons onto flat plains, alluvial fans will form because
   A. the increase in the amount of water from tributary canyons results in deposition
   B. stream velocity decreases due to a widening of the stream channel and a decrease in gradient
   C. stream velocity increases due to a decrease in gradient
   D. all of the above

Hint: Refer to Figure 13.19.

9. What stream feature(s) can develop as a result of regional uplift and erosion?
   A. accelerated downcutting in steams
   B. stream terraces
   C. incised meanders
   D. all of the above

Hint: Refer to Figures 13.9 and 13.20.

10. Which house in the illustration below will ultimately fall into the river channel due to channel bank erosion? Hint: Refer to Figure 13.10.

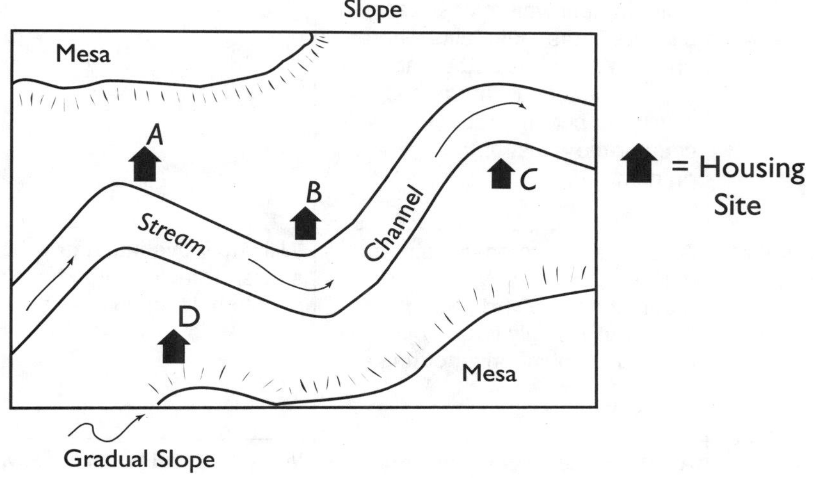

   A. house A
   B. house B
   C. house C
   D. house D

ANSWERS AND EXPLANATIONS FOR EXERCISES AND QUESTIONS

After Lecture

Exercise 1: Stream Velocity

VARIABLE AFFECTING STREAM VELOCITY	RELATIONSHIP OF VARIABLE TO STREAM VELOCITY	ANALOGY
Gradient - the slope of the stream channel	As gradient decreases, stream velocity decreases. Velocity is proportional to the gradient. Note: In headwaters of streams, such as in the mountains where gradient is highest, other factors like decreased discharge and increased channel roughness will counter the effect of the high gradient.	Do you tend to walk faster down a steep slope or a more gradual slope? — faster down a steep slope
Discharge - the amount of water in the stream channel	As discharge increases, stream velocity increases. Velocity is proportional to the discharge. Note: Surprisingly large objects can move down a river during a flood.	Will you move into a new house slower or faster if you have more people helping you? — faster
Sediment load	As the availability of sediment increases, the stream velocity will decrease. Note: Various factors like the bedrock, a landslide, erosion of soil from a burned area and construction can influence the availability of sediment load.	Will you travel faster or slower if you are carrying a heavier load? — slower
Channel characteristics:		
• channel roughness	As channel roughness increases, velocity will decrease.	Cross country hiking without a trail tends to slow one down.
• channel shape	The stream has more contact with the channel surface if the channel shape is very wide or very narrow. More contact with the channel will increase drag and decrease velocity.	When you have more contact with the ground surface, like when you crawl you move slower than if the only contact you have is with the bottom of your feet.

Exercise 3: Relationship between Stream Flow and Groundwater

Why do streams in desert regions typically flow intermittently and streams in more temperate regions, like New England, flow year-round?

During wet period

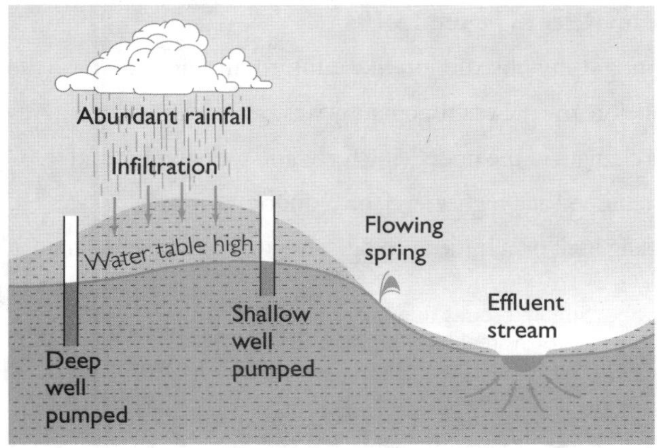

During dry period

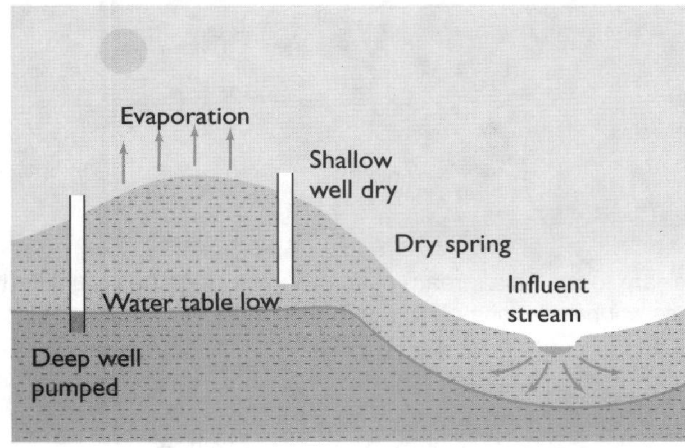

Figure 12:10
Effluent streams are fed by springs from a water table that intersects the stream channel. The water table is well below the channel of an influent stream which flows in response to rainfall, but will quickly dry up as runoff from a storm decreases. Infiltration of stream flow into the channel bottom sediments may help to recharge the groundwater table beneath the channel.

1. D. The Mississippi River is the seventh largest river in its discharge on Earth. Refer to Table 12.1.
2. B. Gravity is the force that drives water down hill.
3. B. Discharge is the volume (typically cubic feet or meters) of water flowing past a point on the stream channel for a given interval of time (typically seconds).
4. A. Velocity is directly proportional to gradient. An increase in slope will increase velocity if all other factors do not counter the change in slope.
5. B. Stream velocity is the most important variable determining the behavior of a stream.
6. A. The change in elevation was 200 feet over the 200 miles canoed. 200feet / 200 miles = 1 foot per mile.
7. B. Refer to Figure 13.23.
8. A. Stream competency is a measure of the size of particles a stream can transport. Stream capacity is a measure of the amount of sediment load a stream can transport.

9. B. Coarser particles like pebbles will settle out before sand and silt.
10. B. The outer bank around a meander is much more likely to be eroded because stream velocity is fastest around the outside of the meandering channel where water depth is greater. On the inside of a meander water depth is lower, velocity slows and the stream is more likely to deposit sediment and form a point bar. Refer to Figure 13.10.
11. Two ways a river can cut through a topography obstruction like a mountain range are:
    A. Antecedence where the river existed before the present topography was created, and it maintained its original course despite changes in the underlying rocks and topography; and
    B. Superposition where the river was established at a higher level, on a uniform surface before eroding down and superimposing itself on a buried geologic structure like an anticline.

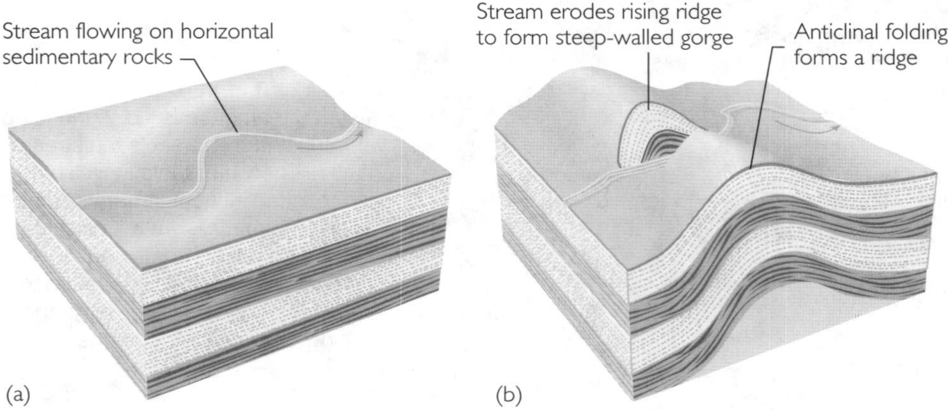

Figure 13.24 (a and b) / An antecedent stream forms as a stream cuts through a ridge being uplifted by tectonic forces. The stream maintains its course despite changes in the underlying rocks and in topography.

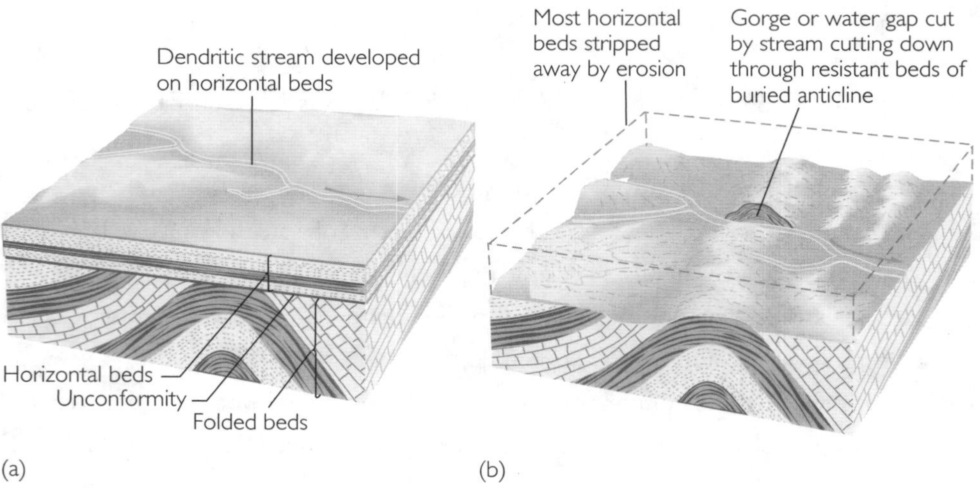

Figure 13.25 (a and b) / The development of a superimposed stream by erosion of horizontal beds overlying folded beds. As the downcutting stream encounters a buried anticline, it erodes a narrow gorge, or water gap, in the beds of the anticline.

Exam Prep

1. C. An increase in rainfall, in stream gradient, or a lowering in base level will increase the stream's velocity and give the stream renewed ability to downcut. Typically, the stream will entrench its channel along its preexisting meandering course.
2. C. Refer to Figure 13.18.
3. C. The regular additions of sediment load is likely to cause the stream velocity in the disturbed stretch to decrease due to the increased availability of load and a reduced gradient. A reduction in velocity may induce deposition.
4. D. The net effect of straightening out the river channel by cutting off a meander bend is to increase the gradient of the stream channel in the cutoff section. An increase in gradient will result in an increase in velocity and increased potential for channel erosion.
5. D. A 50-year flood event has a 2% chance of occurring any year over a 50 year period.
6. D. This may surprise you, but stream velocity typically increases downstream. The reduction of stream gradient is countered by the increase in discharge from tributary channels. The lazy old Mississippi River runs faster past Vicksburg, Mississippi, than up at St. Louis, Missouri.
7. C. A lowering of base level increases the stream gradient, giving the stream renewed ability to erode and transport rock material. Erosion will begin in the vicinity of the change.
8. B. Refer to Figure 13.19 and associated text. Answers A and C are incorrectly stated because velocity would decrease with a decrease in gradient and an increase in discharge would result in erosion not deposition.
9. D. A – C are all possible results of regional uplift which would increase stream gradient and enhance the potential for stream erosion.
10. A. House A is located on the outside of a meander bend which is most susceptible to bank erosion.

Chapter 14
Winds and Deserts

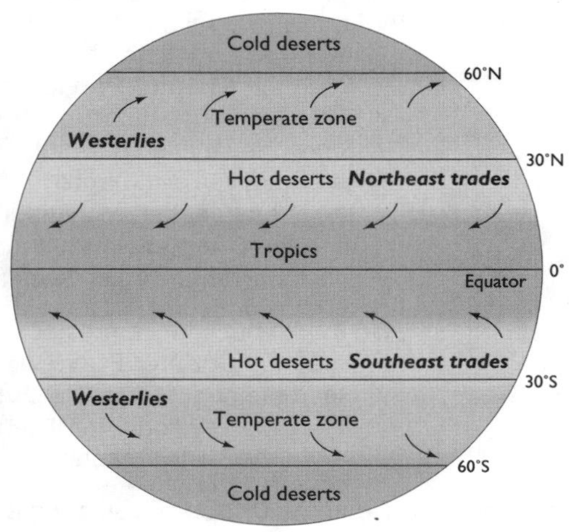

Figure 14.1
Circulation of the atmosphere favors desert conditions between the latitudes of 30° to 15° north and south of the equator because this is generally a region on Earth of warm and dry, descending air.

BEFORE LECTURE

Before you attend lecture be sure to spend some time previewing the chapter. For an efficient preview use the questions below. *Chapter Preview* questions constitute the basic framework for understanding the chapter. Preview works best if you do it just before lecture. With the main points in mind you will understand the lecture better. This in turn will result in a better and more-complete set of notes. How much time should you devote to preview? Obviously, more time is better than less. But even a brief (five or ten minute) preview session just before lecture begins will produce a result you will notice. For a refresher on why previewing is so important see the Appendix: *How to Study Science*.

CHAPTER PREVIEW

- **How do winds erode and transport sand and finer-grained sediments?**
  Brief answer: Turbulent flow and forward motion combine to lift finer particles into the wind and carry them by suspension, sliding, rolling and saltation.

- **How do winds deposit sand dunes and dust?**
  Brief answer: A decrease in wind velocity and gravity causes deposition of sediment being transported by wind. The formation and shape of sand dunes depends on the supply of sand and the strength and variability in wind direction.

- **What factors contribute to the existence of desert regions on Earth?**
  Brief answer: Global atmospheric circulation patterns, distance from water (oceans), topographic barriers like mountains, cold ocean currents, and polar climates may contribute to the formation of desert conditions on Earth.

- **What features are characteristic of desert landscapes?**
  Suggestion: Many diagnostic features of deserts are described in this chapter. One question to ask yourself as you read about these features is, "What features might a geologist look for in the rock record that would serve as evidence for past desert conditions at localities that are no longer deserts?"

TIME MANAGEMENT TIP:	Something is usually better than nothing. Use any available open time before lecture to warm up and preview.
HOW MUCH TIME DO YOU HAVE?	**HOW TO USE IT:**
30 minutes or more	With this much time you can dig deep into the chapter. Do as many of the following as your time allows: ✓ Read the **Chapter Preview** questions and brief answers. ✓ Read the suggestions for **During Lecture**. ✓ Study the key figure for this chapter (always shown at the beginning of the Study Guide Chapter). ✓ Study and annotate any additional figures, hints or suggestions alluded to in the **Chapter Preview**. ✓ Do the **Practice Exercises** and **Study Questions.**
15-20 minutes	Do a brief but intense preview: ✓ Read the **Chapter Preview** questions and brief answers. ✓ Read the suggestions for **During Lecture**. ✓ Study the key figure for this chapter (always shown at the beginning of the Study Guide Chapter).
5-10 minutes	Read the **Chapter Preview** questions and brief answers. Memorize the questions. Then listen for answers during lecture. Even five minutes helps!

The wind grew stronger, whisked under stones, carried up straws and old leaves, and even little clods, marking its course as it sailed across the fields. The air and sky darkened, and through them the sun shone redly, and there was a raw sting in the air.

—John Steinbeck, The Grapes of Wrath, 1939

Vital Information from Other Chapters

- Important information to review in Chapter 6. is Figure 6.18, illustrating a typical desert soil profile, and associated text on *Dry Climates: Pedocals.* Also, review the section entitled *Hydrology and Climate* in Chapter 12.

Website and CD Preview

http://www.whfreeman.com/presssiever

- The Interactive Exercise, *Major Deserts*, at the Website is an excellent preview of where desert regions occur on Earth.

DURING LECTURE

One goal for lecture should be to leave class with a good set of answers to the *Preview Questions*.

- To avoid getting lost in details, keep the "big picture" in mind: Chapter 14 tells two related stories. First, the story of wind: how wind transports sediment and creates sand dunes. Second, the story of deserts: how Earth's circulation patterns produce deserts 15° to 30° away from the equator and how unique features of a desert landscape (e.g. mountain pediments and desert varnish) evolve.
- Focus on understanding Figure 14.1 (atmospheric circulation) and Figure 14.17 (dune types).

AFTER LECTURE

The perfect time to review your notes is right after lecture. The checklist below contains both general review tips and specific suggestions for this chapter.

NOTE REVIEW CHECKLIST

- ✔ All notes legible? (Rewrite so they read easily.)
- ✔ Important points clearly identified? You should now have headers in your notes that tie to each of the questions in the *BEFORE LECTURE/Chapter Preview.*
- ✔ Holes (missing material) filled in from memory?
- ✔ Areas where you don't remember what was said marked for a follow-up session with your instructor, tutor, or study partner?
- ✔ Possible test questions indicated in the margin (TQ)?
- ✔ Additional visual material. *Suggestions for Chapter 14: Sketch a simple figure that integrates the information in Figures 14.1, 14.2, and 14.20. This figure should answer the question: "Why do deserts tend to form between 15° to 30° of the equator?"*
- ✔ Reworked notes into a form that is efficient for your learning style?
- ✔ Created a brief "big picture" overview of this lecture (using a sketch or written outline)?

Intensive Study Session

After this lecture you need to thoroughly master the wind and desert forming processes concepts covered. Schedule at least one hour after lecture for intensive study. Use this time to master key concepts. As always, the majority of your intensive study time should be spent answering questions. Use the Website and CD Activities and Tools suggested below, along with the *Practice Exercises* and *Study Questions*, to insure that you master this chapter. Do as many of these as you have time for during your scheduled study session. Pay particular attention to exercises recommended by your instructor during lecture.

Website and CD Activities and Tools

http://www.whfreeman.com/presssiever

- **Website Q & A Practice Multiple Choice Questions:**
  At the Website complete the *Q & A.* Pay particular attention to the explanations for answers. Flashcards at the Website will help you learn new terms. The *Wind Creating Sand Dunes* animation on your CD illustrates the shape and movement of transverse and barchan dunes. Also look over the dune types and other desert features in the *Photo Gallery* on your CD and at the Website.

PRACTICE EXERCISES AND STUDY QUESTIONS

(Answers and explanations are at the end of this chapter.)

Exercise 1: Sand Dune Types

Complete the table below by filling in the blank spaces. Figure 14.17 and the section in your text on *Dune Types* will be helpful.

DUNE TYPE	CHARACTERISTICS	SAND SUPPLY	WIND DIRECTION/STRENGTH
Barchan			
Transverse			
Blowout		limited to moderate	unidirectional/gusty
Linear			

1. Deserts may be caused by all of the following except
   A. rising air
   B. proximity to cold ocean currents
   C. great distance from the ocean
   D. descending air

2. A dried lake bed with abundant salt deposits and flat enough for the Space Shuttle to land on is called
   A. pediment
   B. desert pavement
   C. a playa
   D. an oxbow lake

3. Loess is
   A. fine dust transported by wind in suspension and deposited on land
   B. dust transported by wind, deposited, and then later eroded and redeposited by water
   C. fine sand transported and deposited by wind
   D. fine-grained salt particles eroded by wind off a playa surface

A Hadley Cell — atmospheric circulation pattern generated by the differential heating of the Earth's surface.

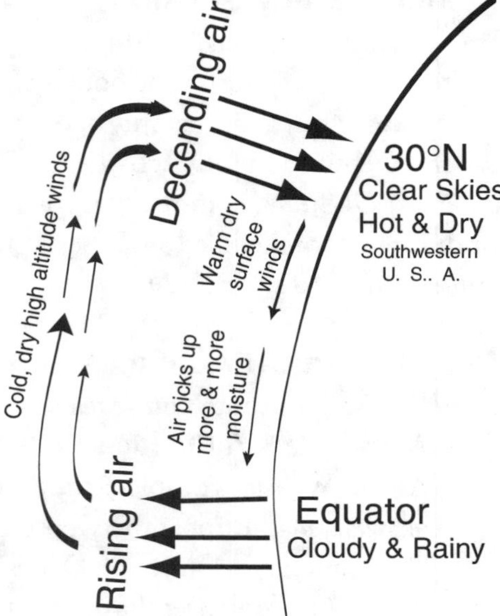

4. Desert pavement may be the product of
   A. deposition of gravel by flash floods
   B. deflation by the wind
   C. deposition of mud by flash floods
   D. desert varnish

Hint: Refer to Figure 14.8.

5. Rock material carried in suspension by wind is mostly the size of
   A. sand
   B. clay
   C. gravel
   D. pebbles

6. Horizontal sedimentary rocks exposed to long periods of weathering and erosion under desert conditions may form
   A. buttes
   B. mesas
   C. anticlines
   D. pediments

7. Deserts such as the Mojave and Great Basin in California and Nevada are formed on the leeward side of mountain ranges, such as the Sierra Nevada, as a result of the
   A. rain shadow effect
   B. great distance from the ocean effect
   C. global circulation effect
   D. cold ocean currents along the California coast

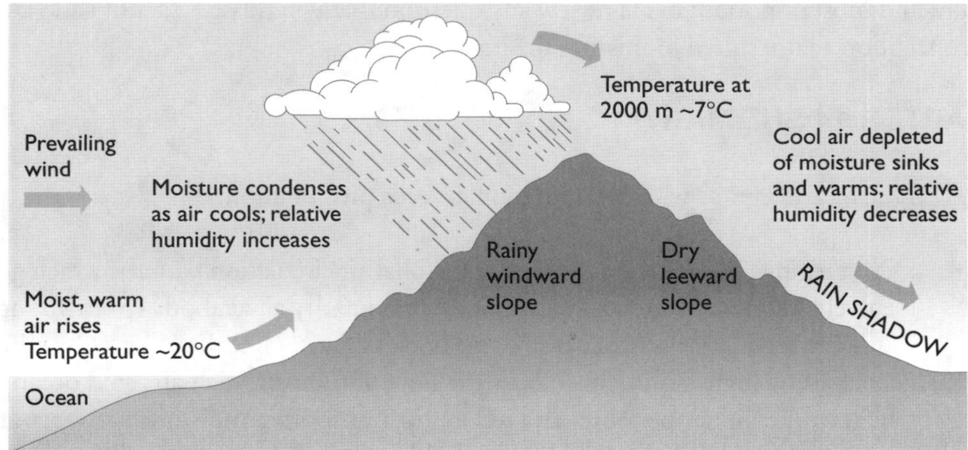

Figure 12.3 / Rain shadows are areas of low rainfall on the leeward (downwind) side of a mountain range. They form as prevailing winds carry moist, warm air to higher elevations where temperatures are cooler. As the rising air cools, precipitation falls on the windward slope, leaving the leeward slope dry.

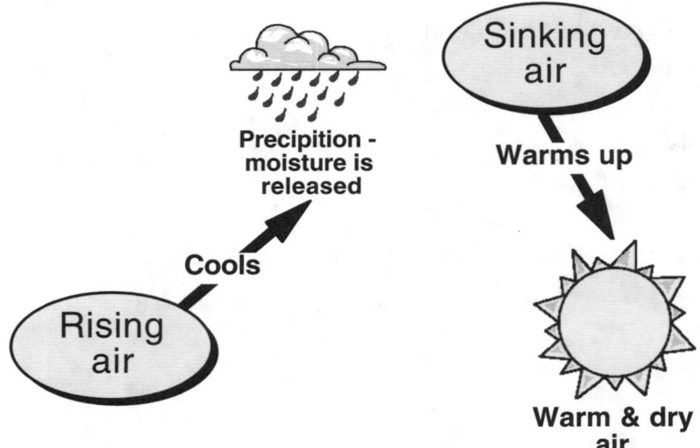

EXAM PREP

Materials in this section are most useful during your preparation for midterm and final exams. For optimal performance midterm preparation should begin about eight days before the exam (see Appendix: *How to Study Geology* for details). The basic idea is a systematic review of material divided into short study sessions.

If you have used regular intensive study sessions to master the material, now you get a pay back! Review for your exam will proceed smoothly and take far less time.

The following *Chapter Summary* and *Practice Exam* should simplify review still further. Read the *Chapter Summary* to begin your session. It provides a helpful overview that should get you back into the material. Next, try the *Practice Exam*. Take it just as you would a midterm: to see how you stand in regard to mastery of this chapter. After you answer the questions, score them. Finally, and most important of all: review any question that you missed. Identify and correct the misconception that resulted in missing the item.

CHAPTER 14 SUMMARY

- Deserts are regions where evaporation exceeds precipitation.

- Desert regions on Earth are the result of: global air circulation patterns which generate a relatively stationary zone of descending, warm and dry air at about 15° to 30° north and south of the equator; distance from large bodies of water like the oceans; the rain shadow generated by high mountains blocks the flow of moisture-rich air; cold ocean currents which reduce air temperatures and reduce the transport of moisture by the air; and polar climates where the air is so cold that it holds very little moisture at all.

- Wind is a major erosional and depositional agent, moving enormous quantities of sand, silt, and dust. Turbulent air flow within wind erodes and transports particles by suspension, sliding, rolling and saltation. As velocity decreases, sediment is pulled out of the air by gravity and deposited as a blanket or dunes of sand and dust.

- Only about 20% of the area of desert regions is covered by sand. The distinctive types of sand dunes are governed by the amount of sand available, the strength of the wind, and the variability in wind direction. Loess, wind blown dust, is another important eolian deposit.

- Geologic and topographic features associated with desert regions include: sand dunes, loess deposits, evaporite (salt) deposits, desert pavement, ventifacts (sandblasted rocks), alluvial fans and alluvial sands and gravels, pediments, mesas, distinctive soils (pedocals, rich in calcium carbonate) and rusty orangish brown colors to weathered rock surfaces. Some of these features are preserved in the rock record and provide geologists clues to ancient desert regions that no longer exist.

Website and CD Activities and Tools
http://www.whfreeman.com/presssiever
You might find looking over the images of dunes and desert features in the *Photo Gallery* a good way to review the characteristics of desert regions.

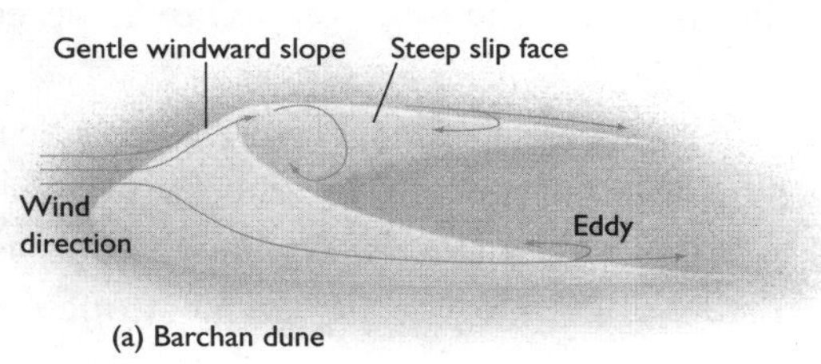

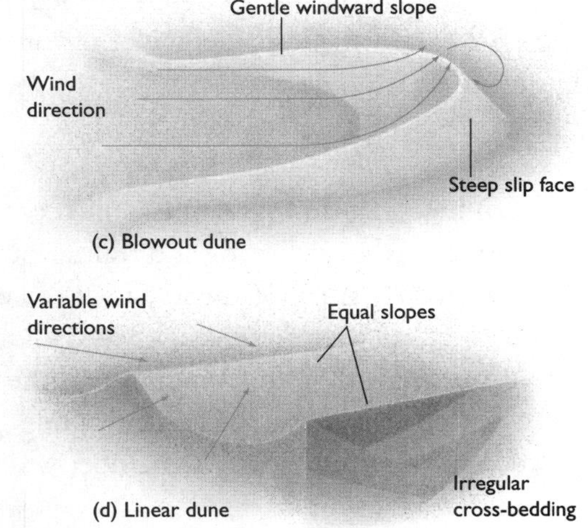

Figure 14.17

Dune types in relation to prevailing winds:

(a) Barchans are crescent-shaped dunes, usually found in groups. The horns of the crescent point downwind.

(b) Transverse dunes are long ridges oriented at right angles to the wind direction.

(c) Blowout dunes are almost the reverse of barchans. The slip face of a blowout dune is convex downwind, whereas the barchan's is concave downwind.

(d) Linear dunes are long ridges of sand whose orientation is parallel to the wind direction.

PRACTICE EXAM QUESTIONS FOR CHAPTER 14

(Answers and explanations are at the end of this chapter.)

1. The best-known loess deposits in North America are derived from
   A. modern deserts in southwestern North America
   B. glacial deposits left barren during the melting of ice sheets at the end of the Pleistocene
   C. ash from volcanoes in Alaska
   D. the Mississippi River Delta

2. The Coriolis Effect deflects winds in the northern hemisphere to the
   A. left (westward)
   B. right (eastward)
   C. north
   D. south

Hint: Refer to the *Coriolis Effect* in Chapter 14.

3. You are lost in the Goblin Desert. The nearest town is due south. Winds blow from the south to the north, but there is no wind blowing today. Furthermore, the sun is totally obscured by clouds. Without a compass, which direction are you going to hike, given the orientation of the barchan dune shown below?
   A. direction A
   B. direction B
   C. direction C
   D. direction D

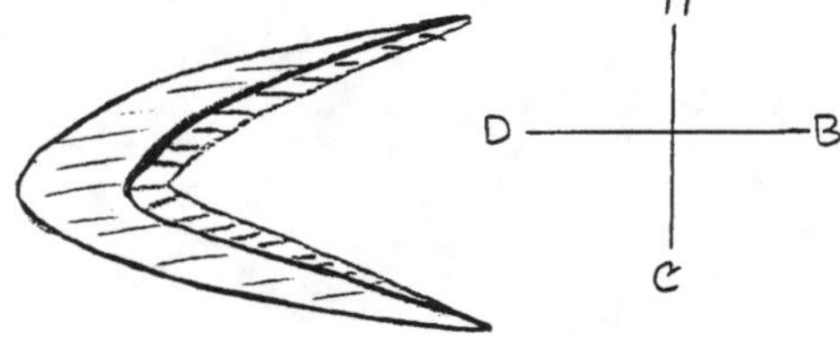

Hint: Refer to Figure 14.17.

4. The Sahara in northern Africa is an example of a
   A. rain-shadow desert in the lee of a large mountain range
   B. cold, coastal-current desert formed due to cold and dry winds off the ocean
   C. desert formed by the flow of the prevailing westerlies
   D. trade wind desert

Hint: Refer to Figure 14.20.

5. While hiking through a dune-filled coastal plain on a windless morning, you become surrounded by dense fog and realize you are lost. You know you are near a shoreline and that the beach will lead you back to camp, but you don't know which direction it is in. You recognize the crescent-shaped dunes that wrap moderate depressions: blow-out dunes! According to your compass, the tapered arms of the dunes point south. Then, remembering that in this region strong, gusty winds come onto the coastal plain off the ocean, you immediately remember which direction to head to get to the beach:
   A. north
   B. south
   C. east
   D. west

Hint: Refer to Figure 14.17.

6. Why are evaporites significant geological deposits?
   A. They are a good paleoenvironmental indicator of ancient desert conditions.
   B. They are a major source of chemicals like borax.
   C. They are a source of salt for the dinner table.
   D. All of the above.

7. Chemical weathering in desert regions is slow due to the lack of water. Therefore desert soils form very slowly and are called ______________________, which are enriched in ______________________________.

8. Petroglyphs, early Native American artwork, are scratched through ________________.

ANSWERS AND EXPLANATIONS FOR EXERCISES AND QUESTIONS

After Lecture

Exercise 1: Sand Dune Types

DUNE TYPE	CHARACTERISTICS	SAND SUPPLY	WIND DIRECTION/STRENGTH
Barchan	Crescent shaped with arms pointing downwind and slip face is concave curve advancing downwind.	limited	unidirectional/strong
Transverse	Long, wavy ridges oriented at right angles to the prevailing wind.	abundant	unidirectional/strong
Blowout	Crescent shaped with arms pointing upwind (into the wind) and the slip face forms the convex curve advancing downwind.	limited to moderate	unidirectional/gusty
Linear	Long, straight ridges more or less parallel to the general direction of the wind.	moderate	variable direction moderate to strong

1. A. Rising air is characteristic of equatorial latitudes where the sun's radiation is more concentrated. As the air rises, it cools and releases abundant rainfall typical of tropical regions. The other three answers all contribute to desert conditions.
2. C. Refer to Figure 14.23. Because it is so flat and smooth, a playa (dried lake bed) in the Mojave Desert east of Los Angles is a regular landing site for the Space Shuttle.
3. A. Loess is fine, wind-blown dust. Cobble-sized rocks can be blown like sailboats across a playa surface. Check out the photo and caption for *Racetrack Playa* in the *Photo Gallery*.

4. B. Refer to Figure 14.8. A photo of an unusual lag deposit of limestone concretions in the Sahara Desert, Egypt, is shown in the *Photo Gallery*.
5. B. Clay is carried in suspension. Sand is typically transported by saltation, sliding, and rolling.
6. B. Refer to Figure 14.26.
7. A. Refer to Figure 12.3.

Exam Prep

1. B. As the Pleistocene ice sheets melted, huge amounts of glacial rock flour were left exposed along the banks of rivers draining the glacial melt water. Wind eroded and transported this glacial dust across North America. Very fertile soils form on loess.
2. B. In the northern hemisphere, the Coriolis Effect deflects air circulation eastward.
3. D. Direction D. The arms of the barchan point down wind. Since the wind is blowing from south to north, the arms are pointing north and the town is in the opposite direction (D).
4. D. Most deserts today are located in the Trade Wind track. Refer to Figure 14.20.
5. B (South). The arms of blowout dunes point into the wind (upwind). Since the wind blows inland from the coast, the arms of the dune point to the beach.
6. D. All of the above answers are appropriate.
7. Pedocal, calcium carbonate. Refer to Figure 6.18.
8. Desert varnish. Refer to Figure 14.21.

Chapter 15
Glaciers: The Work of Ice

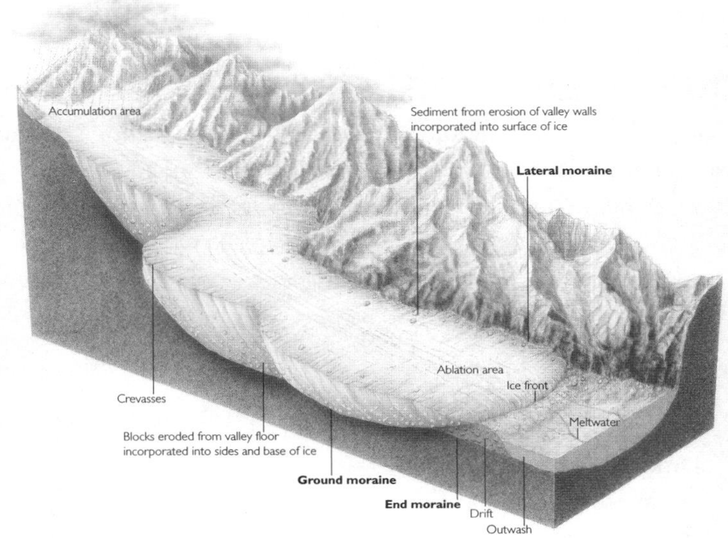

Figure 15.24
Common features associated with a valley glacier.

BEFORE LECTURE

Before you attend lecture be sure to spend some time previewing the chapter. For an efficient preview, use the questions below:

CHAPTER PREVIEW

- **How do glaciers form and how do they move?**
  Brief answer: Glaciers form where snow accumulation exceeds melting (refer to Figure 15.8). Glaciers move by a combination of plastic flow and slip at the base of the ice.

- **How do glaciers erode bedrock, transport and deposit sediments, and shape the landscape?**
  Brief answer: Glaciers erode by scraping, plucking and grinding rock. The rock debris deposited by a glacier is called till. Till may contain rock flour to giant boulders. Glaciers sculpture a distinctive landscape with u-shaped and hanging valleys, arêtes, cirques, moraines, drumlins, kames and other features.

- **What are the ice ages and what causes them?**
  Brief answer: During the Pleistocene many cycles of advance and retreat of continental ice sheets occurred. The ice sheets of the last major advance were gone by about 10,000 years ago, the beginning of the Holocene Epoch. A popular theory for the cause of the Pleistocene ice ages relies on how variations in the Earth's orbit effect the intensity of solar energy reaching Earth.

Vital Information from Other Chapters

- Review Figure 12.1, *Distribution of Water on Earth*, in Chapter 12.

Website and CD Preview

http://www.whfreeman.com/presssiever

Three animations on your CD called *Glaciers: Ablation, Accumulation, and Equilibrium* are recommended viewing before your first lecture on glaciers.

DURING LECTURE

One goal for lecture should be to leave class with a good set of answers to the *Preview Questions*. Note that understanding glacial landscapes is one of the main goals for this lecture. Take notes that will help you understand key landscape features such as the following:

Glacial features formed by the erosive power of glacial ice:

- striations – scratches and grooves – carved in bedrock over which the glacier flowed
- cirque
- u-shaped valley
- hanging valley
- fjord
- arête
- roches moutonée

Glacial features formed by deposition of rock material by glacial ice:

- glacial moraines (the different types are described in Table 15.1)
- drumlin
- esker
- kame
- kettle
- varve

For each feature above, try to sketch essentials and annotate what makes this feature unique. Write figure numbers of slides taken from the text in the margin so you can review them later.

AFTER LECTURE

The perfect time to review your notes is right after lecture. The following *Note Review Checklist* contains both general review tips and specific suggestions for this chapter.

NOTE REVIEW CHECKLIST

✔ All notes legible? (Rewrite so they read easily.)

✔ Important points clearly identified? You should now have headers in your notes that tie to each of the questions in the *BEFORE LECTURE/Chapter Preview.*

✔ Holes (missing material) filled in from memory?

- ✔ Areas where you don't remember what was said marked for a follow-up session with your instructor, tutor, or study partner?
- ✔ Possible test questions indicated in the margin (TQ)?
- ✔ Additional visual material. *Suggestions for Chapter 15: Sketch and label your own version of Figure 15.24. To minimize the artwork, you can shortcut. Instead of drawing mountains, substitute a series of inverted "V's", e.g. /\/\/\/\. Note that simple sketching will work for many of the pictures of glacial features in Chapter 15.*
- ✔ Reworked notes into a form that is efficient for your learning style?
- ✔ Created a brief "big picture" overview of this lecture (using a sketch or written outline)? *Hint: Write a paragraph that starts with "Chapter 15 tells the story of . . ." Use the* Chapter Preview Questions *as a guide.*

Intensive Study Session

Schedule at least one hour after lecture for intensive study. Use this time to master key concepts. By now you know well that mastery is not gained by just reading your text. To master geology you must ask yourself questions and answer them. Use the Website and CD Activities and Tools suggested below along with the *Practice Exercises* and *Study Questions* to insure you master this chapter. Do as many of these as you have time for during your scheduled study session. Pay particular attention to exercises recommended by your instructor during lecture.

STUDY TIP

The pictures in Chapter 15 are designed as a virtual field trip. Use the pictures to master glacial landscape features. Study each picture until you understand what it is and how it differs from other features. (For a list of the landscape features you need to study, refer to the *During Lecture* section.)

Website and CD Activities and Tools

http://www.whfreeman.com/presssiever

- **Website Q & A Practice Multiple Choice Questions:**

  At the Website complete the *Q & A*. Pay particular attention to the explanations for answers. Flashcards at the Website will help you learn new terms. *Glacial Landscapes*, an Interactive Exercise at the Website, reviews some features associated with glaciers and the landscape sculpted by them. The *Photo Gallery* contains many images that provide you with a visual tour of glaciers and their landscapes.

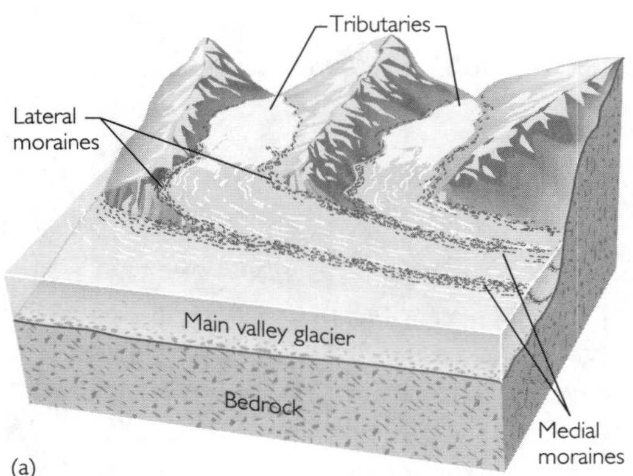

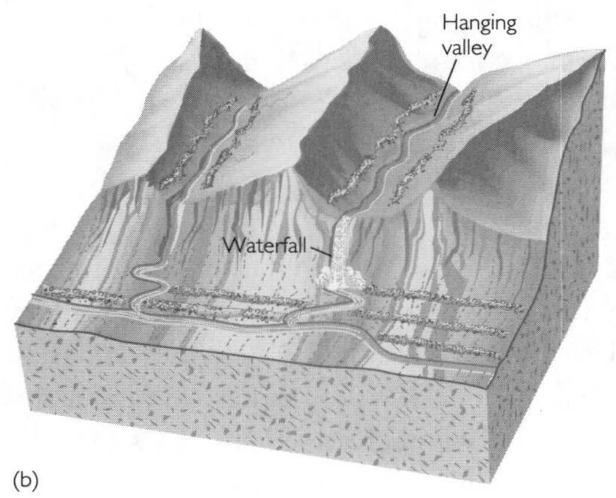

Figure 15.21
The creation of hanging valleys and their waterfalls by tributary valley glaciers.

(a) During glaciation, tributary glaciers enter a major glacier at different floor levels.

(b) After glaciation ends, the ice melts and the region is left with hanging valleys. A stream then forms a waterfall as it hurtles over the steep side of a hanging valley.

PRACTICE EXERCISES AND STUDY QUESTIONS

(Answers and explanations are at the end of this chapter.)

Exercise 1: The Glacially Sculptured Landscape

You are hired as a seasonal ranger at national park with a spectacular glacially sculptured landscape. However, there is not one glacier in the park today. Your job for the summer is to lead natural history hikes along which you interpret the evidence for the past glaciation that helped form the park landscape. Briefly describe five features that you can point out along the nature trail that speak to the past glacial episode in this region.

1. ______________________________

____

2. ______________________________

____

3. ______________________________

____

4. ____

____

5. ____

____

Exercise 2: The Snowline on a Glacier

What is the snowline and what factors control its position?

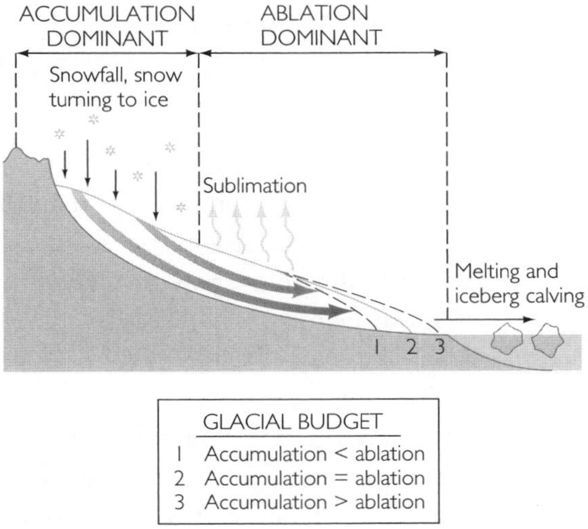

Figure 15.8 / The existence behavior of a glacier depends in large part on the dynamic balance between the accumulation of snow on the upper reaches of the glacier and the ablation of glacial ice from its lower section. The difference between accumulation and ablation is the glacial budget which governs whether a glacier advances or retreats.

1. Glacial ice is a good example of
   A. an igneous rock
   B. a sedimentary rock
   C. a metamorphic rock
   D. a sediment

2. The force which moves glaciers is
   A. recrystallization
   B. melting
   C. lubrication
   D. gravity

3. As snow is transformed into glacial ice, a transitional phase of densely packed granular snow is called
   A. pack ice
   B. firn
   C. alpine ice
   D. crystalline ice

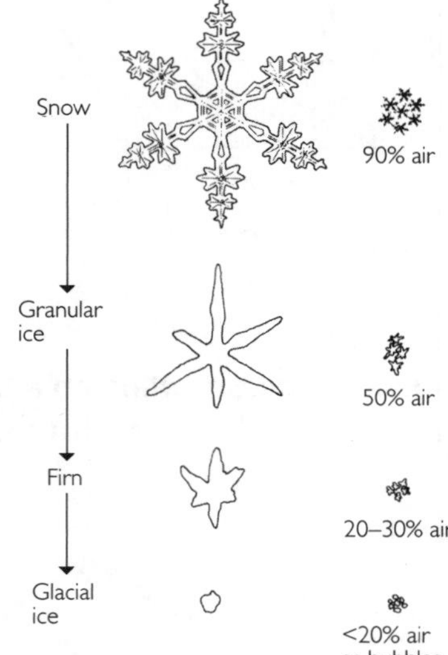

Figure 15.6
Stages in the transformation of snow crystals, first into granular ice, then into firn, and finally into glacial ice. A corresponding increase in density accompanies this transformation as air is eliminated from the crystals.

4. Glaciers retreat when the
   A. accumulation of snow is less than the ablation off the glacier
   B. accumulation of snow exceeds the ablation off the glacier
   C. accumulation of snow is equal to the ablation
   D. snowline moves to lower elevations

5. Moraines are
   A. erosional glacial features carved in bedrock over which the ice flowed
   B. made from glacial outwash sediments
   C. deposits of loess carried by wind from recently glaciated regions
   D. deposits of glacial till

6. A drumlin is a
   A. block of bedrock not quarried away by the bottom of a glacier
   B. sinuous ridge of water-deposited glacial debris
   C. a small depression formed from the melting of a block of ice buried beneath till
   D. streamlined hill constructed of glacial till

7. Ragged, knife-edged ridges are commonly found in glaciated mountains. Such a ridge line is called
   A. a horn
   B. a cirque
   C. a col
   D. an arête

8. The elevation on the glacier above which more snow falls during the winter than can melt in the summer is called the
   A. melt line
   B. snow line
   C. ablation line
   D. evaporation line

9. When continental glaciers advance over the land surface
   A. sea level lowers
   B. sea level rises
   C. plate tectonic processes are especially active
   D. Europe is significantly warmer

10. If all the glacial ice on Earth were to melt, the sea level would be raised by about
   A. 1 m
   B. 10 m
   C. 70 m
   D. 275 m

   Hint: refer to Box 15.1.

11. ______________________ was the first to propose that glacial ice once covered vast areas of North America.
   A. Louis Agassiz
   B. Charles Lyell
   C. Press and Siever
   D. James Hutton

12. Milankovitch calculated the precession, or wobble, of the Earth's rotational axis as a cycle of how many years?
   A. 10,000
   B. 23,000
   C. 100,000
   D. none of the above

   Hint: Refer to Figure 15.34.

EXAM PREP

Materials in this section are most useful during your preparation for midterm and final exams. For optimal performance midterm preparation should begin about eight days before the exam (see Appendix: *How to Study Geology* for details). The basic idea is a systematic review of material divided into short study sessions.

If you have used regular intensive study sessions to master the material, now you get a pay back! Review for your exam will proceed smoothly and take far less time.

The following *Chapter Summary* and *Practice Exam* should simplify review still further. Read the *Chapter Summary* to begin your session. It provides a helpful overview that should get you back into the material. Next, try the *Practice Exam*. Take it just as you would a midterm: to see how you stand in regard to mastery of this chapter. After you answer the questions, score them. Finally, and most important of all: review any question that you missed. Identify and correct the misconception that resulted in missing the item.

CHAPTER 15 SUMMARY

- Glaciers form in cold and snowy climates where snow accumulation exceeds the ablation of ice due to melting, sublimation, wind erosion and iceberg calving. Glacial ice moves by plastic flow and slip along the base, which may be lubricated by melt water. The rate of ice flow varies typically from meters per year to meters per week.
- Glaciers are described as advancing or retreating depending on the balance between snow accumulation and ablation. When ablation exceeds accumulation, the shrinking glacier "retreats" as the toe or terminus migrates up slope. When accumulation exceeds ablation, the expanding glacier "advances" as its toe or terminus migrates down slope.
- Glaciers are powerful agents of erosion and deposition. Glaciers erode by scraping, plucking and grinding rock.
- Landscapes sculptured by ice have distinctive features that have provided geologists with evidence for reconstructing the position of ice sheets during the ice ages and deciphering the existence of ice ages throughout Earth's history. U-shaped and hanging valleys, moraines, arêtes, cirques, drumlins, kames, eskers, striated rock and other features characterize a glacial landscape.
- Ice-laid deposits of rock material are called till, consisting of a heterogeneous mixture of rock, sand and clay. Accumulations of till are called moraines, each type of moraine is named for its position relative to the glacier that formed it. Ancient tills, called tillites, provide evidence for ancient glaciations having occurred numerous times during Earth's history.
- Water-laid deposits from glaciers are called outwash, consisting of sand, gravel, and fine rock flour.
- The ice sheets of the last major advance were gone by about 10,000 years ago, the beginning of the Holocene Epoch. Studies of the geologic ages of glacial deposits on land and sediments of the seafloor show that the Pleistocene glacial epoch consisted of multiple advances (glacial intervals) and retreats (interglacial intervals) of the continental ice sheets. Each advance corresponded to a global lowering of sea level that exposed large areas of continental shelf. During the interglacial intervals, sea level rose and submerged the shelves.
- Although the causes of the ice ages remains uncertain, the general cooling of the Earth leading to glaciation appears to have been the result of plate tectonics that gradually moved continents to positions where they obstructed the general transport of heat from the equator to polar regions.
- A favored explanation for the alternation of glacial and interglacial intervals is the effect of astronomical cycles, by which very small periodic changes in Earth's orbit and axis of rotation alter the amount of sunlight received at the Earth's surface. There is also evidence that decreased levels of carbon dioxide in the atmosphere diminished the greenhouse effect and triggered glaciation.

Website and CD Activities and Tools
http://www.whfreeman.com/presssiever
The *Photo Gallery* contains many images that provide you with a visual tour of glaciers and their landscapes. The images and their captions will reinforce information in the textbook and from lecture.

PRACTICE EXAM QUESTIONS FOR CHAPTER 15

(Answers and explanations are at the end of this chapter.)

Exercise 1: Glacial Advances and Retreats

A glacier advances, halts and retreats. Will the glacier continue to deposit material at its snout while it is halted, and even while it is retreating? Discuss.

1. The relatively mild climate of Europe, despite its northern latitude, is due to
   A. the release of urban and industrial waste heat
   B. greenhouse gases over England's industrialized areas
   C. heat release by ocean currents flowing northeast from equatorial regions
   D. surface winds that have carried heat from the equator

2. Orbital factors that impact the Earth's heat budget primarily affect the
   A. reflectivity of the Earth's upper atmosphere
   B. amount of solar energy reaching our planet
   C. composition of the atmospheric gases
   D. distribution of heat over the globe

3. Which of the following features of a glacial landscape is the product of depositional process, rather than erosion?
   A. fjord
   B. kame
   C. hanging valley
   D. cirque

4. Which of the following is NOT characteristic of glacial till?
   A. sand grains of virtually all quartz
   B. lack of clear stratification
   C. very poor sorting
   D. boulder, cobble, pebble, sand and clay sized rock particles

5. The alternating light and dark layers of fine sediment deposited seasonally in glacial lakes are called
   A. peat
   B. conglomerate
   C. rock flour
   D. varves

Hint: Refer to Figure 15.27.

6. Ablation refers to the
   A. melting, calving, sublimation and erosion of glacial ice
   B. erosion and deposition of glacial sediments
   C. changes in sea level associated with glacial and interglacial periods
   D. glacier advance and retreat

7. The glacial budget is balanced when the rate of accumulation is equal to the rate of
   A. ablation
   B. deposition
   C. advance
   D. retreat

8. The last major advance of continental ice over North America reached
   A. the Gulf of Mexico
   B. California
   C. south of the Great Lakes
   D. Washington, D.C.

Hint: Refer to Figure 15.31.

9. Which of the following may have influenced climate fluctuations during the ice ages?
   A. variations in Earth orbital characteristics
   B. changes in the composition of the atmosphere
   C. plate tectonic movements of the continents
   D. all of the above

ANSWERS AND EXPLANATIONS FOR EXERCISES AND QUESTIONS

After Lecture

Exercise 1: The Glacially Sculptured Landscape

Glacial Erosion and Erosional Landforms, *Glacial Sedimentation and Sedimentary Landforms* and Figure 15.24 in Chapter 15 will be very helpful for completing the brief descriptions of glacial features. Remember, a picture / sketch is worth a "thousand words," so a good way to provide a brief description is with a well-labeled diagram.

A list of possible glacial features that might be found and interpreted by a seasonal ranger include:

Glacial features formed by the erosive power of glacial ice:

- striations—scratches and grooves—carved in bedrock over which the glacier flowed
- cirque
- u-shaped valley
- hanging valley
- fjord
- arête
- roches moutonée

Glacial features formed by deposition of rock material by glacial ice:

- glacial moraine – the different types are described in Table 15.1
- drumlin
- esker
- kame
- kettle
- varve – varves form underwater but may be exposed at the surface if the glacial lake drained or if tectonics movements exposed the lake sediments above water level due to uplift.

All of the above glacial features are described in your textbook and are appropriate answers to Exercise 1.

Exercise 2: The Snow Line on a Glacier

See the section of your text, *How Glaciers Form*. The snow line is that altitude above which snow does not melt in the summer. It varies with latitude and climate. Glaciers can only form above the snow line in climates where adequate snowfall occurs, where they accumulate snow that eventually changes to ice. As glaciers flow to lower elevations, they pass through the snow line and begin to loose ice due to melting, calving, sublimation and wind erosion. This is the zone of ablation. Therefore, the snow line is the line between the zone of accumulation and the zone of ablation on any glacier. See also *Glacial Growth: Accumulation and Glacial Shrinkage: Ablation.*

1. C. Glacial ice fits the general definition of a metamorphic rock. Refer to *Ice as a Rock* in your textbook.
2. D. Gravity is the force that pulls glacial ice downhill. In response to gravity, glacial ice moves by plastic flow and basal slip.
3. B. Refer to Figure 15.6.
4. A. Refer to Figure 15.8.
5. D. Moraines are deposits of till (refer to Table 15.1).

6. D. Drumlins are streamlined hills of till and bedrock. Because drumlins parallel the direction of ice movement, they can be used to reconstruct the direction of movement for the ice sheet.
7. D. Arête.
8. B. Refer to Figure 15.5.
9. A. The movement of water through the hydrologic cycle is altered when glacial ice accumulates on land. Most snowfall that contributes to the growth of the ice sheets ultimately originates as moisture evaporated off the ocean surface. Once trapped as glacial ice, this moisture no longer recycles back to the ocean and the amount of water in the Earth's ocean basins is actually reduced.
10. C. Refer to Box 15.1, *Interpreting Earth and Its System.*
11. A. Louis Agassiz was a pioneer glacial geologist.
12. B. Refer to 15.34.

Exam Prep

Exercise 1: Glacial Advances and Retreats

Rarely does a glacier actually remain stationary. Driven by the force of gravity, glacial ice and the rock material that it carries are moving downhill. The words advancing, retreated and halted are used to describe the movement, or location, of the toe or terminus of the glacier and do not actually refer to the movement of glacial ice within the glacier. The terminus of the glacier will remain stationary (halted), retreat up the valley, or advance down the valley depending on the glacial budget (refer to Figure 15.8). For example, if the snow accumulating in the upper reaches of the glacier equals the loss (ablation) of glacial ice from the lower and warmer reaches of the glacier, the size of the glacier will remain constant and the glacial terminus will remain stationary. Nevertheless, the glacial ice is still flowing down slope with rock material and may pile up a sizeable end moraine (refer to Table 15.1).

1. C. Refer to the section in Chapter 15, *Climate Since the Last Glaciation.*
2. B. Refer to Figure 15.34.
3. B. Kames are small hills of sand and gravel dumped near or at the edge of the ice.
4. A. Till deposits are typically made from a wide variety of rock types depending on the variety of bedrock over which the ice flowed.
5. D. Refer to Figure 15.27.
6. A. Refer to the section in Chapter 15, *Glacial Shrinkage: Ablation.*
7. A. Refer to Figure 15.8.
8. C. Refer to Figure 15.31.
9. D. All of the factors listed play a role in climate change and need to be considered as components of a model that attempts to describe the cause of the ice ages.

Chapter 16
Landscape Evolution

Figure 16.18 / The stages of landscape evolution depend on the balance between uplift and erosion.

BEFORE LECTURE

Before you attend lecture be sure to spend some time previewing the chapter. For an efficient preview, use the questions below.

CHAPTER PREVIEW

- **What are the principal components of landscapes?**
  Brief answer: Landscapes are described in terms of their topography: elevation, the altitude of the surface of the Earth above sea level, and relief (the difference between the highest and the lowest spots in a region). The great variety of landscapes result from sculpturing geologic processes such as erosion and sedimentation by rivers, glaciers, mass wasting, and wind.

- **How do Earth systems and plate tectonics control landscapes?**
  Brief answer: Tectonics, erosion, climate, and the type of bedrock control the evolution of landscapes.

- **How do landscapes evolve?**
  Brief answer: Landscapes go through different phases depending on tectonic activity and climate. For example, a landscape with high relief will form if tectonic activity is high, which in turn stimulates erosion. Erosion will at first enhance relief, but over time water, wind and ice will wear down the high spots and fill in the low spots with sediment.

STUDY TIP
Preview works best if you do it just before lecture. With the main points in mind, you will understand the lecture better. This in turn will result in a better and more complete set of notes.

Vital Information from Other Chapters

- Compared to ice and wind, running water plays the largest role in sculpturing the Earth's land surface today. A review of the last half of Chapter 13, *Streams: Transport to the Oceans*, will greatly reinforce information presented in Chapter 16.

DURING LECTURE

One goal for lecture should be to leave class with a good set of answers to the *Preview Questions*.

- To avoid getting lost in details, keep the "big picture" in mind: Chapter 16 tells the story of how landscapes form. You might expect that a chapter on landscape forming processes would draw on virtually every previous chapter of the book and this is the case.

- Chapter 16 provides a virtual tour of landforms created by geological processes. During lecture be alert to tips that will help you sort out the differences between these new forms.

- Even more important than the landscape features themselves are the processes that form each feature. You already know the processes. Your goal in this chapter is to understand how processes work together to form a particular kind of landscape. Example: the Basin and Range province was created by extensional tectonics (the crust in this region is being stretched by plate movement) interacting with the arid climate.

Learning Styles and Note Taking

Your learning style has a great deal to do with which aspects of lecture you find most helpful.

- Auditory learners learn by listening. They are good at understanding and remembering what the lecturer says. They prefer that directions be spoken out loud.

- Visual learners learn by seeing. Visual images, slides, and flow charts are understood and remembered easily by visual learners.

- Kinesthetic learners learn by moving and experience. They prefer that directions and assignments be written out or presented on slides. Kinesthetic learners do most of their learning and memory work after class. They find it valuable to rewrite notes, redraw pictures, experiment with ideas and concepts.

Which kind of learner are you? To find out, visit the web site of the University Learning Center of the University of Arizona at: **http://www.ulc.arizona.edu/** Select: *Self Assessment*, then *Learning Style Assessment*. You will receive a list of specific study tips and suggestions geared to your learning style.

AFTER LECTURE

The perfect time to review your notes is right after lecture. The checklist below contains both general review tips and specific suggestions for this chapter.

NOTE REVIEW CHECKLIST

- ✔ All notes legible? (Rewrite so they read easily.)
- ✔ Important points clearly identified? You should now have headers in your notes that tie to each of the questions in the *BEFORE LECTURE/Chapter Preview.*
- ✔ Holes (missing material) filled in from memory?
- ✔ Areas where you don't remember what was said marked for a follow-up session with your instructor, tutor, or study partner?
- ✔ Possible test questions indicated in the margin (TQ)?
- ✔ Additional visual material. *Suggestions for Chapter 16: Make some simple sketches to help you learn the features that identify land forms in the chapter. Work some of these into a comparison chart (see Exercise 2).*
- ✔ Reworked notes into a form that is efficient for your learning style?
- ✔ Created a brief "big picture" overview of this lecture (using a sketch or written outline)? *Hint: See Exercise 1.*

Intensive Study Session

After each lecture you need to thoroughly master the processes that form each landscape discussed in the chapter. You should also master the differences between some of the landscape forms discussed in lecture. The best way to accomplish this is by answering questions. Use the Website and CD Activities and Tools suggested below along with the *Practice Exercises* and *Study Questions* to insure you master this chapter. Do as many of these as you have time for during your scheduled study session. Pay particular attention to exercises recommended by your instructor during lecture.

Website and CD Activities and Tools

http://www.whfreeman.com/presssiever

The Interactive Exercise called *Landform Provinces of North America (Face of North America)* at the Website will help you get familiar with the general physiographic provinces on the North American continent. Don't forget to take a look at the *Photo Gallery* photos and their captions on your CD.

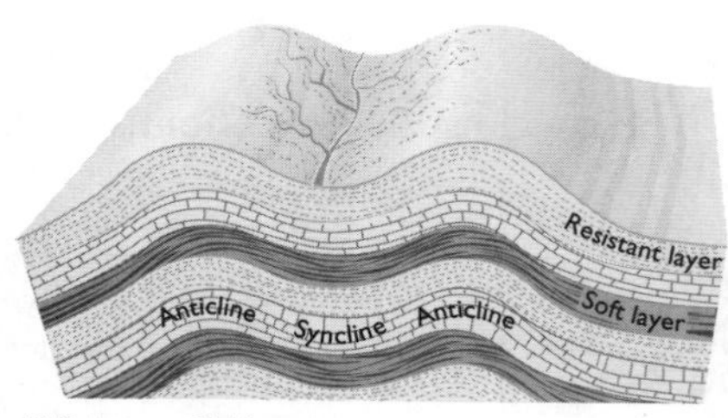

(a) Early stage of folding: ridges over anticlines, streams in synclines

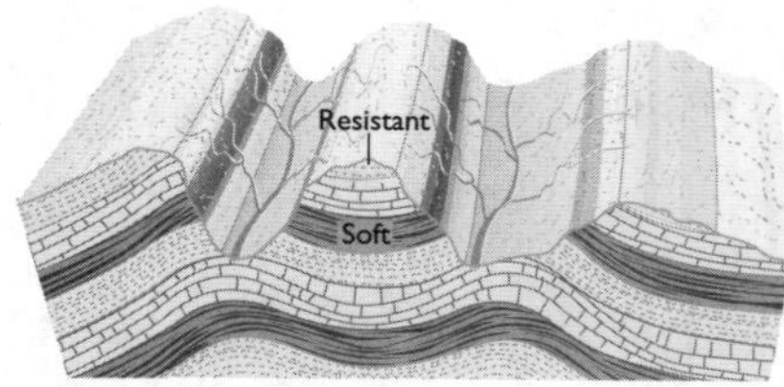

(b) Later stages of erosion: ridges may overlie synclinal axes if capped by resistant beds

Figure 16.10
Two stages in the development of ridges and valleys in folded mountains.

(a) In early stages, ridges are formed by anticlines.

(b) In later stages, the anticlines may be breached and ridges may be held up by caps of resistant rocks as erosion forms valleys in less resistant rock. The resistance of rocks to weather not only depends on their composition but also on the climate. For example, in arid regions such as the Grand Canyon in Arizona, limestone is resistant to weathering and forms cliffs. Since limestone is susceptible to dissolution, in wetter regions limestone tends to weather more rapidly and is not typically a cliff-forming rock.

PRACTICE EXERCISES AND STUDY QUESTIONS

(Answers and explanations are at the end of this chapter.)

Exercise 1: Flow Chart

To review the basic relationships between landscape relief, tectonic activity (uplift), and erosion, fill in the flow chart below. Place the following words in the correct positions on the flow chart:

tectonic activity
high relief
low relief
erosion
physical weathering
chemical weathering

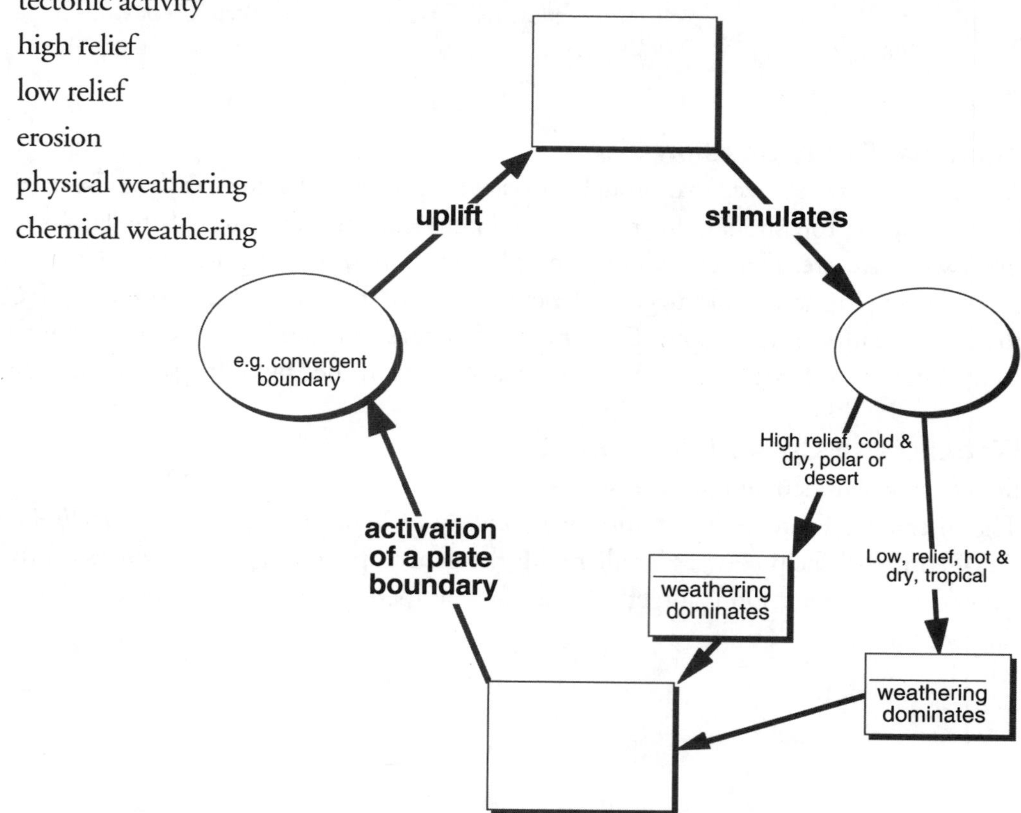

Exercise 2: Comparison of Some Landforms

Chapter 16 provided a virtual fieldtrip experience. To master the landforms discussed in this chapter it will be helpful to develop a comparison chart identifying the important features that distinguish each landform you learned about. Fill in the missing information under "Important Features." Hint: Use the text figures, captions and accompanying text to help you. After you have the features, make a very simple sketch of each landform. Sketching is both a thinking tool and a great kinesthetic learning tool. When you draw, you tap the part of the brain that learns by moving.

LANDFORM	IMPORTANT FEATURE(S)	SKETCH (HINT: KEEP IT SIMPLE)
Mesa (See Fig. 16.6)	A small plateau with ________slopes on all sides. Held up by ____________ ________________________.	
Cuesta (see Fig. 16.7)	A structurally controlled cliff. Somewhat tilted beds, alternating weak and resistance to erosion. Typically undercut and asymmetrical.	
Hogback (see Fig. 16.8)	A structurally controlled cliff with beds that are _______ ______________________. The ridge is more or less ___________________.	
Valley Ridge Topography (see Figs. 16.9 and 16.10)	In young mountains, upfolds (_____________) form ridges and downfolds (_____________) form valleys. As tectonic activity moderates and erosion digs deeper into the structures, the ________________ may form valleys and the synclines ridges.	

1. During the earliest stages of development of a river valley, the valley would have a
   A. simple V-shaped profile
   B. simple U-shaped profile
   C. a low stream gradient
   D. a well-established floodplain

Hint: Refer to Figure 16.12.

2. Elevation is the result of
   A. tectonic activity
   B. the balance between tectonic activity and erosion
   C. erosion and deposition
   D. deposition

3. Relief is the
   A. difference between the highest and lowest point in a region
   B. difference between the highest point and sea level
   C. average height of a landscape
   D. steepness of the slopes

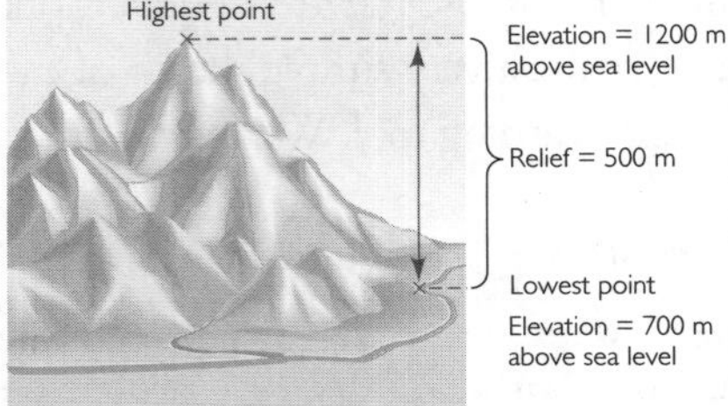

Figure 16.3 / Relief is the difference between the highest and lowest elevations in a regions.

4. Which of the following are important controls on landscape evolution?
   A. tectonics
   B. climate
   C. type of bedrock
   D. all of the above

Exam Prep

Materials in this section are most useful during your preparation for midterm and final exams. For optimal performance, midterm preparation should begin about eight days before the exam (see Appendix: *How to Study Geology* for details). The basic idea is a systematic review of material divided into short study sessions. If you have used regular intensive study sessions to master the material, now you get a pay back! Review for your exam will proceed smoothly and take far less time.

The following *Chapter Summary* and *Practice Exam* should simplify review still further. Read the *Chapter Summary* to begin your session. It provides a helpful overview that should get you back into the material. Next, try the *Practice Exam.* Take it just as you would a midterm: to see how you stand in regard to mastery of this chapter. After you answer the questions, score them. Finally, and most important of all: review any question that you missed. Identify and correct the misconception that resulted in missing the item.

CHAPTER 16 SUMMARY

- Landscapes are described in terms of their topography: elevation, the altitude of the surface of the Earth above sea level; relief, the difference between the highest and the lowest spots in a region; and the varied landforms produced by erosion and sedimentation by rivers, glaciers, mass wasting, and wind. Elevation is a balance between tectonic activity and erosion rate.
- Tectonics (uplift and subsidence), erosion, climate, and the type of bedrock control the evolution of landscapes. Water, wind, and ice act to erode and transport rock material from the high spots and deposit it in the low spots. Over time, relief is subdued by both erosion and sedimentation.

- Landscapes go through different phases depending on tectonic activity and climate. For example, a landscape with high relief will form if tectonic activity is high, which in turn stimulates erosion. Erosion will at first enhance relief, but over time water, wind and ice will wear down the high spots and fill in the low spots with sediment.
- Current views of landscape evolution emphasize the balance between erosion and tectonic uplift. If uplift is faster, the mountain will rise; if erosion is faster, the mountains will be lowered. When tectonics dominates, mountains are high and steep, and they remain so as long as the balance is in favor of tectonics. When erosion exceeds uplift, slopes become lower and more rounded.
- Mountain building processes dominate convergent plate boundaries. Various feedback mechanisms may influence evolution of a mountain system at a convergent boundary. As the mountains become higher, glaciers can form. Ice is a very effective agent of erosion. A negative feedback then develops where, as mountains get higher, glaciers become bigger and more numerous, and the faster the ice can erode. Over the long term (tens to hundreds of millions of years), this feedback mechanism probably speeds up the wearing down of high mountains. However, over the short term (thousands to millions of years) erosion may promote uplift and actually result in the highest peaks rising even higher. For more information on this refer to Chapter 19 and the Geology Essay, *The Himalayas and Tibetan Plateau.*

Website and CD Activities and Tools
http://www.whfreeman.com/presssiever
Don't forget to take a look at the *Photo Gallery* photos and their captions on your CD.

PRACTICE EXAM QUESTIONS FOR CHAPTER 16

(Answers and explanations are at the end of this chapter.)

1. The Earth has two fundamental levels on its surface. They are the
   A. land and oceans
   B. crust and mantle
   C. mountains and trenches
   D. continental crust and ocean basins

2. Mountain belts are found commonly in association with
   A. hotspots
   B. convergent boundaries
   C. transform faults
   D. mid-ocean ridges

3. The Appalachian Valley and Ridge province is characterized by a landscape controlled by
   A. a series of regional faults that were active millions of years ago
   B. glacial erosion and deposition since a continental ice sheet once covered the entire region
   C. wind erosion and deposition partly constrained by zones of dense vegetation
   D. an intricate series of anticlines and synclines

Hint: Refer to Figure 16.11.

TEST-TAKING TIP:
Key-Word the Stem
If you are unsure of the answer to a question, it is sometimes helpful to underline the key words in the stem. What are the key words in the stem for item 3? How does key-wording help you get the item? For more about this, see the answers at the end of the chapter.

4. The Basin and Range province is a region of alternating mountain ranges and elongated basins (valleys) that lies between the Colorado Plateau and the Sierra Nevada. What factors played a major role in the evolution of the Basin and Range landscape?
   A. thrust faulting associated with a convergent plate boundary
   B. large rivers cutting into a series of anticlines and synclines
   C. large rivers cutting into very thick accumulation of wind deposited loess
   D. recent tectonic activity interacting with a semiarid to arid climate

Hint: Refer to Figure 14.25.

5. An example of a negative-feedback process in landscape evolution is how
   A. tectonic uplift provokes an increase in erosion rate
   B. tectonic uplift may provoke additional uplift
   C. rivers wash sediments out of subsiding basins
   D. interactions at convergent boundaries result in low relief

Hint: Refer to Figure 16.15.

6. At which locality will physical weathering be the most dominant?
   A. tropical rainforests
   B. polar regions
   C. jungle regions
   D. swampy areas of low relief

Hint: Refer to *Latitude Mimics Elevation* in Chapter 16.

The physical landscape is baffling in its ability to transcend whatever we would make of it.
—Barry Lopez, Arctic Dreams

ANSWERS AND EXPLANATIONS FOR EXERCISES AND QUESTIONS

After Lecture

Exercise 1: Flow Chart

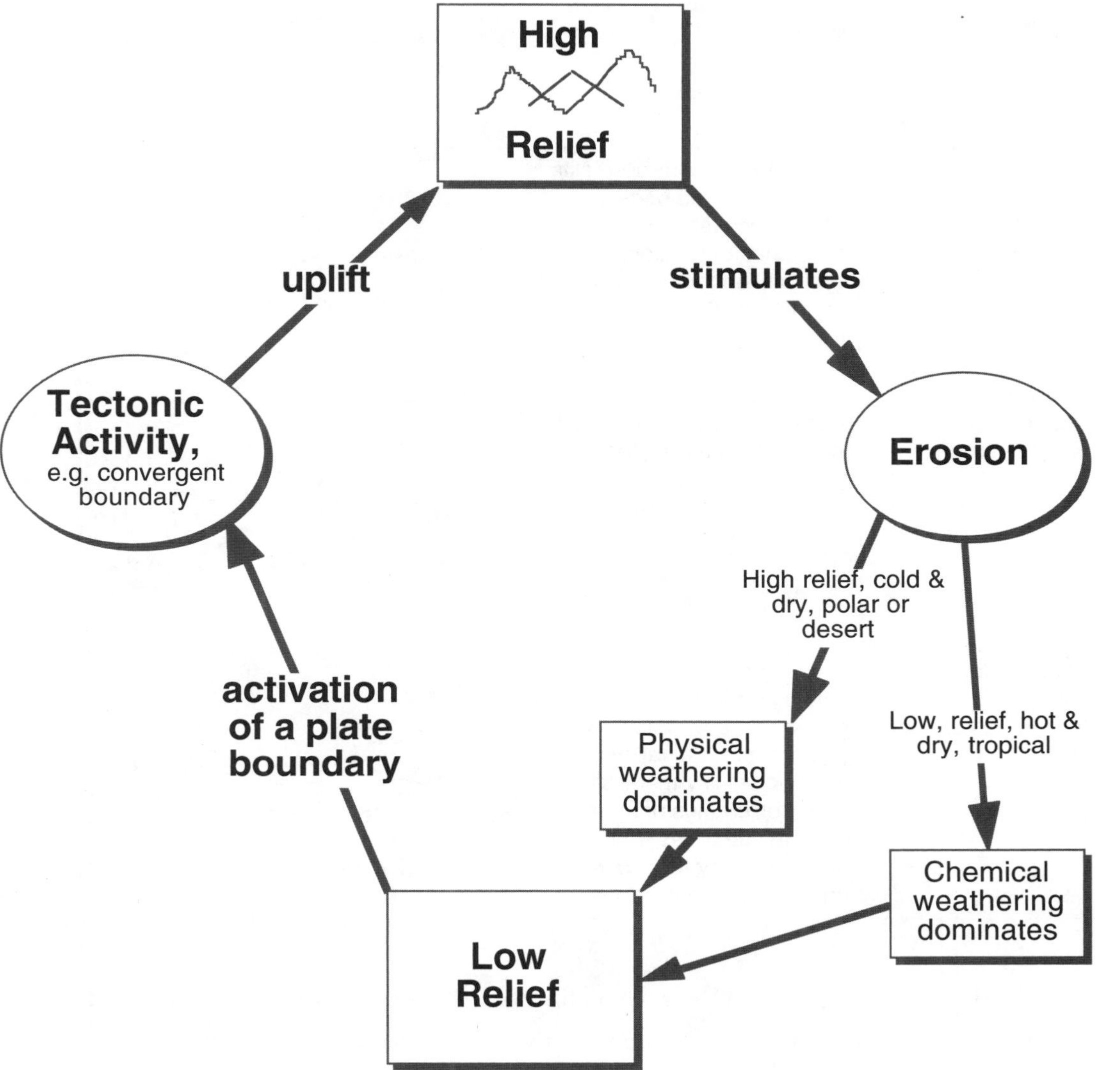

Exercise 2. Comparison of Some Landforms

LANDFORM	IMPORTANT FEATURE(S)	SKETCH (HINT: KEEP IT SIMPLE)
Mesa (See Fig. 16.6)	A small plateau with steep slopes on all sides. Held up by undeformed sedimentary layers or lava flows.	Mesa
Cuesta (see Fig. 16.7)	A structurally controlled cliff. Somewhat tilted beds, alternating weak and resistance to erosion. Typically undercut and asymmetrical.	S.S. Shl. S.S.
Hogback (see Fig. 16.8)	A structurally controlled cliff with beds that are steeply dipping to vertical. The ridge is more or less symmetrical.	S.S. S.S. Side view; S.S. S.S. S.S. end view
Valley Ridge Topography (see Figs. 16.9 and 16.10)	In young mountains, upfolds (anticlines) form ridges and downfolds (synclines) form valleys. As tectonic activity moderates and erosion digs deeper into the structures, the anticlines may form valleys and the synclines ridges.	RIDGE anticline VALLEY syncline RIDGE anticline

1. A. River valleys begin with a V-shaped profile. As the sides of the steep valley retreat and the floor of the valley widens, a floodplain and relatively flat-floored valley will evolve. The gradient in a youthful river system is steep, but decreases though time unless other geologic events are superimposed on the history of the drainage. Glaciers carve U-shaped valleys. Refer to Figure 16.12.
2. B. Elevation is the result of the balance between tectonic activity (uplift and subsidence) and erosion.
3. A. Refer to Figure 16.3.
4. D. Tectonic activity, erosion, climate, and type of bedrock are important controls on landscape evolution.

Exam Prep

1. D. Earth's surface has two fundamental levels, the continents which on average are about a half a mile above sea level and the ocean floor which on average is about 2.5 miles below sea level.
2. B. Mountains typically form at convergent plate boundaries.
3. D. Refer to Figure 16.11.

TEST-TAKING TIP:
Key Word the Stem

Here's the stem with key words underlined:
"The Appalachian Valley and Ridge province is characterized by a landscape controlled by…"

Asking yourself, "What are the key terms in this stem?" may help you focus on the concept. Once you think Valley and Ridge you are likely to remember that in young mountains, valleys often form in synclines and ridges on anticlines. In more eroded mounain ranges like the Appalachians, the anticlines may form valleys and the synclines ridges (refer to Figure 16.10). This is also a good technique for kinesthetic learners because the movement of drawing may unlock your kinesthetic memory.

4. D. The Basin and Range province is a result of extensional tectonics — the crust in this region is being pulled apart.
5. A. Refer to Figure 16.15.
6. B. Physical weathering will dominate in polar regions because glaciers that are typically found in polar regions are powerful agents of physical erosion and the very low temperatures will slow down or stop chemical weathering reactions. The wet and warm conditions of the other regions will enhance chemical weathering over physical weathering.

Chapter 17
The Oceans

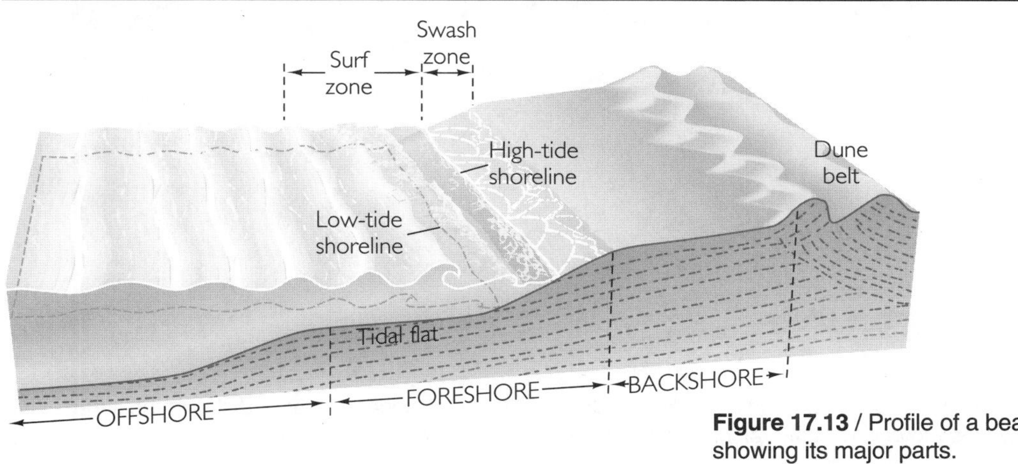

Figure 17.13 / Profile of a beach, showing its major parts.

. . . over all the face of Earth
Main ocean flowed, not idle; but, with warm
Prolifick humour softening all her globe . . .
—John Milton
Paradise Lost, Book VII, l 278-280

BEFORE LECTURE

Before you attend lecture be sure to spend some time previewing the chapter. For an efficient preview use the questions below. For a refresher on why previewing is so important, see the Appendix: *How to Study Science*.

CHAPTER PREVIEW

- **What processes shape the shoreline?**
  Brief answer: Waves and tides shape the shoreline.

- **What are the major components of the continental margins and adjacent ocean floor?**
  Brief answer: Continental margins are flooded portions of the continent. The continental slope marks the edge of the continent and a transition to deeper water and the ocean floor. Turbidity currents transport fine sediments off the continental shelf and onto the adjacent abyssal ocean floor.

- **How is the deep seafloor formed?**
  Brief answer: The deep seafloor is constructed by volcanism at the mid-oceanic ridges and at oceanic hot spots.

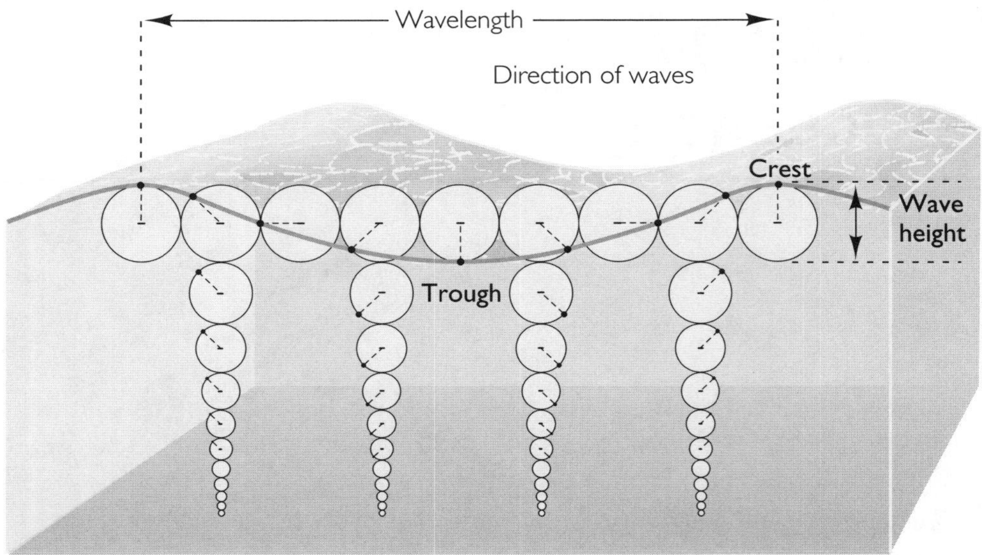

Figure 17.3 / The orbital movement of water particles as a wave advances. Water particles at the surface and beneath waves move in circular vertical orbits. Orbits decrease in radius with depth.

Vital Information from Other Chapters

- Review Chapter 7: *Sediments and Sedimentary Rocks*. Pay particular attention to the sections on *Marine Environments and Sediments*.

Website and CD Preview

http://www.whfreeman.com/presssiever

- The Interactive Exercise called *Parts of a Beach* at the Website will familiarize you with the terms used to describe a beach before lecture.

DURING LECTURE

One goal for lecture should be to leave class with a good set of answers to the *Preview Questions*. To avoid getting lost in details, keep the "big picture" in mind: Chapter 17 tells the story of the ocean depths. You will learn about various landform features of the ocean and the geological processes that create them.

NOTE-TAKING TIP

There is a lot of new terminology in this chapter. Because the unit is so terminology-rich, your lecturer may use terms you are not familiar with. Mark these terms in your notes so that you can check them out later. Put the abbreviation *def.* (define) in the margin to remind yourself to do this.

AFTER LECTURE

The perfect time to review your notes is right after lecture. The following checklist contains both general review tips and specific suggestions for this chapter.

NOTE REVIEW CHECKLIST

- ✔ All notes legible? (Rewrite so they read easily.)
- ✔ Important points clearly identified? You should now have headers in your notes that tie to each of the questions in the *BEFORE LECTURE/Chapter Preview.*
- ✔ Holes (missing material) filled in from memory?
- ✔ Areas where you don't remember what was said marked for a follow-up session with your instructor, tutor, or study partner?
- ✔ Possible test questions indicated in the margin (TQ)?
- ✔ Additional visual material. *Suggestions for Chapter 17: The most important figures for this chapter are Figures 17.13 (Beach Profile), 17.7 (Wave Refraction), and the figure you will complete in Exercise 1 below (Shoreline Profile).*
- ✔ Reworked notes into a form that is efficient for your learning style?
- ✔ Created a brief "big picture" overview of this lecture (using a sketch or written outline)?

Intensive Study Session

After each lecture you need to thoroughly master the concepts covered. You need to do this before you attend the subsequent lecture. The ideas of geology are like a stack of boxes. Each new idea rests on all ideas (boxes) stacked beneath it.

STUDY TIP

Now is the time to look up those terms you didn't understand. Skim the margin of your notes (you did mark them "def", didn't you?). Your text provides two helpful aids for dealing with terms you don't know. *Key Terms and Concepts* (at the end of each text chapter) lists all new terms and handily provides the page number of a term so you can look it up. Alternatively, use the *Glossary* at the end of the text.

Schedule at least one hour after lecture for intensive study. Use this time to master key concepts. As always, spend the majority of your study time answering questions. Read the text selectively. Use it to help you understand why you missed a question or to work out exercises.

The person who does not read has no advantage over the person who cannot read.

Use the Website and CD Activities and Tools suggested below, along with the *Practice Exercises* and *Study Questions,* to insure that you master this chapter. Do as many of these as you have time for during your scheduled study session. Pay particular attention to exercises recommended by your instructor during lecture.

Website and CD Activities and Tools
http://www.whfreeman.com/presssiever
Complete the Q & A at the Website. Pay particular attention to the explanations for the answers. Also at the Website are Flashcards to help you learn new terms, and an Interactive Exercise called *The Oceans and the Seafloor.*

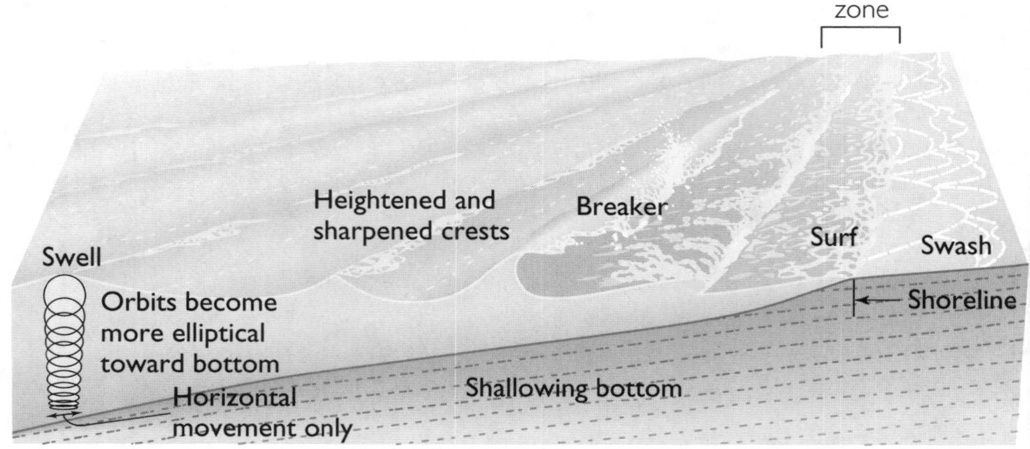

Figure 17.4 / Formation of a breaking wave as swell meets a shallowing bottom. Waves become more closely spaced until, heightened and sharpened, they break in the surf zone.

PRACTICE EXERCISES AND STUDY QUESTIONS

(Answers and explanations are at the end of this chapter.)

Exercise 1: Profile from the Shoreline to the Ocean Floor

Fill in the blanks to correctly label this profile. Hint: Refer to Figure 17.29.

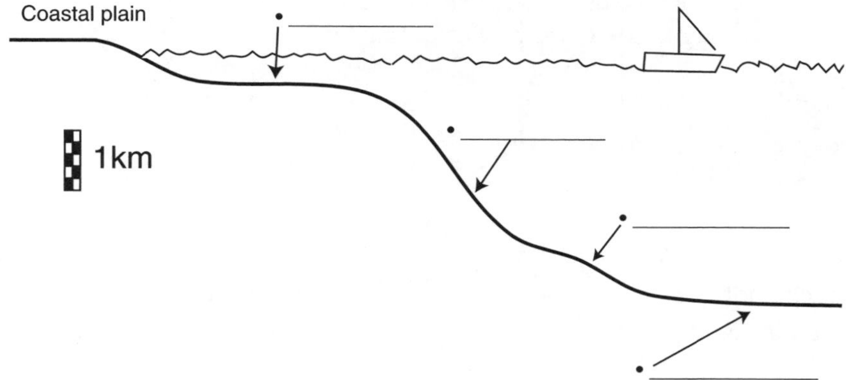

Does the profile above illustrate an active or passive continental margin? ________________

Exercise 2: Passive vs. Active Continental Margins

Characterize each locality described below as either a passive or an active continental margin. Figures 17.22, 17.27, 17.28 and 17.29 are useful references.

A. far from a crustal plate boundary ______________________________

B. east coast of North America _________________________

C. west coast of South America ____________________________

D. very broad, featureless continental shelf ____________________________

E. California coast along the San Andreas Fault ____________________________

F. continental margin with no volcanic activity for millions of years ______________________

1. Ocean waves are mostly generated by
   A. tides
   B. ships
   C. the wind
   D. earthquakes

2. Geologically, the edge of a continent is considered to be
   A. on the ocean side of the sea-floor trenches
   B. the continental shelf
   C. the continental rise and slope
   D. the shoreline

3. Which of the following processes is most important for building the ocean floor?
   A. volcanism
   B. metamorphism
   C. precipitation of carbonate rocks
   D. deposition of sediment derived from the land

4. The bending of waves as they approach shore is called
   A. wave reflection
   B. wave erosion
   C. longshore drift
   D. wave refraction

5. An erosional coast is characterized by
   A. sea cliffs and stacks
   B. barrier islands
   C. coral reefs
   D. estuaries

6. Ocean floor rock is made up of
   A. basalt and pelagic sediments
   B. granite and gneiss
   C. obsidian and sand
   D. rhyolite and carbonate sediments

7. The rock that makes up a seamount is
   A. basalt
   B. limestone
   C. granite
   D. marine sedimentary rocks

8. Sea stacks are
   A. piles of sedimentary rocks near the shore
   B. the erosional remnants of sea cliffs
   C. formed where a river drains onto the coastline
   D. formed by rapidly growing corals on a reef

9. The deep abyssal ocean plain lies at a water depth of
   A. between 100 to 500 meters
   B. 600 to 1000 meters
   C. 2000 to 3000meters
   D. 4000 to 6000 meters

10. Recent precise satellite measurements of the sea surface shows that
    A. Sea level is rising a few millimeters per year.
    B. Sea level is not changing.
    C. Sea level is dropping a few millimeters per year.
    D. Sea level rises and falls with the seasons.

11. What is sediment-charged water which flows rapidly down the continental slope called?
    A. turbidity current
    B. longshore current
    C. tidal current
    D. tsunami

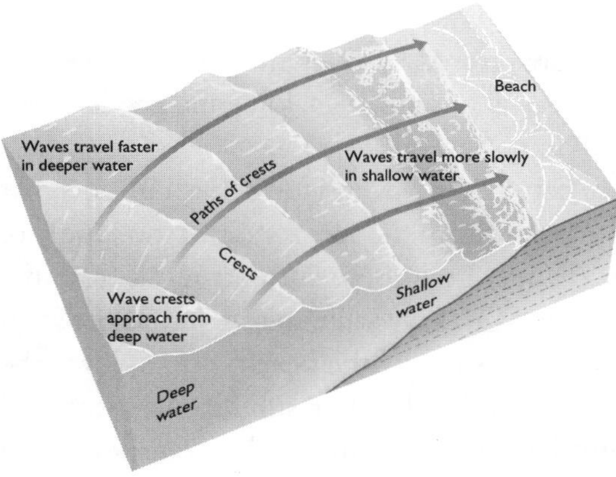

Figure 17.6 / Wave refraction is the bending of lines of wave crests as they approach the shore at an angle to the shoreline.

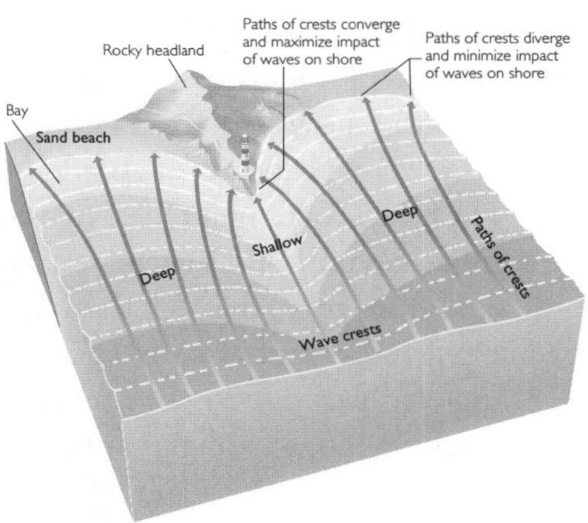

Figure 17.7 / Wave refraction around a headland and bay. Wave energies are concentrated at headlands and dispersed in the bays.

EXAM PREP

Materials in this section are most useful during your preparation for midterm and final exams. For optimal performance, midterm preparation should begin about eight days before the exam (see Appendix: *How to Study Geology* for details). The basic idea is a systematic review of material divided into short study sessions.

If you have used regular intensive study sessions to master the material, now you get a pay back! Review for your exam will proceed smoothly and take far less time.

The following *Chapter Summary* and *Practice Exam* should simplify review still further. Read the *Chapter Summary* to begin your session. It provides a helpful overview that should get you back into the material. Next, try the *Practice Exam*. Take it just as you would a midterm: to see how you stand in regard to mastery of this chapter. After you answer the questions, score them. Finally, and most important of all: review any question that you missed. Identify and correct the misconception that resulted in missing the item.

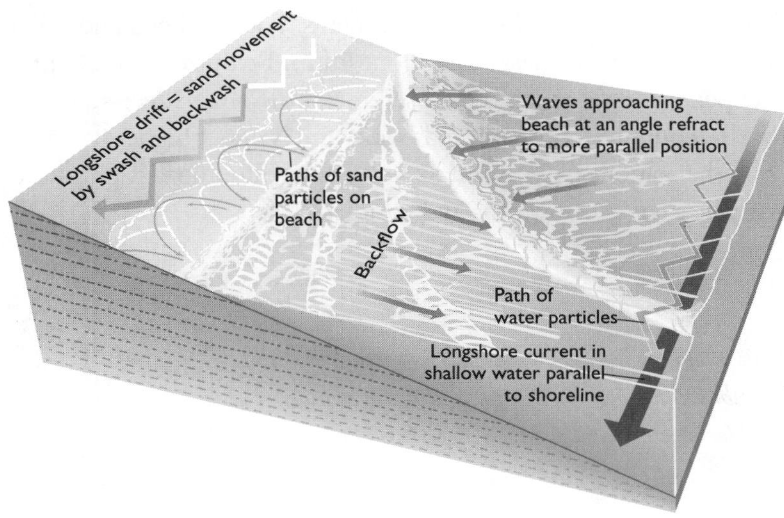

Figure 17.8
Longshore drift is the zigzag movement of sand grains thrown up on the beach face by waves hitting the shore at one angle and washing back at a different angle.

CHAPTER 17 SUMMARY

- Waves and tides shape the shoreline. Waves are created by the wind blowing over the surface of the water. Ocean tides on Earth are a result of centrifugal force and gravitational forces acting between the Earth, Moon and Sun.
- The beach is a result of the dynamic balance between waves and longshore currents that erode and transport rock material along the coast within the surf zone and the supply of sand from rivers that deliver sand to the surf zone. Longshore currents result from the zigzag movement of water on and off the beach. Waves typically splash onto shore at an angle, in part due to wave refraction. The backwash—off the beach—runs down the beach slope at a small but opposite angle to the swash. The net results of this swash and backwash of water on and off the beach slope is a longshore current that transports sand parallel to the beach within the surf zone.
- The topography along the shoreline is a product of tectonic forces elevating or depressing the Earth's crust and changes in sea level.
- Continental margins are flooded portions of the continent. The continental slope and rise mark the edge of the continent and a transition to deeper water and the ocean floor.
- Passive continental margins form where rifting and seafloor spreading carry continental margins away from plate boundaries. Active continental margins form where oceanic lithosphere is subducted beneath a continent, or a transform fault coincides with the continental margin. Continental shelves are broad and relatively flat at passive continental margins, and are narrow and uneven at active margins.
- Turbidity currents transport fine sediments off the continental shelf and onto the adjacent abyssal ocean floor. Turbidity currents can both erode and transport sediments. Submarine canyons and fans are formed by turbidity currents.
- The deep seafloor is constructed by volcanism at the mid-oceanic ridges and at oceanic hot spots. Sedimentation plays a secondary role in constructing the deep seafloor.
- Coral reefs and atolls are constructed by coral and other marine organisms. Reef construction plays an important role in modulating wave energy and creating a favorable environment for shallow marine life.
- In the deep sea, fine-grained pelagic, terrigenous, and biochemically precipitated sediments settle to the seafloor. Foraminiferal oozes, composed of tiny foraminiferal shells, are the most abundant biochemical component of pelagic sediments. Foraminiferal and other carbonate oozes are abundant at depths less than about 4 km. Below a certain depth, called the carbonate compensation depth, carbonate sediments dissolve in deep seawater. Deep ocean water is colder, contains more carbon dioxide, and is under higher pressure. All these factors increase the solubility of carbonate sediments. Silica ooze is produced by sedimentation of the silica shells of diatoms and radiolaria.

. . . though we know the sea to be an everlasting terra incognita,
so that Columbus sailed over numberless unknown worlds. . . .
—Hermann Melville, Moby-Dick (1851)

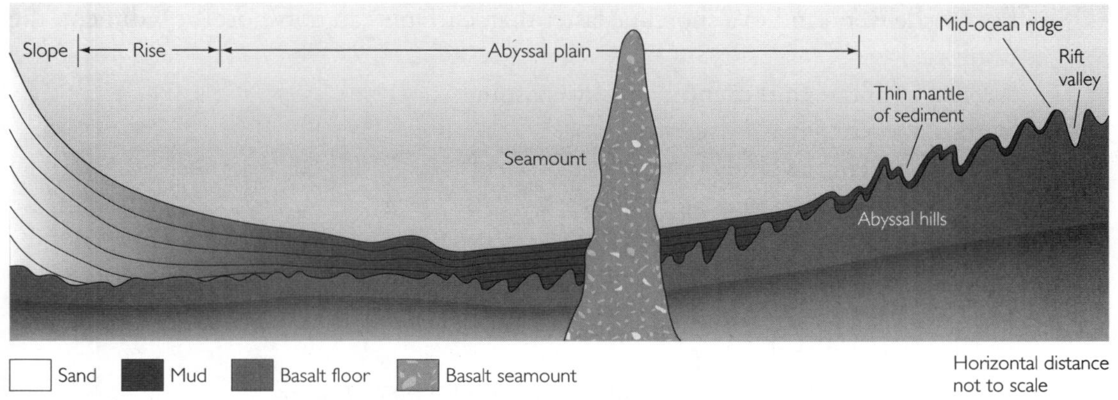

Figure 17.26 / Profile from the continental rise to the Mid-Atlantic Ridge.

Website and CD Activities and Tools
http://www.whfreeman.com/presssiever
At the Website is a Geology Essay entitled *Sea Level*, an interesting article about sea level change from *Scientific American*. Also review the many photos with descriptions of coastal and oceanic features in the *Photo Gallery* on your CD.

PRACTICE EXAM QUESTIONS FOR CHAPTER 17

(Answers and explanations are at the end of this chapter.)

1. The highest tides occur when
   A. the Sun, Moon, and Earth are all aligned
   B. the Sun and Moon are at right angles to the Earth
   C. the Earth is closest to the Sun
   D. the Moon is in either its first or last quarters

   Hint: Refer to Figures 17.9 and 17.10.

2. You live on a beach front. Your up-current neighbors are planning to build a groin to halt erosion and increase the width of their beach. What effect will this have on your beach?
   A. The groin will likely cause your beach to grow.
   B. It is likely that your beach will remain unchanged.
   C. The groin will likely cause your beach to erode.
   D. None of the above.

   Hint: Refer to 17.1, *Earth Policy*.

3. Headlands and points experience greater erosion than bays and inlets because of
   A. longshore drift
   B. wave refraction
   C. wave reflection
   D. the more resistant rock in the headland

4. If a river delivers sand to a shoreline faster than currents can transport the sediment, the result is
   A. erosion of the sand to form a rocky coastline
   B. formation of a marine terrace
   C. expansion of the sandy beach
   D. formation of sea stacks and pillars

5. If you accidentally get caught in a rip current that is carrying you out to sea what should you do?
   A. Swim parallel to the shore.
   B. Rest and float with the rip far out to sea and then swim back.
   C. Swim to shore because this is the shortest distance.
   D. Scream for help.

6. Coral atolls form in tropical oceans by the
   A. upward growth of coral from deep, submarine mountains
   B. upward growth of coral on a continental shelf that is gradually subsiding
   C. upward growth of coral on subsiding volcanic islands
   D. upward growth of coral on exposed sections of the oceanic ridge system

Hint: Refer to Figure 17.34.

7. What is the motion of individual water molecules as a wave travels?
   A. The molecules travel along with the wave.
   B. The molecules follow a straight up and down path.
   C. The molecules follow a roughly circular path.
   D. The molecules travel laterally along with longshore currents.

8. A result of wave refraction is that
   A. wave energy is concentrated on headlands and points
   B. wave energy is concentrated in the bays
   C. sediment is deposited in the vicinity of headlands, making them larger
   D. wave energy is largely dissipated uniformly along the coastline

9. Submarine canyons are formed by
   A. rifting of the lithosphere
   B. erosion due to rivers
   C. turbidity flows
   D. glacial erosion

10. On the diagram below, draw an arrow indicting the direction of the longshore current which will be produced by the ocean waves hitting the beach.

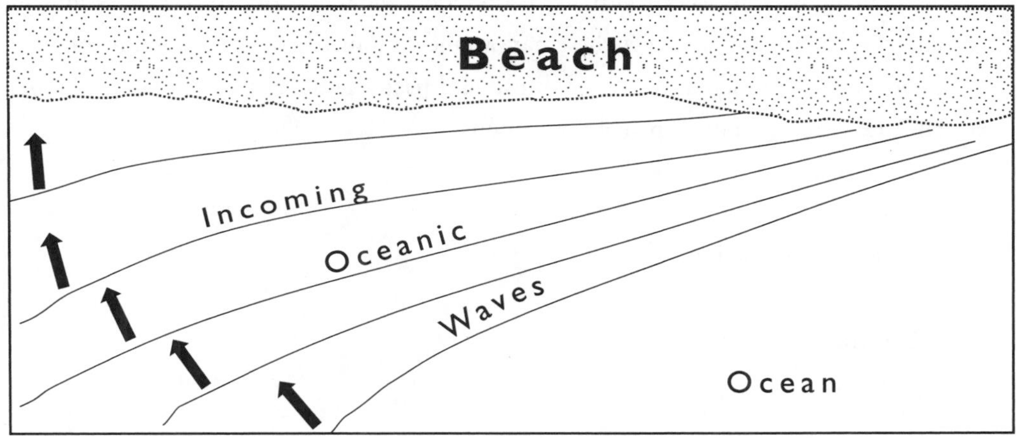

ANSWERS AND EXPLANATIONS FOR EXERCISES AND QUESTIONS

After Lecture

Exercise 1: Profile from the Shoreline to the Ocean Floor

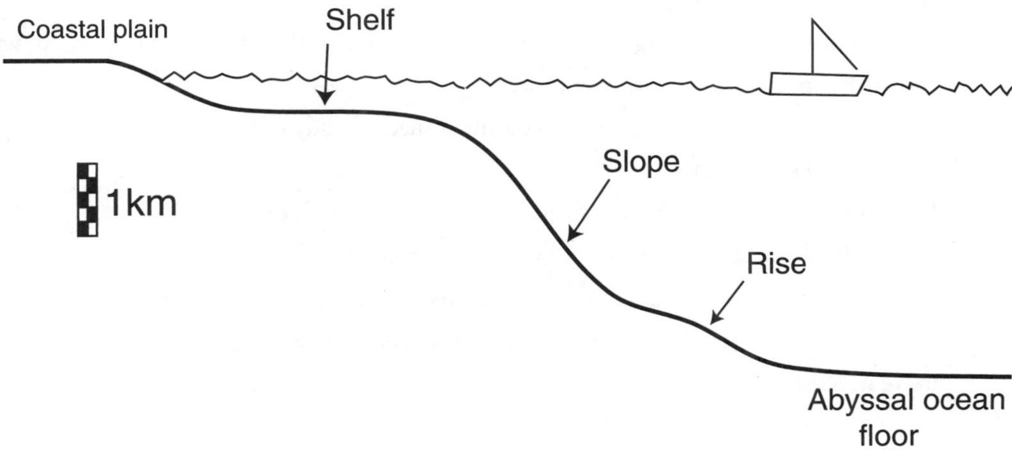

Does the profile above illustrate an active or passive continental margin? passive

Exercise 2: Passive vs. Active Continental Margins

A passive margin is a continental borderland far from an active plate boundary. In contrast, active margins are associated with subduction zones and transform faults. The volcanic activity and frequent earthquakes give these narrow and tectonically deformed continental margins their name. Continental shelves are broad and relatively flat at passive continental margins and are narrow and uneven at active margins.

A. passive
B. passive
C. active (subduction)
D. passive
E. active (transform)
F. passive

1. C. The wind generates most waves in the oceans.
2. C. The edge of the continent is rarely if ever the shoreline. It is the continental slope and rise which may be hundreds of kilometers from the shoreline.
3. A. Mafic volcanism at the oceanic spreading centers generate new ocean crust.
4. D. Wave refraction is the tendency of a wave to bend into the shoreline that it approaches. Waves bend as the wave bottom "drags" on the shallowing bottom before the shoreline, slowing that portion of the wave down. See Figures 17.4, 17.6 and 17.7.
5. A. Refer to Figure 17.15.
6. A. Mafic volcanism at divergent plate boundaries generates the basaltic sea floor crust, which is buried in pelagic sediments over time.
7. A. Seamounts are extinct submarine volcanoes made mostly of basalt.
8. B. Refer to Figure 17.15.
9. D. 4000 to 6000 meters is the depth of the abyssal ocean floor.
10. A. Sea level appears to be rising about 4 millimeters per year.
11. A. Turbidity currents, often triggered by earthquakes, plunge down the continental slopes. They are flowing masses of turbid muds suspended in water, denser than the clear water above it. See Figures 17.30 and 17.31.

Exam Prep

1. A. Refer to Figures 17.9 and 17.10.
2. C. The groin will block the drift of sand past your beach in the surf zone. Robbed of a continuous supply of sand, longshore currents will carry sand from your property, which will not be replenished by the sand supply up current. Your sandy beach is likely to disappear in a few years.
3. B. Wave energy is focused on points and headlands by wave refraction.
4. C. Refer to Figure 17.14 illustrating the sand budget—the dynamic balance between inputs and outputs of sand along the shoreline.

5. A. Rip currents run perpendicular to the beach. Therefore, the best way to escape from the grip of a rip tide is to swim parallel to the beach—unless you are up for some long distance swimming.
6. C. Reef forming corals require sunlight to grow and do not thrive in water much deeper than 20 meters. Therefore, atolls begin as fringing reefs around a volcanic island. Over time, as the volcanic island slowly subsides, the coral reef grows upward, maintaining a shallow marine environment.
7. C. Refer to Figure 17.3.
8. A. Refer to Figure 17.7.
9. C. Turbidity currents (flows) are both agents of erosion and deposition of sediment along the continental slope and adjacent ocean floor (refer to Figure 17.31).
10. The longshore current will flow parallel to the beach in the surf zone and to the left, since the waves are coming in at an angle from the lower right. The best answer has the arrow close to the beach-front in the surface zone. See Figure 17.8.

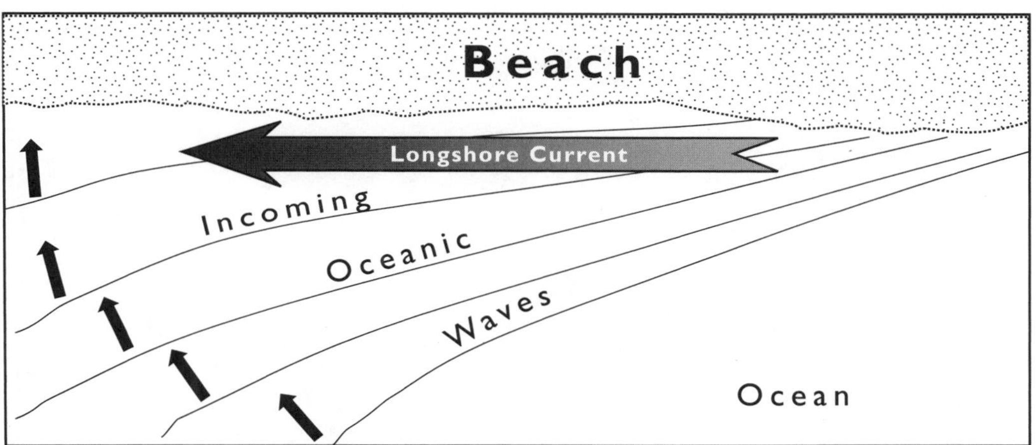

Chapter 18
Earthquakes

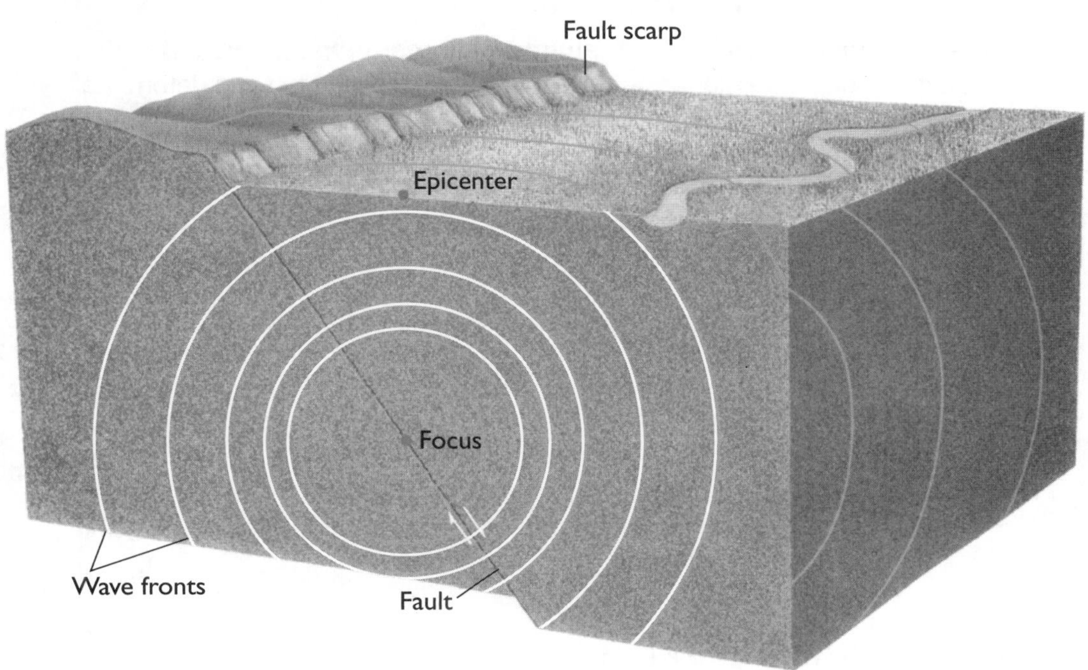

Figure 18.3 / Seismic waves radiate from the focus of an earthquake.

BEFORE LECTURE

Before you attend lecture be sure to spend some time previewing the chapter. For an efficient preview use the questions below.

CHAPTER PREVIEW

- **What is an earthquake?**
  Brief answer: An earthquake is a shaking of the ground caused by seismic waves that emanate from a fault that moves suddenly (refer to Figure 18.1).

- **What are the three types of seismic waves?**
  Brief answer: P (primary/compressional) waves, S (secondary/shear) waves, and surface waves (refer to Figure 18.8).

- **What is earthquake magnitude and how is it measured?**
  Brief answer: Earthquake magnitude is a measure of the size of the earthquake. The Richter magnitude is determined from the amplitude of the ground motion. The moment magnitude is closely related to the amount of energy radiated by the earthquake (refer to Figure 18.11).

(cont'd on next page)

CHAPTER PREVIEW (cont'd)

- **Where do most earthquakes occur?**
  Brief answer: Most earthquakes occur along crust plate boundaries, but not all.
- **What governs the type of faulting that occurs in an earthquake?**
  Brief answer: Tensional, compressional, and shear stresses determine the type of faulting.
- **What causes the destructiveness of earthquakes?**
  Brief answer: Ground motion and avalanches, fires, and tsunamis triggered by the earthquake can cause widespread destruction.
- **What can be done to mitigate the damage of earthquakes?**
  Brief answer: Building codes and restrictions and emergency planning and preparedness can minimize the consequences of a severe earthquake.
- **Can scientists predict earthquakes?**
  Brief answer: Scientists can characterize the degree of risk in a region, but they cannot consistently predict earthquakes.

Vital Information from Other Chapters

- A careful review of Chapter 10: *Folds, Faults, and Other Records of Rock Deformation* will provide you with important prerequisite information. Chapter 10 emphasizes brittle and plastic styles of rock deformation. Earthquakes are thought to be the result of the elastic behavior of solid rocks, analogous to the snap-back from a rubber band when it breaks.

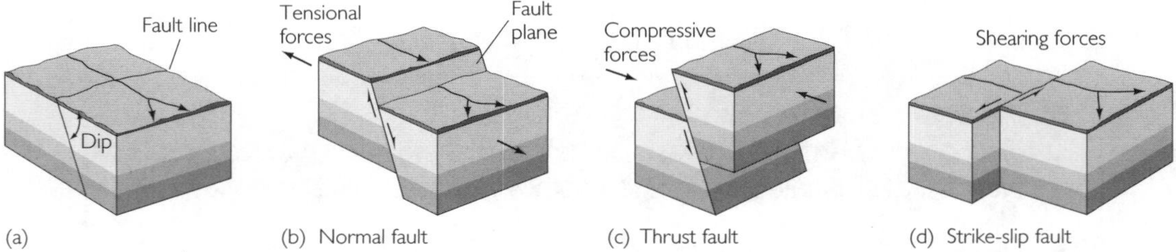

Figure 18.12 / The three main types of fault movements that initiate earthquakes and the stresses that cause them: (a) situation before movement takes place; (b) normal fault due to tensile (tensional) stress; (c) thrust (reverse) fault due to compressive stress; (d) strike-slip fault due to shear stress.

Website and CD Preview

http://www.whfreeman.com/presssiever

- Three animations entitled *Normal Fault*, *Reverse Fault*, and *Strike Slip Fault* are worth checking out before your first lecture on earthquakes.

DURING LECTURE

One goal for lecture should be to leave class with a good set of answers to the *Preview Questions.*

- To avoid getting lost in details, keep the "big picture" in mind: Chapter 18 tells the story of earthquakes: how earthquake activity is measured (by seismic waves) and how earthquake activity is driven by plate tectonics (the plate boundary sdetermines the kind of fault that is formed by an earthquake).
- Focus on understanding the differences between the three kinds of seismic (P, S, Surface) waves.

AFTER LECTURE

The perfect time to review your notes is right after lecture. The checklist below contains both general review tips and specific suggestions for this chapter.

NOTE REVIEW CHECKLIST

✔ All notes legible? (Rewrite so they read easily.)

✔ Important points clearly identified? You should now have headers in your notes that tie to each of the questions in the *BEFORE LECTURE/Chapter Preview.*

✔ Holes (missing material) filled in from memory?

✔ Areas where you don't remember what was said marked for a follow-up session with your instructor, tutor, or study partner?

✔ Possible test questions indicated in the margin (TQ)?

✔ Additional visual material. *Key Visual Material for Chapter 18: Figures 18.1* Elastic Rebound, *18.3* Seismic Waves, *and the figure you will develop in Exercise 1 (*Practice Exercises and Study Questions*). Be sure to add sketches of each of these figures to your notes.*

✔ Reworked notes into a form that is efficient for your learning style?

✔ Created a brief "big picture" overview of this lecture (using a sketch or written outline)? *What is the most interesting thing you learned about earthquakes in this chapter?*

Intensive Study Session

After each lecture you need to thoroughly master the differences between the different kinds of seismic waves (see *Exercise 2*) as well as the way faults at an earthquake site depend on plate tectonic locations.

Use the Website and CD Activities and Tools suggested below, along with the *Practice Exercises* and *Study Questions,* to insure you master this chapter. Do as many of these as you have time for during your scheduled study session. Pay particular attention to exercises recommended by your instructor during lecture.

In times of change, learners inherit the earth, while the learned find themselves beautifully equipped to deal with a world that no longer exists.
—Eric Hoffer

Website and CD Activities and Tools

http://www.whfreeman.com/presssiever

Complete the Q & A at the Website. Pay particular attention to the explanations for the answers. Also at the Website are Flashcards to help you learn new terms. The Interactive Exercises *Earthquakes and Plate Boundaries* and *Earthquakes at Convergent Plate Boundaries* will help you link the terms with actual features. Don't forget to view the three animations illustrating an earthquake along three major types of faults. Check out the Geology Essay *New Madrid Earthquake* at the Website.

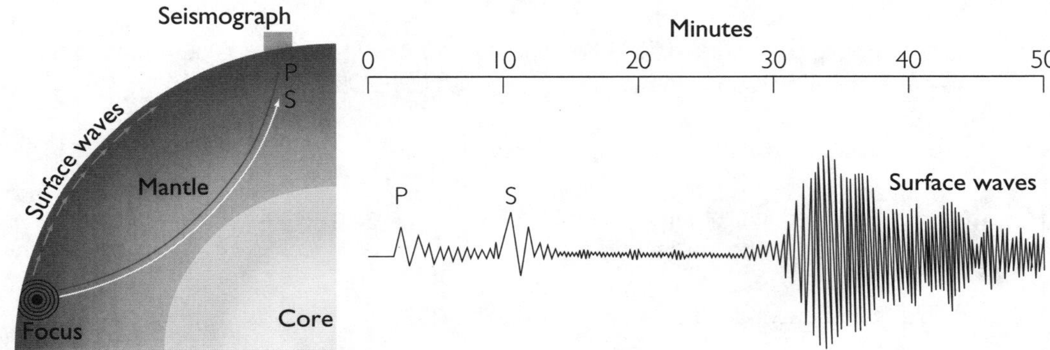

Figure 18.6 / Seismographic recording of P, S, and surface waves from a distant earthquake. The cross section shows the paths followed by the three types of waves.

PRACTICE EXERCISES AND STUDY QUESTIONS

(Answers and explanations are at the end of this chapter.)

Exercise 1: Fault zone (star marks the location from which fault movement propagated)

Label the diagram below by filling in the blanks using the following terms: fault scarp, earthquake focus, fault zone, earthquake epicenter. Use arrows to show the relative motion along the fault zone.

What type of fault is this? ___________________________

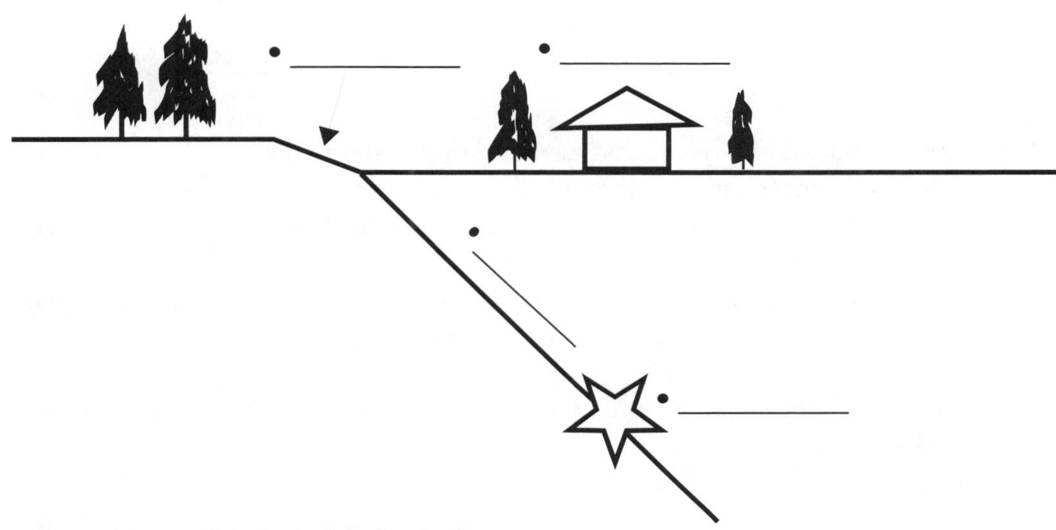

Exercise 2: Characteristics of Seismic Waves

Complete the table below by filling in the blank boxes. Figure 18.7 and the section of the textbook on *Seismic Waves* will be helpful.

SEISMIC WAVES	P (PRIMARY) WAVES	S (SECONDARY) WAVES	SURFACE WAVES
Relative speed		second fastest	
Motion of material through which wave propagates			rolling/elliptical and sideways motion
Medium through which wave will propagate			confined to the Earth's surface
Analogy with common wave forms		S wave propagaton is difficult to visualize. It is somewhat analogous to the way cards in a deck of playing cards slide over eachother as you shuffle the deck.	

Exercise 3: Factors that Influence the Damage Caused by an Earthquake

Briefly discuss five factors that can influence the damage caused by an earthquake.

1. ______________________________

2. ______________________________

3. ______________________________

4. ______________________________

5. ______________________________

1. Elastic rebound theory says that earthquakes are produced when
   A. rocks deform plastically along a fault to produce anticlines and synclines
   B. rocks abruptly slip past each other after an extended period during which elastic deformation is built up in the rocks
   C. magma within the Earth abruptly begins to flow, elastically deforming surrounding rocks
   D. abrupt movement along faults is caused by tidal forces

2. The actual rupture-point within the crust that results in an earthquake is called
   A. the tsunami
   B. the focus
   C. the epicenter
   D. static release

3. The order of arrival of seismic waves at a recording station is
   A. P waves, S waves, surface waves
   B. S waves, surface waves, P waves
   C. P waves, surface waves, S waves
   D. all waves arrive simultaneously

4. In order to locate an earthquake epicenter, a minimum of ______ seismic stations is required.
   A. one
   B. two
   C. three
   D. between five and twelve, depending on the location of the earthquake

Hint: Refer to Figure 18.9

5. Richter Scale for earthquake magnitude measures the
   A. damage caused by the earthquake
   B. amount of energy released by the earthquake
   C. amount of ground motion
   D. duration of the earthquake

EXAM PREP

Materials in this section are most useful during your preparation for midterm and final exams. For optimal performance midterm preparation should begin about eight days before the exam (see Appendix: *How to Study Geology* for details). The basic idea is a systematic review of material divided into short study sessions.

If you have used regular intensive study sessions to master the material, now you get a pay back! Review for your exam will proceed smoothly and take far less time.

The following *Chapter Summary* and *Practice Exam* should simplify review still further. Read the *Chapter Summary* to begin your session. It provides a helpful overview that should get you back into the material. Next, try the *Practice Exam*. Take it just as you would a midterm: to see how you stand in regard to mastery of this chapter. After you answer the questions, score them. Finally, and most important of all: review any question that you missed. Identify and correct the misconception that resulted in missing the item.

It often happens that the wave flees the place of its creation,
while the water does not;
like the waves made in a field of grain by the wind.
Where we see the waves running across the field while the grain remains in place.
—Leonardo de Vinci

CHAPTER 18 SUMMARY

- An earthquake is a shaking of the ground caused by seismic waves that emanate from a fault that moves suddenly. When the fault moves, the strain built up over years of slow deformation by tectonic forces is released in a few minutes as seismic waves.
- Elastic rebound theory explains why earthquakes occur. Over a period of time the application of stress causes rock to slowly deform (bend) elastically until its breaks and the rock snaps back as the fault moves. This is analogous to stretching a rubber band until it breaks and snaps back to sting your hand.
- There are three major types of seismic waves. Two types of waves travel through the Earth's interior: P (primary / compressional) waves, which move through all forms of matter and move the fastest, and S (secondary / shear) waves, which move through solids only and move at about half the speed of P waves. Surface waves need a free surface like the Earth's surface to ripple. They move more slowly than the interior waves, but cause most of the destruction associated with earthquakes.
- Earthquake magnitude is a measure of the size of the earthquake. The Richter magnitude is determined from the amplitude of the ground motion. The moment magnitude is closely related to the amount of energy radiated by the earthquake. The moment magnitude scale depends on the product of the slip of the fault when it broke, the area of the fault break, and the rigidity or stiffness of the rock.
- Most earthquakes occur along crust plate boundaries, but not all. Earthquakes at divergent plate boundaries are usually shallow, lower magnitude and a consequence of tensional stress. Convergent plate boundaries produce shallow and deep earthquakes of low to

high magnitudes and commonly due to compressive stress. Transform faults produce shallow to moderately deep earthquakes of low to high magnitude and usually in response to shear stress.

- The destructiveness of an earthquake does not depend on the magnitude alone. In addition to ground motion, the duration of the earthquake, avalanches, fires, liquefaction, tsunamis, proximity to population centers, and the construction design of buildings all help to determine the destructiveness of an earthquake.
- The damage caused by earthquakes can be mitigated by: regulating the construction design of buildings in earthquake zones; bolting houses to their foundation; in your home, securing appliances connected to gas lines and tall furniture to walls and keeping heavy items at low levels; having a community plan for dealing with emergencies generated by earthquakes. Geologists generate seismic-risk maps to aid public authorities with their evaluations. Scientists can characterize the degree of risk in a region, but they cannot consistently predict earthquakes.

Website and CD Activities and Tools

http://www.whfreeman.com/presssiever

The *Photo Gallery* on your CD and at the Website has many pictures with informative captions of earthquakes and related features. If you are a visual learner, the *Photo Gallery* will be an especially useful resource for you.

PRACTICE EXAM QUESTIONS FOR CHAPTER 18

(Answers and explanations are at the end of this chapter.)

1. What is the factor of ground motion intensity between a Richter magnitude 4 earthquake and a Richter magnitude 8 earthquake?
   A. factor of 2
   B. factor of 100
   C. factor of 1000
   D. factor of 10,000

2. Earthquakes are the result of
   A. elastic deformation building up to a point where rocks break
   B. elastic deformation that is relieved by movement along a fault
   C. elastic deformation that is not relieved by fault creep
   D. all of the above

3. Primary seismic waves (P waves)
   A. travel only through solid material
   B. travel only through liquids and gas
   C. travel through solid, liquid, or gas
   D. are the slowest seismic waves

4. Secondary seismic waves (S waves)
   A. travel parallel to P waves (parallel waves)
   B. can travel only through solid material
   C. can travel through solid, liquid, or gas
   D. are the fastest seismic waves

5. A significant finding that supports the theory of plate tectonics is that most earthquakes occur
   A. randomly in the middle of crustal plates
   B. at all active plate boundaries
   C. only at plate boundaries that move toward each other
   D. only at plate boundaries that slide past each other

6. You just start a job as a county planner in Colorado when the Board of Supervisors mandates earthquake risk assessment. Your first task is to assess the potential for a major seismic event in an area that has experienced only a few minor earthquakes. So, you decide to
   A. install a state-of-the-art seismic recording station to monitor earthquake activity
   B. develop a seismic risk map showing the likelihood of an earthquake based on the number that have occurred in the past
   C. develop an earthquake protection plan with local and state officials
   D. investigate the records for tsunamis

7. Given four structures all built identically, which one would sustain the MOST damage during an earthquake when located the same distance from its epicenter?
   A. one built on a hillside composed of unfractured granite
   B. one built on a quartz-cemented sandstone formation
   C. one built on solid, unfractured granite bedrock
   D. one built on a water-saturated stream delta

Hint: Refer to *How Earthquakes Cause Their Damage* in your textbook.

8. Your seismograph has just recorded an earthquake. You are curious whether the earthquake occurred in North America or somewhere else in the world. Given the arrival time of 5 minutes for the first P waves and 10 minutes for the first S waves, you determine that the approximate distance between you and the earthquake is
   A. 100 kilometers
   B. 1000 kilometers
   C. 3000 kilometers
   D. 7000 kilometers

Hint: Use the graph in Figure 18.9 (b) to determine the correct distance.

9. Which of the following does NOT characterize a divergent plate boundary?
   A. shallow-focus earthquakes
   B. basalt eruptions
   C. deep-focus earthquakes
   D. a rift valley

Hint: Refer to Figure 18.15.

10. Of the states listed below, which one has the lowest potential for seismic hazard?
    A. Washington
    B. Utah
    C. Texas
    D. New York

Hint: Refer to Figure 18.21.

11. If a Richter 7 magnitude earthquake occurred at each one of the four locations below, which one would be felt the furthest from its epicenter?
    A. Seattle, Washington
    B. Salt Lake City, Utah
    C. Los Angeles, California
    D. New York, New York

Hint: Start by considering which locality is underlain by the oldest, coldest, most rigid bedrock. Seismic waves will travel more efficiently through rocks that are colder and less fractured — more rigid.

12. The first motions of an earthquake as recorded from four different stations are displayed on the strike slip fault shown below. Using this first motion data, and Figure 18.13, determine the direction of relative motion along the fault at the time of the earthquake.

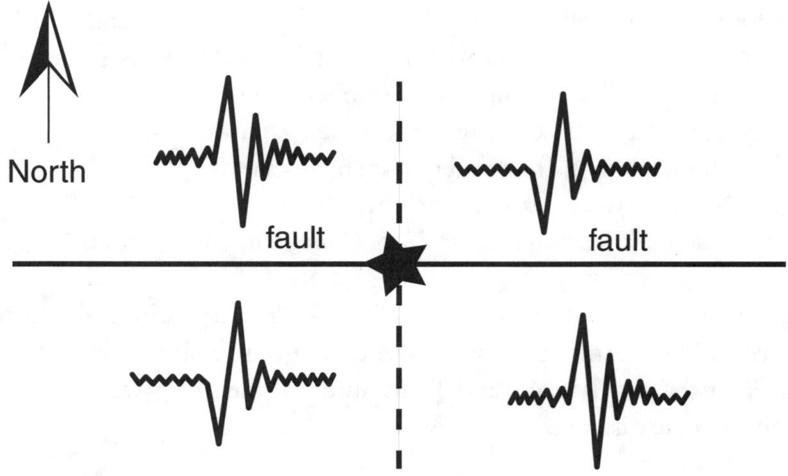

Bird's eye view of a fault zone with first motion data for P waves arriving at seismograph stations during an earthquake. The star marks the epicenter. Dashed line is a north-south reference line plotted on this map with the first motion.

A. north side moved east (right) and south side moved west (left)
B. north side moved west (left) and south side moved east (right)
C. north side moved down and the south side moved up
D. north side moved up and the south side moved down

ANSWERS AND EXPLANATIONS FOR EXERCISES AND QUESTIONS

After Lecture

Exercise 1: Fault zone (star marks the location from which fault movement propagated)

What type of fault is this? <u>normal fault</u>

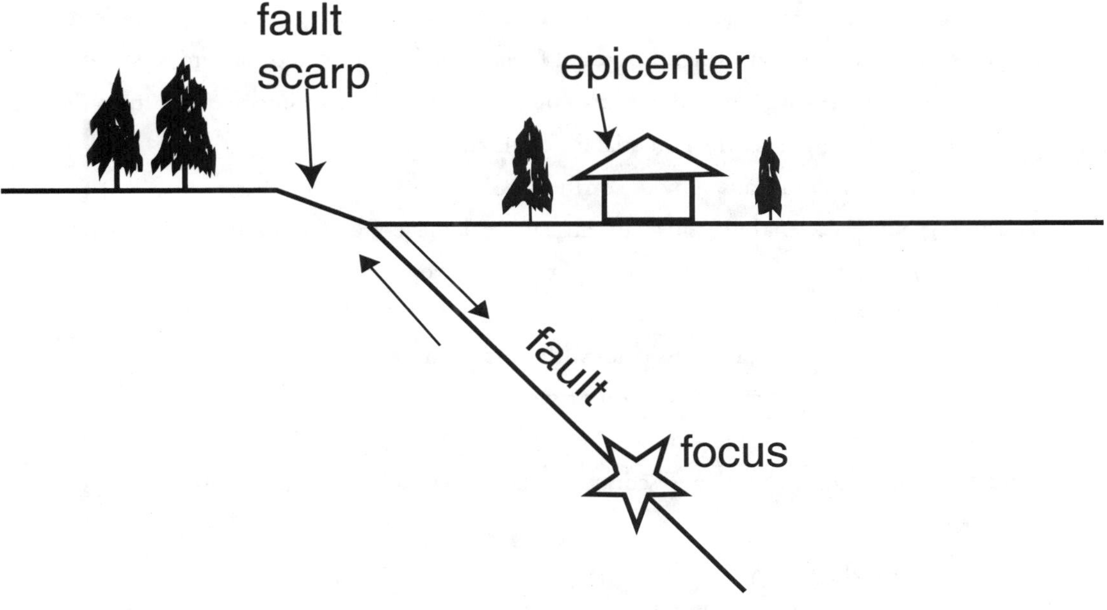

Exercise 2: Characteristics of Seismic Waves

Figure 18.7 and the section of the textbook on *Seismic Waves* will be helpful.

SEISMIC WAVES	P (PRIMARY) WAVES	S (SECONDARY) WAVES	SURFACE WAVES
Relative speed	fastest	second fastest	slowest
Motion of material through which wave propagates	compressional push / pull	shearing	rolling/elliptical and sideways motion
Medium through which wave will propagate	solids liquid gas	solids only	confined to the Earth's surface
Analogy with common wave forms	sound waves	S wave propagaton is difficult to visualize. It is somewhat analogous to the way cards in a deck of playing cards slide over eachother as you shuffle the deck.	ocean waves

Exercise 3: Factors that Influence the Damage Caused by an Earthquake

Factors that influence the damage caused by an earthquake include:

- The magnitude of the earthquake.

  Obviously, a larger magnitude earthquake tends to be more destructive, but many other factors besides magnitude can contribute to the potential for destruction.
- The duration of the earthquake.

  Earthquakes that last longer usually cause more damage. Consider this analogy: the longer you push someone on a swing, the greater their swing backward and forward. Likewise, if structures experience a longer duration of pushes and pulls by seismic waves, the more they will sway and the greater the potential for damage.
- The ground acceleration caused by the earthquake.

  Faster ground acceleration is likely to cause more damage. Rapid ground acceleration may have contributed to the destructiveness of the Northridge, California, earthquake in 1994.
- The depth to the focus of the earthquake.

  The energy of seismic waves dissipates with distance from the focus.
- Proximity to a coastline where a tsunami may hit.
- Fires ignited by ruptured gas and power lines.
- Liquefaction of water-rich soils and sediments that are typically associated with coastal regions and developments.
- Preexisting building designs and codes for the community.

You may come up with other reasonable factors not listed above.

1. B. Refer to Figure 18.1.
2. B. An earthquake propagates from the focus of the earthquake. The epicenter is the location on the surface directly above the focus.
3. A. P waves travel fastest and surface waves travel with the slowest velocities.
4. C. Three (refer to Figure 18.9).
5. C. Richter Scale measures the amount of ground motion.

Exam Prep

1. D. The Richter Scale for earthquake magnitude is exponential. With every unit of increase in magnitude, such as from 4 magnitude to 5 magnitude, the ground motion increases by a factor of 10.
2. D. All answers are correct.
3. C. P waves, like sound waves, travel through solids, liquids and gases.
4. B. S waves travel only though solids.
5. B. Earthquakes are strongly associated with all active crustal plate boundaries.
6. B. Generating a seismic risk map would be a first step in earthquake risk assessment.
7. D. Water-saturated stream delta sediments exhibit liquefaction in response to the shaking from an earthquake.

8. C. 3000 kilometers based on the seismic travel-time curves in Figure 18.9 (b).
9. C. Deep focus earthquakes do not occur in association with divergent boundaries. How do you explain this observation?
10. C. Texas has the lowest seismic risk compared to the other regions listed. See Figure 18.21.
11. D. New York. Seismic waves propagate more efficiently through cold, less fractured, more rigid rocks. Los Angeles and southern California are riddled with faults. This broken ground will dampen seismic waves. The same applies to Utah and the northwestern United States. There is also a higher geothermal gradient in the northwestern United States because of the active volcanic arc associated with subduction beneath the North American plate. Hotter rocks tend to propagate seismic waves less efficiently.
12. B. This would be a left-lateral strike slip fault.

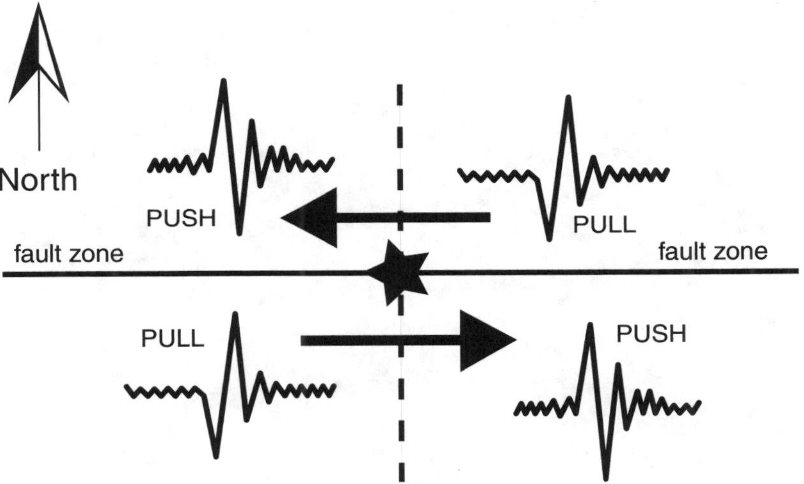

Bird's eye view of a fault zone with first motion data for P waves arriving at seismograph stations during an earthquake. The star marks the epicenter. Dashed line is a north-south reference line plotted on this map with the first motion.

Chapter 19
Exploring Earth's Interior

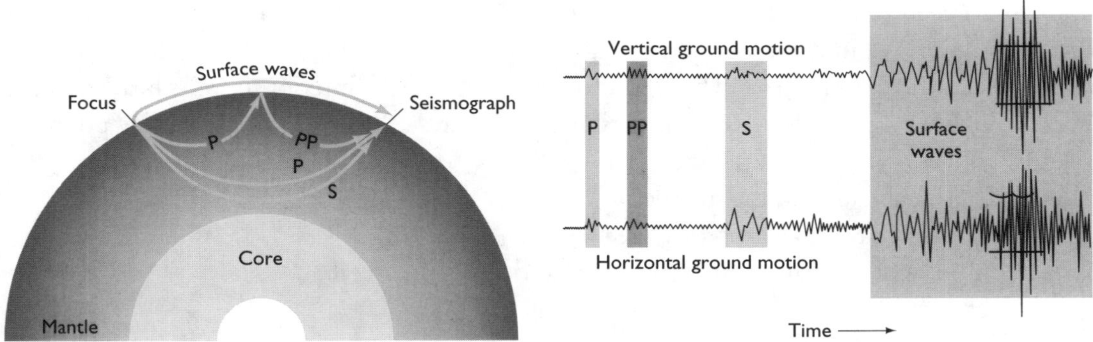

Figure 19.4 / Seismographic recording of P, PP, S and surface waves from a distant earthquake. Seismic waves are used to 'image' the Earth's interior in much the same way X-rays are used in medicine.

BEFORE LECTURE

CHAPTER PREVIEW

- **What do seismic waves reveal about the Earth's interior?**
  Brief answer: Seismic waves reveal that the Earth has a concentrically zoned internal structure. The silicate-rich crust lies on a denser mantle consisting of iron-rich silicates. A portion of the upper mantle, called the asthenosphere is partially molten. The liquid outer core and solid inner core are mostly iron.

- **What is the principle of isostasy and what evidence supports it?**
  Brief answer: The principle of isostasy proposes that continents float on the denser mantle, supported by buoyant roots that project into the mantle. Seismic waves reveal the existence of these roots.

- **Where does the energy that drives geological processes come from?**
  Brief answer: Earth heated up in the process of becoming a planet, and its temperature was increased further by the heat released by the decay of radioactive elements. The Sun's energy is responsible for climate, wind, and rain.

- **What is paleomagnetism and what is its importance?**
  Brief answer: The Earth's magnetic field flips back and forth over geologic time. Preserved in some rocks is a record of past changes in the orientation of Earth's magnetic field.

Vital Information from Other Chapters

- It is very important to review the information on seismic waves presented in *Studying Earthquakes* at the beginning of Chapter 18. A quick review of Chapter 1 will also give you a head start on Chapter 19.

DURING LECTURE

One goal for lecture should be to leave class with a good set of answers to the *Preview Questions.* To avoid getting lost in details, keep the "big picture" in mind: Chapter 19 tells the story of the interior of the earth, its shape and composition, and how Earth's interior supplies heat energy to drive geological processes. Focus on understanding Figures 19.4.19.5, and 19.6. Key points:

- Concentrically zoned structure.
- Continents float on the mantle.
- Mantle behaves like a viscous fluid.
- P & S waves reveal a liquid outer core and solid inner core.
- Heat transfer occurs via convection.
- Earth's magnetic field is generated by electrical fields in a moving liquid, iron-rich core.

NOTE-TAKING TIP

We all have moments when we don't understand a point being made in lecture. When momentarily confused, it is important to continue taking notes. Hopefully, insight will come. But if it does not, the notes you take will provide a clue to what you need to investigate further in your text or in a conversation with your instructor.

AFTER LECTURE

The perfect time to review your notes is right after lecture. The checklist below contains both general review tips and specific suggestions for this chapter.

NOTE REVIEW CHECKLIST

- ✔ All notes legible? (Rewrite so they read easily.)
- ✔ Important points clearly identified? You should now have headers in your notes that tie to each of the questions in the *BEFORE LECTURE/Chapter Preview.*
- ✔ Holes (missing material) filled in from memory?
- ✔ Areas where you don't remember what was said marked for a follow-up session with your instructor, tutor, or study partner?
- ✔ Possible test questions indicated in the margin (TQ)?
- ✔ Additional visual material. *Suggestions for Chapter 19: Figures 19.4.19.5, and 19.6, 19.7.*
- ✔ Reworked notes into a form that is efficient for your learning style?
- ✔ Created a brief "big picture" overview of this lecture (using a sketch or written outline)?

Intensive Study Session

Schedule at least one hour after lecture for intensive study. Use this time to master key concepts. By now you know well that mastery is not gained by just reading your text. To master geology you must ask yourself questions and answer them. Use the Website and CD Activities and Tools suggested below, along with the *Practice Exercises* and *Study Questions*, to insure you master this chapter. Do as many of these as you have time for during your scheduled study session. Pay particular attention to exercises recommended by your instructor during lecture.

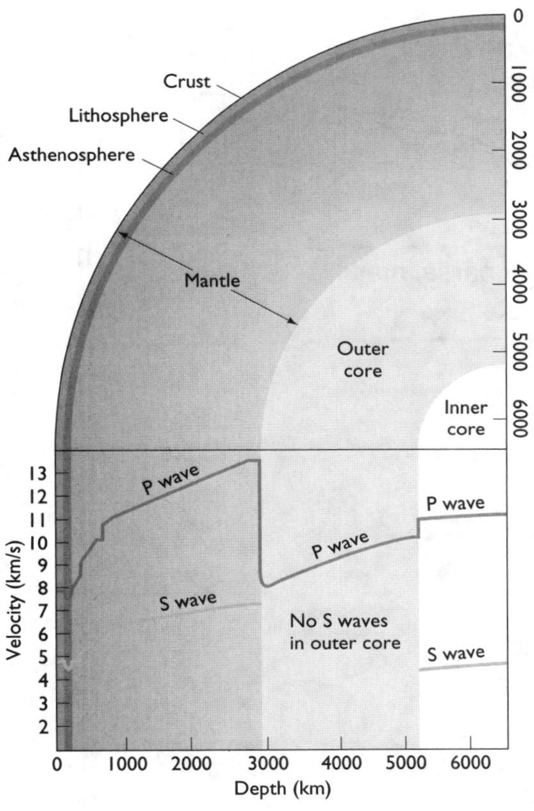

Figure 19.5 / Changes in P- and S-wave velocities with depth in the Earth reveal the sequence of layers that make up Earth's interior.

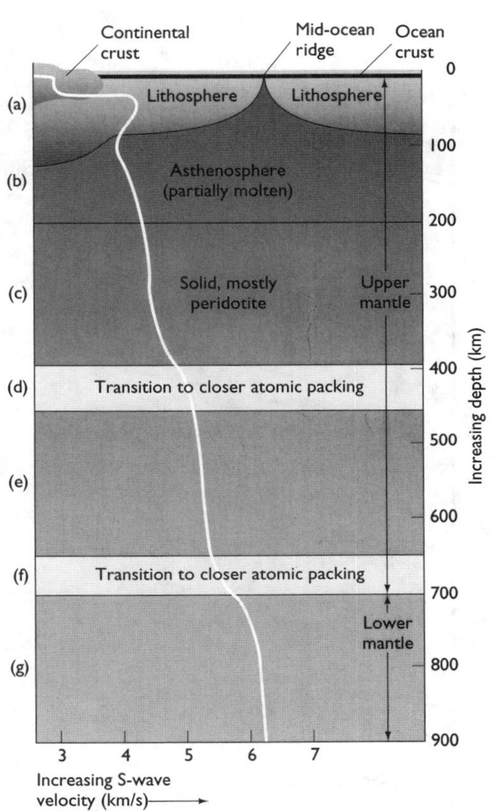

Figure 19.7 / The structure of the upper mantle, the outermost 900 km of the Earth, is illustrated by a plot of S-wave velocity against depth. Changes in velocity mark the strong lithosphere, the weak asthenosphere, and two zones in which changes occur because increasing pressure forces a rearrangement of the atoms into denser or more compact crystalline structures.

Website and CD Activities and Tools

http://www.whfreeman.com/presssiever

Complete the Q & A at the Website. Pay particular attention to the explanations for the answers. Also at the Website are Flashcards to help you learn new terms.

PRACTICE EXERCISES AND STUDY QUESTIONS

(Answers and explanations are at the end of this chapter.)

Exercise 1: The Earth's Interior Layers

Fill in the blanks. Use Figures 19.5, 19.6, and 19.7 as references, in addition to the section *The Composition and Structure of the Interior.*

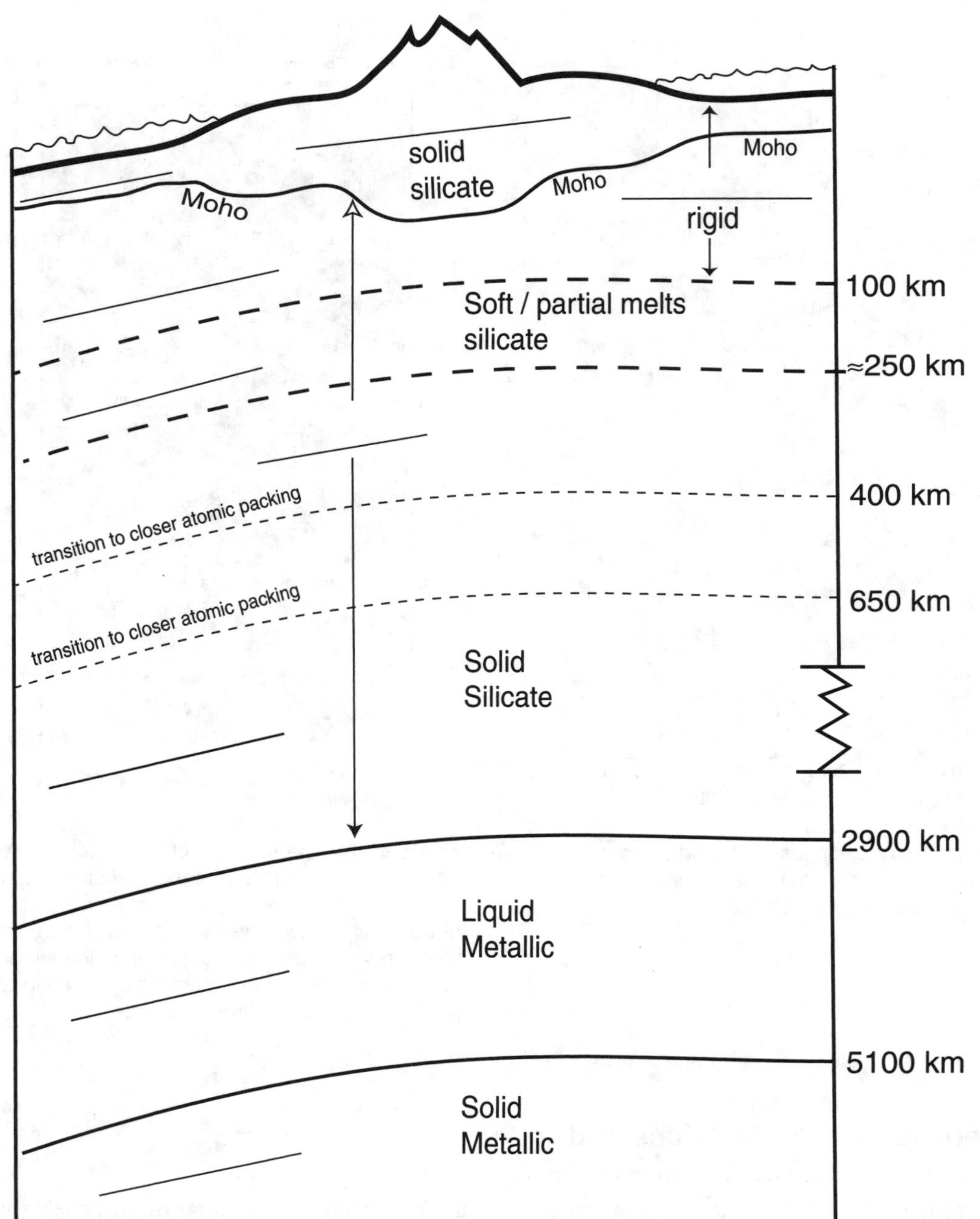

Exercise 2: The Characteristics of Earth's Internal Layers: Complete the table below by filling in the blanks and blank boxes.

LAYERS	VOLUME % of total	MASS % of total	DENSITY g/cm^3	PHYSICAL STATE	COMPOSITION	OBSERVATIONS AND EVIDENCE THAT SUPPORT THE CHARACTERISTICS OF THE LAYER
CRUST	0.60	0.42				
Continental Crust	0.44	0.25	2.7	solid		Very heterogeneous. 40-65 km thick. Formed at convergent boundaries by orogenic processes.
Ocean Crust	0.16	0.17	3.0			Very homogeneous. About 10 km thick. Formed at ________________ from ______ magmas and lavas.
MANTLE	83.02	67.77	3.3 - 5.7			
Mantle Lithosphere	(no data)	(no data)	(no data)			Crust and mantle lithosphere makes up Earth's _______________. Thickness ranges from 0 km at spreading centers to 200 km beneath continents. S wave velocities _________ through it.
Asthenosphere	(no data)	(no data)	(no data)			Weak zone. ______________ melting. Reaches close to the surface at spreading centers and deepens under older seafloor. S waves __________and are partially absorbed.
Lower Mantle	(no data)	(no data)	(no data)			Abrupt _________ in S wave velocities at 400 and ____ km mark changes in mantle structure - collapse of the ____________________ of minerals
CORE		31.79				
Outer Core	15.68	(no data)	9.9 - 12.2			P waves slow down. S waves are _______________. Iron-nickel composition is consistent with bulk density of the ______.
Inner Core	0.70	(no data)	12.6 - 13.0			P waves suddenly __________ at 5100 km. S waves are ___________ through inner core. Composition also consistent with the natural abundance of iron and meteorites.

1. Earth's core has a radius that is about __________ of the Earth's radius.
   A. 1/8
   B. 1/4
   C. 1/2
   D. 3/4

Hint: Refer to Figure 19.5.

2. The thickness of Earth's crustal plates is
   A. the same on the continents as under the oceans
   B. at its thinnest under the oceans
   C. at its thinnest in the continents
   D. completely unknown

Hint: Refer to Figure 19.6.

3. The likely composition of the upper mantle is
   A. felsic
   B. mafic
   C. ultramafic
   D. diamonds

4. Which of the following constitutes the rigid, outer layer of Earth's crustal plates?
   A. asthenosphere
   B. lithosphere
   C. crust
   D. mantle

5. Continental crust has an overall composition corresponding closely to that of
   A. ultramafic
   B. mafic
   C. felsic to intermediate
   D. peridotite

6. The lithosphere is a ________________ layer, as opposed to the asthenosphere.
   A. plastic
   B. fluid
   C. weak
   D. rigid

7. The inference that Earth's outer core is liquid is supported by the observation that
   A. P-waves do not pass through it
   B. S-waves do not pass through it
   C. P-waves travel more rapidly through it
   D. S-waves travel more slowly through it

8. The highest-density component of the Earth is
   A. the mantle
   B. the continental crust
   C. the core
   D. the whole Earth

9. Earth's North Magnetic Pole is located
   A. at the North Geographic Pole
   B. in Alaska
   C. between Greenland and Baffin Island
   D. in China

Hint: Refer to an Atlas; Figure 19.11 offers a good description also.

10. The Earth's magnetic field is thought to be generated by
    A. permanent magnetism of minerals within the mantle
    B. permanent magnetism of the solid iron-rich inner core
    C. electrical currents generated by movement of the liquid outer core
    D. electrical currents generated by convection in the asthenosphere

11. Earth's internal heat is generated by
    A. planet formation
    B. large scale impacts of meteorites and comets
    C. decay of radioactive elements
    D. all of the above

12. The asthenosphere is a ________________ layer, as opposed to the lithosphere.
    A. brittle
    B. plastic
    C. molten
    D. rigid

EXAM PREP

Materials in this section are most useful during your preparation for midterm and final exams. For optimal performance, midterm preparation should begin about eight days before the exam (see Appendix: *How to Study Geology* for details). The basic idea is a systematic review of material divided into short study sessions.

While there is no substitute for preparation, the grade you earn also depends on test-taking skill. Use your learning style to improve your performance on the next exam.

TEST-TAKING TIPS

Test Taking and Learning Styles

VISUAL LEARNERS

- Use written directions.
- When you get stuck on an item, close your eyes and picture flow charts, pictures, field experiences or text.

AUDITORY LEARNERS

- Pay attention to verbal directions.
- Repeat written directions quietly to yourself (moving your lips is often enough).
- When you get stuck, remember your lecturer's voice covering this section.

KINESTHETIC LEARNERS

There are a variety of things kinesthetic learners find helpful when they get stuck on a test item. Try some of these:

- When you get stuck, move in your chair or tap your foot to trigger memory.
- Feel yourself doing a lab procedure.
- Sketch a flow chart to unlock memory of a process.
- Stuck on a geology problem? Sketch what is being described to get you started.

If you have used regular intensive study sessions to master the material, now you get a pay back! Review for your exam will proceed smoothly and take far less time.

The following *Chapter Summary* and *Practice Exam* should simplify review still further. Read the *Chapter Summary* to begin your session. It provides a helpful overview that should get you back into the material. Next, try the *Practice Exam*. Take it just as you would a midterm: to see how you stand in regard to mastery of this chapter. After you answer the questions, score them. Finally, and most important of all: review any question that you missed. Identify and correct the misconception that resulted in missing the item.

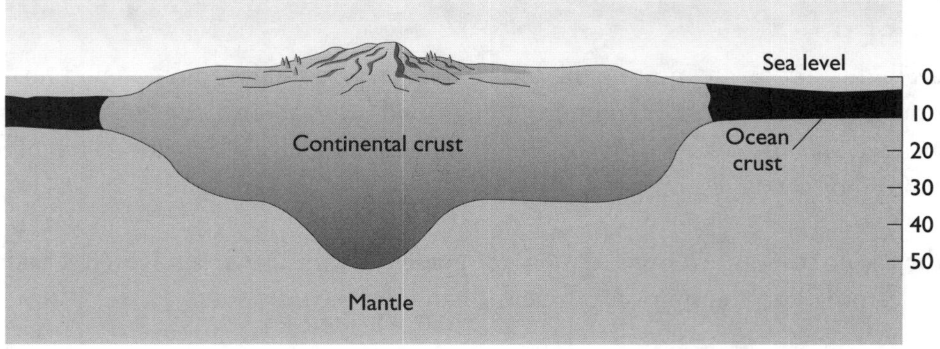

Figure 19.6 / Seismic waves reveal the boundary between the crust and underlying mantle and variations in the thickness of the crust. Relatively light continental crust projecting into the denser mantle serves as a buoyant root providing "flotation" for the continent. The root is deeper under mountains, where flotation is required to support the heavier load, in accordance with the principle of isostasy.

CHAPTER 19 SUMMARY

- Seismic waves reveal that the Earth has a concentrically zoned internal structure. The silicate-rich crust lies on a denser mantle consisting of iron-rich silicates. The Moho or Mohorovicic discontinuity in seismic wave velocities marks the boundary between the crust and the mantle. Earth's tectonic plates are large fragments of the lithosphere which include the crust and upper-most, rigid mantle. Below the lithosphere in the upper mantle is a partially molten zone called the asthenosphere. Abrupt increases in seismic wave velocities coupled with laboratory studies on high pressure minerals suggest that there are zones at progressively greater depths within the mantle where the crystal structure of minerals collapse under the intense pressure to form more compact atomic structures and, therefore, different minerals.
- The principle of isostasy proposes that continents float on the denser mantle, supported by buoyant roots that project into the mantle. Seismic waves reveal the existence of these roots. The principle of isostasy implies that, over longer periods, the mantle has little strength and behaves like a viscous fluid when it is forced to support the weight of continents and mountains.
- P wave and S wave shadow zones reveal a liquid outer core and a solid inner core. P wave velocities in the core, the natural abundance of iron in nature, the existence of iron-nickel meteorites, the Earth's strong magnetic field, and the need for a very dense core to account for the overall mass of the Earth all support an iron-nickel composition for the Earth's core.
- Earth heated up in the process of becoming a planet, and its temperature was increased further by the heat released by the decay of radioactive elements. Heat transfer within the Earth is thought to be largely a result of convection. Convection occurs when a heated fluid expands and rises because it becomes less dense than the surrounding material. Convection can occur in solids that "flow" over longer periods, similar to the way a ball of Silly Putty will spread out into a pancake shape under its own weight over a relatively short period. Convection is driven by the force of gravity and density differences between hotter and cooler materials. Plate-tectonic movements can be explained by convection in the upper mantle.
- The average increase in temperature measured for the upper crust is about 2° to 3° C per 100 meters. The rate of increase of Earth's internal temperature must decrease rapidly within the mantle, otherwise the entire Earth would be hot enough to melt. Seismic waves tell us that most of the Earth is solid.
- The main component of the Earth's magnetic field is thought to be generated by electrical fields within a moving, liquid, iron-rich outer core. The Earth's magnetic field flips back and forth over geologic time. Reversals in the Earth's magnetic field are thought to last for periods of between 1000 and 6000 years. Preserved in some rocks is a clear record of past changes in the orientation of Earth's magnetic field. The chronology of magnetic field reversals has been worked out so that the direction of remnant magnetization of a rock formation is often an indicator of its stratigraphic age.

Website and CD Activities and Tools

http://www.whfreeman.com/presssiever

Exploring Earth's Interior is an Interactive Exercise at the Website that provides you with a brief overview of some key concepts in Chapter 19.

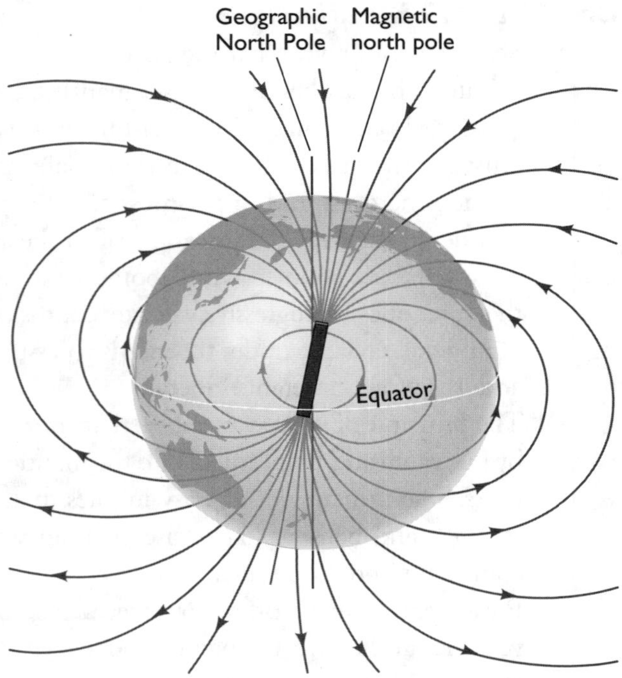

Figure 19.11
The axis of Earth's magnetic field is slightly inclined (11°) from the axis of rotation. Today, magnetic lines of force emerge from the polar region of the southern hemisphere and converge on the polar region of the northern hemisphere. Earth's magnetic field episodically reverses direction over geologic time. The high temperatures of the Earth's interior would destroy permanent magnetism. The flow of electrical currents in the Earth's hot, fluid, metallic, dynamic outer core probably generates the magnetic field.

Geographic North Pole
Magnetic north pole
Equator

PRACTICE EXAM QUESTIONS FOR CHAPTER 19

(Answers and explanations are at the end of this chapter.)

1. Except for where associated with descending crustal plates, deep earthquakes do not occur in the asthenosphere because
   A. the asthenosphere doesn't move anywhere except for at subduction zones
   B. the asthenosphere is composed of peridotite or related material
   C. the asthenosphere is a rigid plate that cannot rupture
   D. the asthenosphere is weak, and deforms plastically

2. Key elements in a reasonable model to explain what drives lithospheric plates across the Earth's surface would include
   A. the tides, the Coriolis effect, and the lithosphere
   B. heat transfer by convection, gravity, and the asthenosphere
   C. the shrinking Earth hypothesis and isostatic adjustments
   D. the electro-dynamo theory for the generation of the Earth's magnetic field

3. Lava beds become magnetized by the Earth's magnetic field when
   A. they are struck a sharp blow by a meteorite
   B. cooling iron-rich minerals become magnetized parallel to the Earth's field
   C. the Earth's magnetic field reverses itself
   D. electricity from lightning strikes passes through the lava beds

Hint: Refer to Figure 19.14.

4. Which layer of the Earth experiences the most rapid increase in temperature with increasing depth?
   A. the lithosphere
   B. the asthenosphere
   C. the mantle
   D. the liquid outer core

Hint: Refer to Figure 19.10.

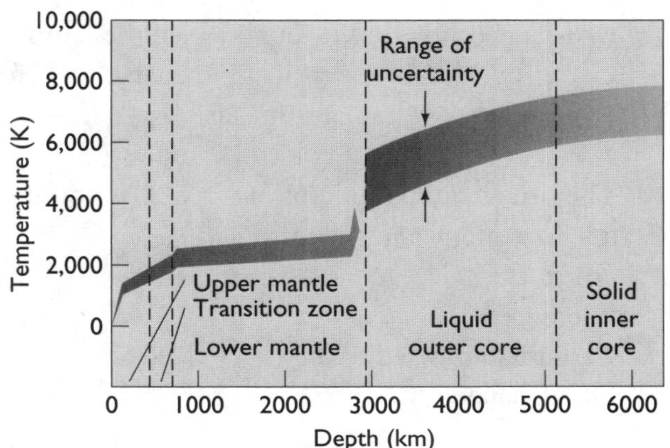

Figure 19.10 / Estimated increase in temperature with depth in the Earth is inferred from studies of volcanoes, speed of seismic waves, laboratory experiments, and theory. The width of the curve shows the range of uncertainty within which the actual geotherm lies. Temperature is given in kelvins (K=°C + 273°).

5. Between which boundary in the Earth's interior does the greatest change in composition occur?
   A. the lithosphere-asthenosphere boundary
   B. the crust-mantle boundary
   C. the mantle-core boundary
   D. the boundary between the outer and inner core

6. The crust is typically thickest beneath
   A. high mountain ranges and plateaus on the continents
   B. ocean spreading centers
   C. continental interiors like the Great Plains in North America
   D. beneath passive margins of continents where topography is very flat

Hint: Refer to Figure 19.6.

7. Significant increases in S wave velocity at 400 and 650 kilometers depth are explained by
   A. changes in the chemical composition of the mantle
   B. a collapse of the crystal structures to more close packed forms
   C. changes in the degree of partial melting within the mantle
   D. a rapid increases in temperature

Hint: Refer to Figure 19.7.

8. The Earth's core is inferred to be composed of mostly iron because
   A. iron is naturally very abundant
   B. most meteorites representing the interstellar matter from which the planets formed are rich in iron
   C. iron is very dense, so its presence in the core would account for the Earth's average density
   D. all of the above

9. The velocity at which seismic waves leave the focus of an earthquake does not depend on the
   A. depth at which the earthquake occurs
   B. density of the rock through which the seismic waves travel
   C. composition of the rock along the fault where the earthquake occurred
   D. magnitude of the earthquake

10. The time required for erosion to level a mountain range by erosion is ____________ due to isostatic adjustments.
   A. lengthened
   B. reduced
   C. dramatically reduced
   D. unchanged

11. Without any knowledge of what seismic waves tell us about the Earth's interior, why is it unreasonable to assume that a large portion of the lower mantle is molten?
   A. Actually, it could be molten. We just don't see evidence for it because silicate magmas are trapped within the Earth due to the confining pressures.
   B. Direct measurements show that the temperature at the core / mantle boundary is not high enough to melt the lower mantle.
   C. The magnetic field strength would be greatly reduced if more of the Earth's interior was molten.
   D. Silicate magmas are less dense and would rise to the surface, so we should observe widespread volcanic activity across the Earth's surface. Whereas, the molten iron-rich outer core is too dense to rise.

12. As rocks experience increased pressure with depth, P-waves in general will ______________as they migrate through them.
   A. travel faster
   B. travel slower
   C. travel at the same velocity
   D. rapidly die out

13. When a reversal of the Earth's magnetic field occurs, the
   A. sense of rotation of the Earth is also reversed
   B. Earth flips over in its orbit
   C. magnetic polarity of the Earth reverses, such that the north end of a magnetic compass needle points toward the south geographic pole
   D. almost all the igneous sedimentary rocks of the ocean floor reverses in magnetization to match the new orientation of the magnetic field

14. The Moho, or Mohorovicic discontinuity between the crust and the mantle, was first detected based on
   A. the abrupt decrease in seismic velocities as they cross it
   B. the abrupt increase in seismic velocities as they cross it
   C. the S wave shadow zone through which S waves do not pass
   D. the observation that no earthquakes occur below the Moho

ANSWERS AND EXPLANATIONS FOR EXERCISES AND QUESTIONS

After Lecture

Exercise 1: The Earth's Interior Layers

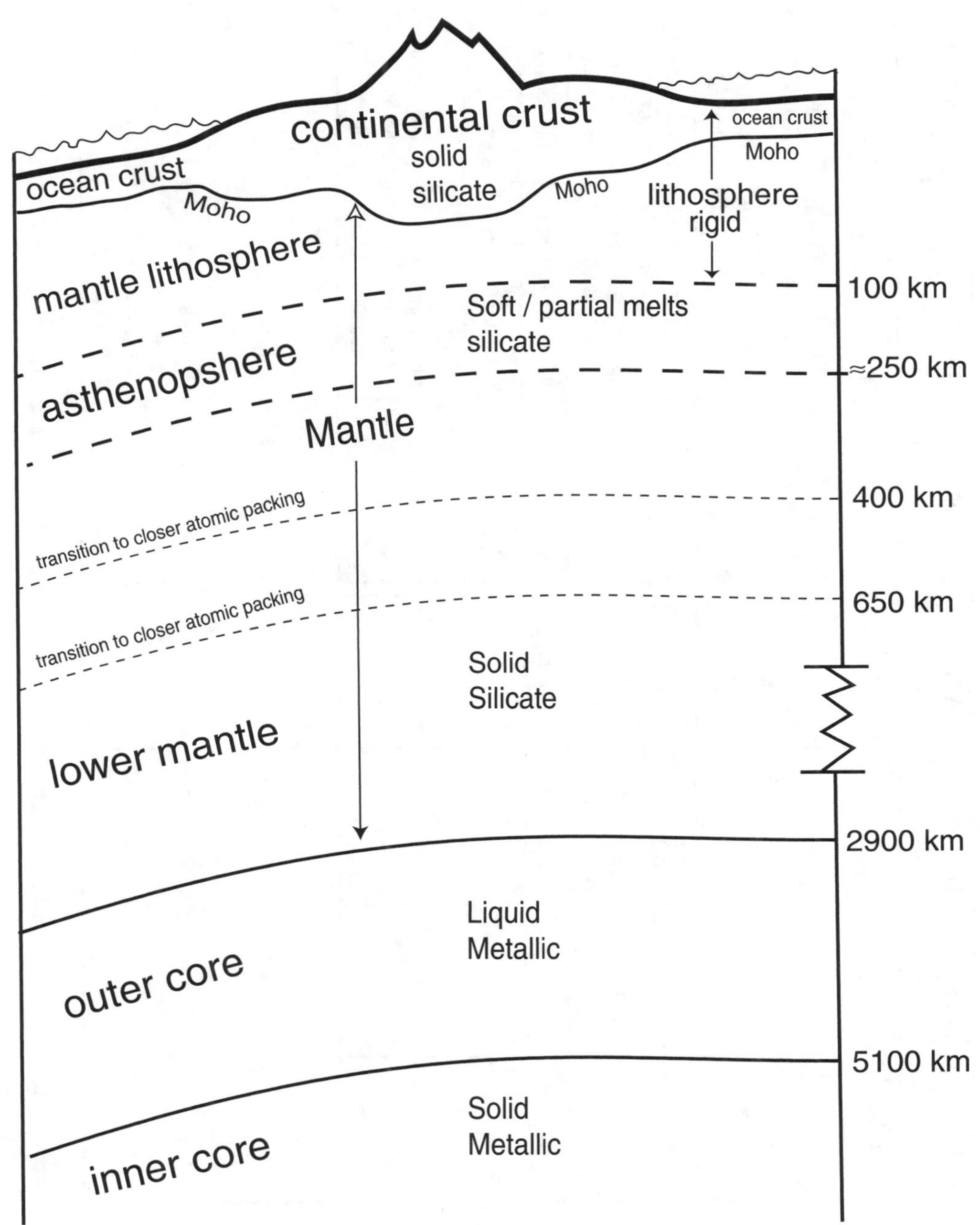

Exercise 2: The Characteristics of Earth's Internal Layers

LAYERS	VOLUME % of total	MASS % of total	DENSITY g/cm³	PHYSICAL STATE	COMPOSITION	OBSERVATIONS AND EVIDENCE THAT SUPPORT THE CHARACTERISTICS OF THE LAYER
CRUST	0.60	0.42				
Continental Crust	0.44	0.25	2.7	solid	mostly felsic - intermediate silicate material	Very heterogeneous. 40-65 km thick. Formed at convergent boundaries by orogenic processes.
Ocean Crust	0.16	0.17	3.0	solid	mafic silicate minerals	Very homogeneous. About 10 km thick. Formed at divergent boundaries from mafic magmas and lavas.
MANTLE	83.02	67.77	3.3 - 5.7			
Mantle Lithosphere	(no data)	(no data)	(no data)	rigid solid	ultramafic silicates - peridotite	Crust and mantle lithosphere makes up Earth's tectonic plates. Thickness ranges from 0 km at spreading centers to 200 km beneath continents. S wave velocities speed up through it.
Asthenosphere	(no data)	(no data)	(no data)	soft solid or plastic	ultramafic silicates - peridotite	Weak zone. A few percent melting. Reaches close to the surface at spreading centers and deepens under older seafloor. S waves slow down and are partially absorbed.
Lower Mantle	(no data)	(no data)	(no data)	solid	ultramafic silicates - peridotite	Abrupt increases in S wave velocities at 400 and 650 km mark changes in mantle structure - collapse of the crystalline structure of minerals
CORE		31.79				
Outer Core	15.68	(no data)	9.9 - 12.2	liquid	mostly iron and nickel	P waves slow down. S waves are completely blocked. Iron-nickel composition is consistent with bulk density of the Earth.
Inner Core	0.70	(no data)	12.6 - 13.0	solid	mostly iron and nickel	P waves suddenly speed up at 5100 km. S waves are transmitted through inner core. Composition also consistent with the natural abundance of iron and meteorites.

1. C. Earth's core is about one-half the radius of planet Earth. See Figure 19.5: in the illustration, Earth's center is at 6,000 km depth, and the outer core is at 3,000 km depth
2. B. Refer to Figure 19.6. The ocean basins exist on Earth because the ocean crust is thinner and more dense than the continental crust. Therefore, the ocean crust sits lower.
3. C. Ultramafic or peridotite.
4. B. Earth's tectonics plates are large fragments of the lithosphere, a rigid layer consisting of the crust and upper most mantle.
5. C. Continental crust is very heterogeneous in composition. Nevertheless, its average composition is felsic to intermediate.
6. D. The lithosphere is thought to be a rigid layer that rests on a plastic, weak asthenosphere.
7. B. S waves do not pass through the outer core.
8. C. The metallic core has the highest density (refer to Exercise 2).
9. C. Earth's north magnetic pole is located between Greenland and Baffin Island.
10. C. The electro-dynamo theory explains the Earth's magnetic field as a consequence of electrical currents moving in the fluid, metallic outer core. The silicate minerals in the asthenosphere are not nearly as good conductors of electricity. Therefore, flow of electrical currents and the associated magnetic field is probably much weaker in the asthenosphere.
11. D. Earth's internal heat comes from all sources listed.
12. B. The asthenosphere is thought to be a soft, plastic, zone within the upper mantle that may be partially melted.

Exam Prep

1. D. Earthquakes with foci between 100 and 700 km occur only in association with subduction zones. It is thought that earthquakes are generated within the cooler, more rigid and brittle subducting slab. Below about 700 km the Earth's mantle appears to become too plastic for earthquakes to occur.
2. B. Heat transfer by convection helps to create the high and low spots that mark tectonic plate boundaries. Gravity pulls the plates apart at the spreading centers and the old, cold, denser, subducting slab sinks into the mantle due to gravity. The soft, plastic asthenosphere decouples the tectonic plates from the rest of the mantle and allows the rigid lithosphere to move laterally with less drag from mantle materials.
3. B. Refer to Figure 19.14.
4. A. Refer to Figure 19.10.
5. C. The mantle – core boundary is marked by the most significant change in composition.
6. A. The thickest continental crust on Earth lies beneath the Tibetan Plateau, Himalaya Mountains and portions of the Andes Mountains (refer to Figure 19.6).
7. B. Refer to Figure 19.7.
8. D. All of the above.

9. D. The magnitude of the earthquake does not affect seismic wave velocities. However, depth with the Earth, density and composition do influence seismic wave velocities.
10. A. Refer to Figure 19.6. As a mountain range is eroded, the buoyant root of less dense crustal rocks within the mantle continues to rise and replace some of the materials removed by erosion.
11. D. If a significant portion of the lower mantle were molten, we should see lots more volcanic activity on the Earth's surface in addition to tectonic plate boundary volcanism.
12. A. Seismic wave velocities depend on the elasticity, rigidity, and density of materials. These properties are dependent on the composition, physical state, and compactness of the atomic structure of the material.
13. C. Refer to Figure 19.12.
14. B. The Moho was first detected by an abrupt increase in seismic wave velocities due to the change from lower density crustal rocks to the more dense ultramafic rocks within the mantle.

Chapter 20
Plate Tectonics: The Unifying Theory

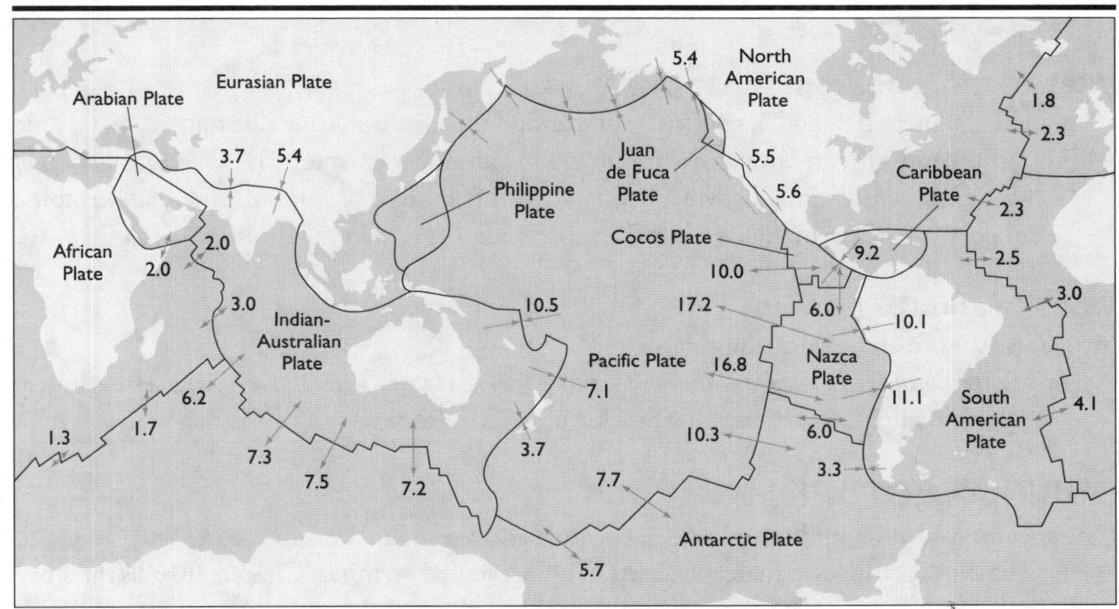

Figure 20.12 / Relative velocities (in centimeters per year) and directions of plate separation and convergence. Opposed arrowheads indicate convergence. Diverging arrowheads indicate plate separation at ocean ridges. Parallel arrowheads, as along the San Andreas fault in California, indicate transform faults, where plates slide past each other. Spreading is fastest between the Pacific and Nazca plates and slowest between the North American and Eurasian plates. [*Data from C. Demets, R. G. Gordon, D. F. Argus, and S. Stein, Model Nuvel-1, 1990.*]

BEFORE LECTURE

Before you attend lecture be sure to spend some time previewing the chapter. For an efficient preview use the questions below.

CHAPTER PREVIEW

- **What is the theory of plate tectonics?**
  Brief answer: The Earth's lithosphere is broken into about a dozen moving plates. Volcanoes, earthquakes and crustal deformation are concentrated along the active plate boundaries.
- **What are some of the geologic characteristics of plate boundaries?**
  Brief answer: Mountains typically form along convergent and transform plate boundaries.Where divergent plate boundaries are exposed on land, subsiding basins and mafic volcanism are typical.
- **How can the age of the seafloor be determined?**
  Brief answer: Magnetic anomaly bands in the seafloor and the chronology of magnetic reversals deciphered from lavas and sediments on land are used to determine the approximate age of the seafloor.
- **How do lithospheric plates form and move?**
  Brief answer: Driven by Earth's internal heat, convection of hot and cold matter within the mantle, the force of gravity and the existence of an asthenosphere are important factors in any model for the driving mechanism of plate tectonics.

From time to time in the history of science, a fundamental concept appears that unifies a field of study by pulling together diverse theories and explaining a large body of observations. Such a concept in physics is the theory of relativity; in chemistry, the nature of the chemical bond; in biology, DNA; in astronomy, the Big Bang;
and in geology, plate tectonics.

—Press and Siever

Vital Information from Other Chapters

- In Chapters 1, 3, and 5, review the figures and their captions that illustrate plate tectonic boundaries. If you have any extra time, look ahead to Chapter 21, *The Deformation of the Continental Crust.* It will provide you with more examples of how plate-tectonic theory integrates many facets of the geosciences.

Website and CD Preview

http://www.whfreeman.com/presssiever

- *Plate Boundary Descriptions* and *Plates of the World* are Interactive Exercises definitely worth completing before your first lecture on this topic.

DURING LECTURE

One goal for lecture should be to leave class with a good set of answers to the *Preview Questions.*

- To avoid getting lost in details, keep the "big picture" in mind: Chapter 20 tells the story of *Plate Tectonics*: Earth's moving plates and the geological features associated with converging and diverging plate boundaries. Plate tectonics underlies and explains modern geology. In that sense, this chapter provides a summary of your entire geology course.
- Focus on understanding Figure 20.12, *Plate Separation.* Also study Figures 20.18, 20.19 and 20.20 to understand the geological features associated with diverging and converging plate boundaries. Hint: It will be helpful to have these figures handy during lecture. Annotate with comments made by your instructor.

AFTER LECTURE

The perfect time to review your notes is right after lecture. The checklist below contains both general review tips and specific suggestions for this chapter.

NOTE REVIEW CHECKLIST

✔ All notes legible? (Rewrite so they read easily.)

✔ Important points clearly identified? You should now have headers in your notes that tie to each of the questions in the *BEFORE LECTURE/Chapter Preview.*

✔ Holes (missing material) filled in from memory?

✔ Areas where you don't remember what was said marked for a follow-up session with your instructor, tutor, or study partner?

✔ Possible test questions indicated in the margin (TQ)?

- ✔ Additional visual material. *Suggestions for Chapter 20: Figures 20.12 Plate Separation. Figures 20.18, 20.19 and 20.20 are also essential aids to understand the geological features associated with diverging and converging plate boundaries.*

- ✔ Reworked notes into a form that is efficient for your learning style?

- ✔ Created a brief "big picture" overview of this lecture (using a sketch or written outline)? *Hint: An overview of this chapter is really an overview of the entire course. Exercise 1 in Practice Exercises and Study Questions is your best overview strategy for this chapter. It is also an essential aspect of preparing for your final exam, so be sure to complete Exercise 1.*

Intensive Study Session

After each lecture you need to thoroughly master the concepts covered. You need to do this before you attend the subsequent lecture. The ideas of geology are like a stack of boxes. Each new idea rests on all ideas (boxes) stacked beneath it.

Schedule at least one hour after lecture for intensive study. Use this time to master key concepts. By now you know well that mastery is not gained by just reading your text. To master geology, you must ask yourself questions and answer them. Use the Website and CD Activities and Tools suggested below, along with the *Practice Exercises* and *Study Questions*, to insure you master this chapter. Do as many of these as you have time for during your scheduled study session. Pay particular attention to exercises recommended by your instructor during lecture.

Website and CD Activities and Tools

http://www.whfreeman.com/presssiever

Complete the Q & A at the Website. Pay particular attention to the explanations for the answers. Also at the Website are Flashcards to help you learn new terms. On your CD there are five animations illustrating the four types of plate boundaries and the tectonic history of passive continental margins. As you watch these brief animations, be sure to write down any questions you have about them. Then, check in your textbook and with your instructor for answers.

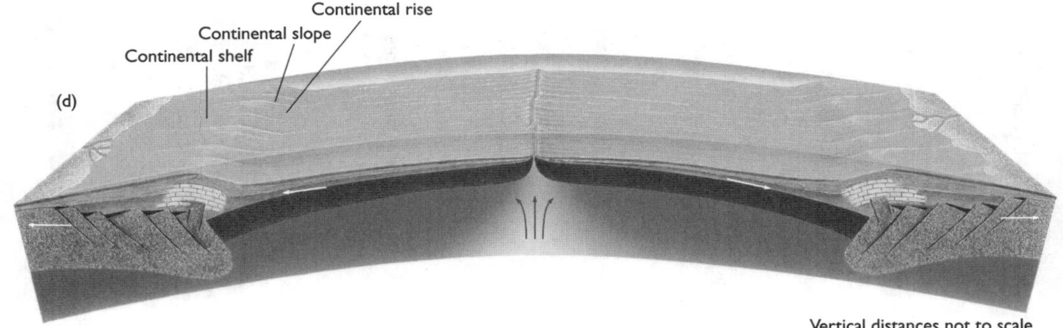

Figure 20.17 (d) / The Atlantic margins of Europe, Africa, and North America. The opening of the Atlantic Ocean Basin coincided with the breakup of Pangaea. Rifting of Pangaea began during the Triassic. Since then, the Atlantic has evolved into a mature ocean basin with a mid-oceanic ridge.

PRACTICE EXERCISES AND STUDY QUESTIONS

(Answers and explanations are at the end of this chapter.)

Exercise 1: Characteristics of Active Tectonic Plate Boundaries

Complete the table below by filling in the blanks. Chapters 3, 4, 5, 18, 19 and 20 are useful resources.

CHARACTERISTICS	DIVERGENT	CONVERGENT			TRANSFORM
		OCEAN/OCEAN	OCEAN/ CONTINENTAL	COLLISION	
EXAMPLES	Mid-Atlantic Ridge	Marianas and Tonga Trenches			
TOPOGRAPHY		trench, volcanic island arc, forearc			offset of topographic and geologic features
EARTHQUAKE FOCI				shallow to intermediate	shallow
EARTHQUAKE MAGNITUDE	low				
DEFORMATION STYLE		compressional - thrusting and folding		compressional - intense thrusting and folding	
VOLCANISM	basaltic lava flows, pillow lavas, hydrothermal vents	composite volcanoes, lava flows, ash flows, explosive eruptions		not characteristic	not characteristic
MAGMA TYPE		mafic to felsic		possible felsic plutons generated	none
ASSOCIATED ROCK ASSEMBLAGES AND GEOLOGIC FEATURES	pillow lavas and ophiolite suite form at divergent margins	mafic volcanics and associated sediments		suture zone with ophiolite suite, highly folded and faulted sediments	
IMPLICATIONS		formation of new land, possible microplates, raw materials for the growth of continental crust	deforms leading edge of continent, continental crust is thickened		redistribution of continental crust, formation of microplates

1. What is the most important process in building ocean floor?
   A. volcanism
   B. metamorphism
   C. precipitation of carbonate rocks
   D. deposition of sediments onto continental shelves

2. The youngest ocean crust is located
   A. along the oceanic ridges
   B. in the oceanic trenches
   C. around hot spots
   D. on the abyssal sea floor

3. Rates of seafloor spreading today range from
   A. 0.01 to 0.1 cm/year
   B. 0.1 to 1.0 cm/year
   C. 1.0 to 15 cm/year
   D. 15 to 100 cm/year

Hint: Refer to Figure 20.12.

4. The motion found along transform boundaries is where plates
   A. are being forced together so as to produce a mountain system
   B. move apart to create a widening rift valley
   C. are being consumed by subduction
   D. move horizontally past each other

5. The ______________________ is an example of a divergent plate margin.
   A. East African Rift
   C. Japan Trench
   D. Himalayas
   A. San Andreas fault

6. Of the following, which does NOT represent a plate boundary?
   A. San Andreas Fault
   B. Red Sea
   C. Hawaiian Islands
   E. Peru-Chile Trench

7. The following are associated with active plate boundaries, EXCEPT
   A. Atlantic coast of North America
   B. Gulf of California
   C. Himalayas
   D. Northwestern North America

8. The supercontinent that broke up in the Mesozoic era is called
   A. Laurasia
   B. Atlantis
   C. India
   D. Pangaea

9. According to plate tectonic theory, mid-ocean ridges are
   A. locations of plate convergence
   B. pull-apart zones where new oceanic crust is produced
   C. places where oceanic crust is consumed
   D. transform faults

10. The ______________________________ are an example of a collision zone between two pieces of continental crust riding on converging lithospheric plates.
   A. Himalaya Mountains between India and Asia
   B. islands of Japan
   C. Aleutian Islands in Alaska
   D. Andes Mountains in South America

11. The Hawaiian Islands are located on a
   A. subduction zone
   B. spreading center
   C. transform fault
   D. hot spot

12. Of the following crustal plate features, which is least likely to see volcanoes form?
   A. hot spots
   B. convergent boundaries
   C. divergent boundaries
   D. transform boundaries

13. During the late Paleozoic
   A. India collided with Asia to form the Himalaya Mountains
   B. the supercontinent, Rodinia, formed
   C. continents assembled into Pangaea
   D. North America collided with Pangaea

Hint: Refer to Figure 20.23.

14. The breakup of Pangaea began in the ______________ period of geologic time.
   A. Cambrian
   B. Permian
   C. Triassic
   D. Tertiary

Hint: Refer to Figures 20.23 and 20.24.

EXAM PREP

It's time to get ready for final exam week. As with any big project, devoting some time up front to getting organized will pay big dividends. Use the following worksheet as a guide to help you get organized. Modify the suggestions to fit your personal situation and needs.

FINAL EXAM PREP
WORKSHEET

(to be completed three weeks prior to exam)

Adapted with permission from the University Learning Center, University of Arizona.

1. **COURSE SHEETS:** In your notebook, set up four or five separate sheets of paper—one sheet for each course you are taking. At the top of each sheet list the course and the grade you presently have (be realistic, not hopeful).

2. **DATE:** List the date of the final exam under each course name.

3. **COMPREHENSIVE FINALS:** Mark with a "C": each course with a comprehensive final.

4. **EXAM FORMAT:** Identify the format of the exam (multiple choice, essay, etc.) under the date of the final for each course.

5. **TASK:** Identify the levels of thinking expected. Hint: Previous midterms are your ultimate resource on this question. List all kinds of questions. Estimate what percent of total points will be devoted to each kind of thinking.
    - Application to Real World Situations
    - Problem Solving
    - Critical Thinking
    - Understanding Principles
    - Memory of Basic Facts

6. **RANK FINALS IN ORDER OF IMPORTANCE:** In the upper right-hand corner of each sheet, rank in order the most critical and important final to the least important final—the final that will make the least difference in your grade. (Be aware of how much impact your final exam has on your overall class grade.)

7. **LIST WHAT THE TEST WILL COVER:** For each course on each sheet, list everything the test will cover—remember which exams are comprehensive.
    - Handouts??
    - Chapters?? (which ones??)
    - Lectures??
    - Discussions??
    - Other??

    Check your syllabi to be sure you have not left anything out.

8. Draw a line beneath this list. Then list what you still have left to do for that particular course.
    - Which chapters do you still have to preview?
    - Which lecture notes do you need to review and update?
    - Which *Practice Exercises* and *Study Questions* do you need to complete?
    - Which labs do you still have to hand in?
    - What papers do you still have to write?

9. Draw another line. Now list the test preparation strategies you will use to study for the exam—study groups or study patterns, self-questioning using the annotations, mapping, charting, questions and answers, concept cards, going over old tests and quizzes, making up your own problems.

10. Now fill in the calendar, identifying exams, finals and when papers are due. Each day you need to do something from No. 8, but you will also need to study and review for the finals at least two hours a day. Be sure to use all of your available time—weekends, waiting time, etc.

Work Toward These Goals:

- Finish all work under No. 8. Review *Practice Exercises* and *Study Questions*, etc. one week prior to your first final. Review all lecture notes by asking yourself the questions out loud or by having someone quiz you five days prior to your first final (allow two to three hours).
- Divide the work that remains so that you do an *Eight-Day Study Plan* for each course that you assigned a high priority in No. 6.
- Remove the distractions from your life. This is not the week to be captured by TV or other addictions. Stick to your priorities. Tell friends and family that you need to focus all your energy on your finals until they are over.
- Avoid burnout. Build into your schedule time for adequate sleep, relaxation and exercise.

EXAM PREP

The following *Chapter Summary* and *Practice Exam* should simplify review of Chapter 20. Read the *Chapter Summary* to begin your session. It provides a helpful overview that should get you back into the material. Next, try the *Practice Exam*. Take it just as you would a midterm: to see how you stand in regard to mastery of this chapter. After you answer the questions, score them. Finally, and most important of all: review any question that you missed. Identify and correct the misconception that resulted in missing the item.

CHAPTER 20 SUMMARY

- For over the last century some geologists have argued for the concept of continental drift based on the jigsaw-puzzle fit of the coasts on both sides of the Atlantic, the geological similarities in rock ages and trends in geologic structures on opposite sides of the Atlantic, fossil evidence suggesting that continents were joined at one time, and the distribution of glacial deposits as well as other paleoclimatic evidence.
- In the last half of the Twentieth Century, the major elements of the plate tectonic theory were formulated. Starting in the 1940's, ocean floor mapping began to reveal major geologic features on the ocean floor. Then the match between magnetic anomaly patterns on the seafloor with the paleomagnetic time scale revealed that the ocean floor had a young geologic age and was systematically older away from the oceanic ridge systems. The concepts for seafloor spreading, subduction and transform faulting evolved out of these and other observations.
- According to the theory of plate tectonics, the Earth's lithosphere is broken into about a dozen moving plates. The plates slide over a partially molten, weak asthenosphere, and the continents, embedded in some of the moving plates, are carried along.
- There are three major types of boundaries between lithospheric plates: divergent boundaries where plates move apart; convergent boundaries where plates move together and one plate often subducts beneath the other; and transform boundaries where plates slide past each other. Volcanoes, earthquakes and crustal deformation is concentrated along the active plate boundaries. Mountains typically form along convergent and transform plate boundaries. Where divergent plate boundaries are exposed on land, subsiding basins and mafic volcanism are typical.
- Various methods have been used to estimate and measure the rate and direction of plate movements. Seafloor spreading rates vary between 2 to 17 cm per year today.
- Distinct assemblages of rocks characterize each type of plate boundary. Finding these diagnostic rock assemblages embedded in continents, geologists have been able to reconstruct plate tectonic events and plate configurations in the geologic past.
- Accretion of microplate terranes is now recognized as an important process in the formation of mountains and the growth of continents along active continental margins.
- Driven by Earth's internal heat, convection of hot and cold matter within the mantle, the force of gravity and the existence of an asthenosphere are important factors in any model for the driving mechanism of plate tectonics.

Website and CD Activities and Tools

http://www.whfreeman.com/presssiever

Use Chapters 20 and 21 and *Photo Gallery* images and their captions on your CD as a virtual tour of active plate boundaries around the world. Deformation at active plate boundaries generates much of the spectacular topography and scenery found on Earth.

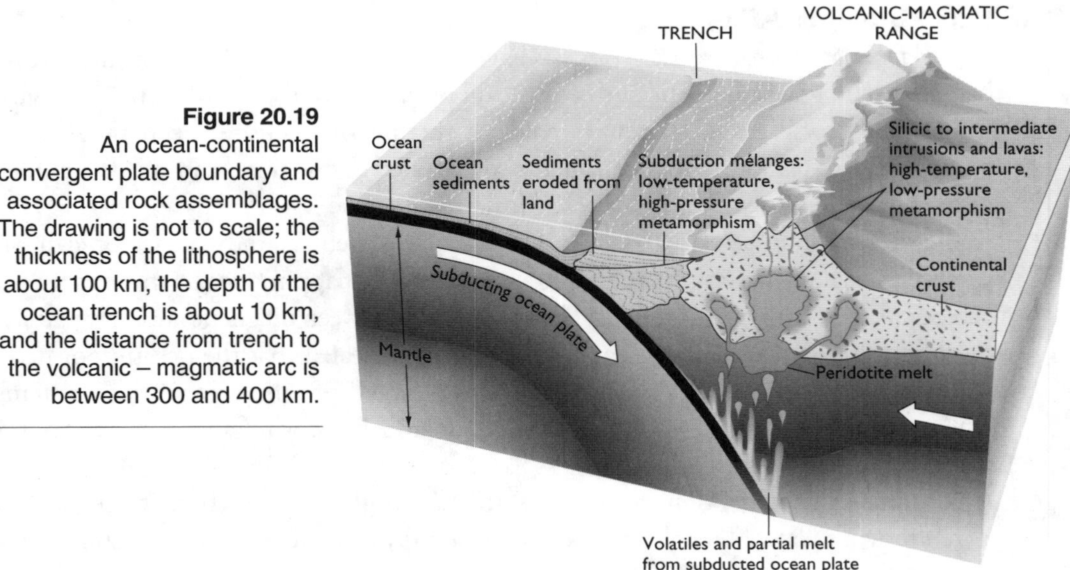

Figure 20.19 An ocean-continental convergent plate boundary and associated rock assemblages. The drawing is not to scale; the thickness of the lithosphere is about 100 km, the depth of the ocean trench is about 10 km, and the distance from trench to the volcanic – magmatic arc is between 300 and 400 km.

PRACTICE EXAM QUESTIONS FOR CHAPTER 20

(Answers and explanations are at the end of this chapter.)

1. All of the following are observed at mid-oceanic ridges EXCEPT
   A. shallow-focus earthquakes
   B. basalt eruptions
   C. deep-focus earthquakes
   D. a rift valley

2. Areas of continents that experience crustal plate separation are characterized by
   A. mountain building
   B. transform faults
   C. long, downfaulted valleys
   D. deep-focus earthquakes

3. Which of the following localities are not associated with an active plate boundary?
   A. West coast of northern Europe
   B. Southern California
   C. Japan Islands
   D. Himalaya Mountains in Asia

4. The magnetic anomaly patterns discovered in the seafloor were significant because they
   A. established an estimated age for the seafloor once they were correlated with the known magnetic reversal chronology
   B. represented absolute proof that the sea floor was spreading apart
   C. allowed for a reconstruction of the ancient supercontinent Rodinia
   D. revealed the driving force for plate tectonics in mantle convection

5. The oldest rocks on the sea floor are about
   A. 20 million years old
   B. 175 million years old
   C. 500 million years old
   D. one billion years old

6. Oceanic crust
   A. ranges in age from Jurassic to Precambrian
   B. is progressively older further from the oceanic ridges
   C. is progressively younger further from the oceanic ridges
   D. is the same age virtually everywhere

7. An assemblage of rock consisting of deep-sea sediments, pillow lavas, gabbro, and peridotites that represents fragments of ocean lithosphere is called
   A. an ophiolite suite
   B. a mélange of metamorphosed sediments
   C. a Benioff suite
   D. a Franciscan suite

8. Passive and active continental margins differ from one another in that active margins
   A. are characterized by strong tides, with large differences between high and low water
   B. lie adjacent to regions of sea floor spreading
   C. lie adjacent to regions of subduction
   D. are characterized by broad gently sloping continental shelves

9. Volcanic island arcs like the Aleutian islands are associated with
   A. convergent boundaries
   B. divergent boundaries
   C. transform boundaries
   D. a chain of hot spots

Hint: Refer to Figure 20.18.

10. Rift valleys are associated with
   A. convergent boundaries
   B. divergent boundaries
   C. transform boundaries
   D. active continental margins

Hint: Refer to Figures 20.4 and 20.17.

11. Rift valleys are associated with what form of tectonic activity?
   A. subduction
   B. continental collision
   C. continental rupture
   D. movement on a transform fault

12. The Atlantic coast of North America is a passive margin along which is being deposited
    A. shales, sandstones, limestones
    B. basalts and pillow lavas
    C. mélange sediments and ophiolites
    D. volcanic ash and felsic lava flows

13. The Franciscan mélange found in the California coastal ranges represents
    A. rocks associated with and metamorphosed in a subduction zone
    B. reef and passive continental margin sediments
    C. river and lake sediments deposited during the last ice age
    D. red sandstones and siltstone associated with an ancient desert

14. Magnetic anomalies in the seafloor are caused by
    A. magnetic reversals recorded by lavas erupted at oceanic spreading centers
    B. changes in the atomic structure of minerals in response to changing ocean depth
    C. the metamorphism of deep-sea sediments
    D. the heating up of subducting oceanic lithosphere as it plunges deeper into the mantle

15. Microplate terranes are typically accreted to other terranes or continents at
    A. hot spots
    B. rift zones
    C. convergent plate boundaries
    D. divergent boundaries

Hint: Refer to Figures 20.21 and 20.22.

16. Sediments in oceanic basins are all Cretaceous in age or younger. Why?
    A. Any older sediments that existed have been dissolved because they are deposited at great depths on the ocean floor.
    B. Older sediments are covered by younger sediments and are not therefore accessible for study.
    C. Older sediments have been subducted or accreted.
    D All the older sediments have been metamorphosed.

17. Granitic plutons are forming today
    A. at oceanic ridges
    B. beneath hot spots like Hawaii
    C. within volcanic arcs associated with convergent plate boundaries
    D. along transform faults which provide fracture along which magmas can move

Hint: Refer to Figure 20.19.

18. Continents grow at
    A. divergent plate boundaries
    B. convergent plate boundaries
    C. transform plate boundaries
    D. hot spots

ANSWERS AND EXPLANATIONS FOR EXERCISES AND QUESTIONS

After Lecture

Exercise 1: Characteristics of Active Tectonic Plate Boundaries

CHARACTERISTICS	DIVERGENT	CONVERGENT			TRANSFORM
		OCEAN/OCEAN	OCEAN/ CONTINENTAL	COLLISION	
EXAMPLES	Mid-Atlantic Ridge	Marianas and Tonga Trenches			
TOPOGRAPHY		trench, volcanic island arc, forearc			offset of topographic and geologic features
EARTHQUAKE FOCI				shallow to intermediate	shallow
EARTHQUAKE MAGNITUDE	low				
DEFORMATION STYLE		compressional - thrusting and folding		compressional - intense thrusting and folding	
VOLCANISM	basaltic lava flows, pillow lavas, hydrothermal vents	composite volcanoes, lava flows, ash flows, explosive eruptions		not characteristic	not characteristic
MAGMA TYPE		mafic to felsic		possible felsic plutons generated	none
ASSOCIATED ROCK ASSEMBLAGES AND GEOLOGIC FEATURES	pillow lavas and ophiolite suite form at divergent margins	mafic volcanics and associated sediments		suture zone with ophiolite suite, highly folded and faulted sediments	
IMPLICATIONS		formation of new land, possible microplates, raw materials for the growth of continental crust	deforms leading edge of continent, continental crust is thickened		redistribution of continental crust, formation of microplates

1. A. Mafic volcanism at the oceanic ridges builds the seafloor.
2. A. Refer to Figures 20.10 and 20.11.
3. C. Refer to Figure 20.12.
4. D. The typical plate motion along most transform faults is horizontal slip (shearing). However, along curves in the transform fault, transextension (forming a depression) and transcompression (forming mountains) may be generated.
5. A. Refer to Figures 20.4 and 20.5.
6. C. The Hawaiian islands formed over a hotspot in the middle of the Pacific ocean plate. Some hotspots such as in Iceland are located coincidentally adjacent to a spreading center. Most hotspots on Earth are not directly associated with plate boundaries.
7. A. The Atlantic coast of North America is a passive continental margin, which is not associated with an active plate margin.
8. D. Pangaea began to breakup during the Triassic Period of the Mesozoic (refer to Figure 20.24).
9. B. Mid-oceanic ridges or spreading centers are divergent boundaries where the crust is extending (pulling apart) and mafic magmas are intruding upward from the asthenosphere to feed basaltic volcanism that is building new ocean floor.
10. A. Refer to Figure 20.20.
11. D. Hawaiian islands are located on a hot spot in the middle of the Pacific plate.
12. C. Volcanism is not characteristic of transform plate boundaries. Some volcanism, typically minor amounts, may occur in association with transform faults where transextension is occurring.
13. C. Refer to Figure 20.23.
14. C. Pangaea began to rift apart during the Triassic (refer to Figures 20.23 and 20.24).

Exam Prep

1. C. Deep foci earthquakes do not occur in association with the ocean ridge systems. The asthenosphere essentially reaches close to the ocean floor along the ridge systems. The soft, partially melted rocks within the asthenosphere are too warm and soft to exhibit the brittle or elastic behavior required to generate earthquakes. Earthquakes along the ridge system are, therefore, restricted to the colder, older, and more rigid oceanic lithosphere.
2. C. Long downfaulted valleys, or rift valleys, are classic features of continental rifting or separation. See Figures 20.4 (b) and 20.5. At least two areas of North America display evidence of ancient rifting. Where could you expect to see such evidence? See Figure 20.17 to think about one area.
3. A. Like the east coast of North America, the west coast of northern Europe is a passive continental margin.
4. A. Refer to Figures 20.9 and 20.10, and the section *Inferring Seafloor Spreading Ages and Spreading Velocity*.

5. B. Refer to Figure 20.11.
6. B. Refer to Figure 20.11.
7. A. Refer to Figures 20.14 and 20.15.
8. C. Active continental margins lie adjacent to convergent and transform plate boundaries. The rifting of a continent, as illustrated in Figure 20.17, results in the formation of a passive continental margin on opposite sides of the rift basin.
9. A. Refer to Figure 20.18.
10. B. Refer to Figure 20.17.
11. C. Refer to Figure 20.17.
12. A. Refer to *Rock Assemblages at Intracontinental Rifts* and Figure 20.17.
13. A. Refer to Figure 20.20 (a).
14. A. Refer to Figure 20.10.
15. C. Refer to Figure 20.21 and 20.22.
16. C. The oldest ocean floor in the ocean basins today is Jurassic in age. Based on the principle of superposition, sediments deposited on top of the ocean floor can not be any older than Jurassic and most will be younger. The Cretaceous period of the Geologic Time Scale occurs immediately after the Jurassic within the Mesozoic era.
17. C. Felsic (granitic) to intermediate magmas are typically generated by subduction along ocean – continental convergent plate margins (refer to Figure 20.19).
18. B. The processes acting at convergent plate margins typically thicken and add new materials to the overriding margin of the continent (refer to Figures 20.20, a and b, 20.21 and 20.22). The geologic history of the continents is covered in the next chapter.

Chapter 21

Deformation of the Continental Crust

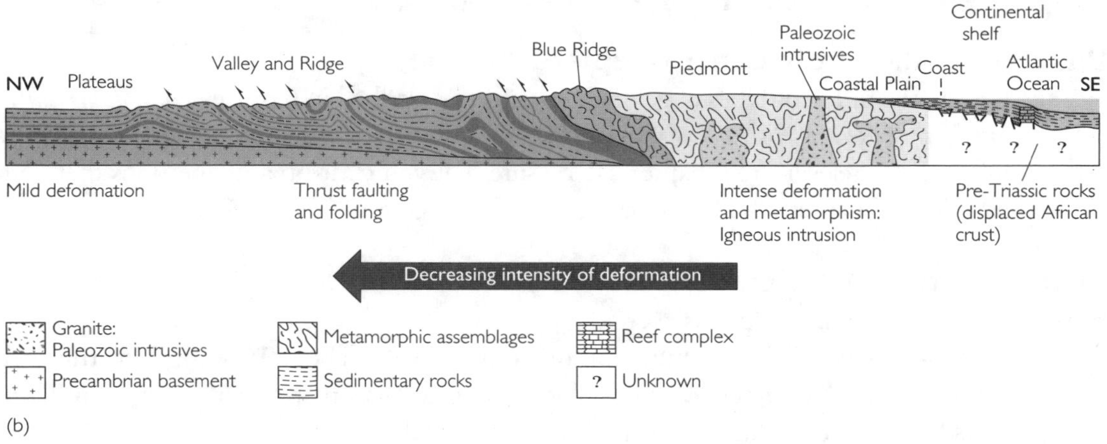

Figure 21.7 (b) / An idealized cross-section of the Appalachian Mountain region of the south and central eastern United States. Multiple plate collisions in the Paleozoic folded sediments of the Valley and Ridge province and transported them westward as large-scale thrust sheets. The modern coastal plain and shelf developed after the Triassic-Jurassic rifting of North America from Africa and the opening of the modern North Atlantic Ocean. [*After S. M. Stanley, Earth System History, New York, W. H. Freeman, 1999.*]

BEFORE LECTURE

Before you attend lecture be sure to spend some time previewing the chapter. For an efficient preview use the questions below. *Chapter Preview* questions constitute the basic framework for understanding the chapter. Preview works best if you do it just before lecture. With the main points in mind you will understand the lecture better. This in turn will result in a better and more-complete set of notes. How much time should you devote to preview? Obviously, more time is better than less. But even a brief (five or ten minute) preview session just before lecture begins will produce a result you will notice. For a refresher on why previewing is so important see the Appendix: *How to Study Science*.

CHAPTER PREVIEW

- **What are the major belts of deformation of North America?**
  Brief answer: Most of the continental crust can be divided into belts that have been deformed during various geologic periods (refer to Figures 21.2 and 21.3).
- **What events typify an orogeny (mountain building episode) caused by plate convergence?**
  Brief answer: Plate convergence initiates the crustal deformation and other processes that build mountains.
- **What are epeirogeny and orogeny?**
  Brief answer: Regional movements of the crust such as postglacial isostatic rebound that are simply up-and-down displacements without severe deformation are called epeirogenic events. Episodes of intense deformation such as the extensive and complex folding and faulting in the Appalachians, Alps and Himalayas are called orogenic events.

STUDY TIP

When confronted with new scientific terminology, sometimes the dictionary can help. For example, if you look up the terms epeirogeny and orogeny you would find the following origins for the roots of these terms:

epeirogeny — Gk. Epeiros, continent

orogeny — Gk.oros, mountain

Vital Information from Other Chapters

- Review the different types of geologic structures described in Chapter 10, *Folds, Faults, and Other Records of Rock Deformation*; Chapter 16, *Landscape Evolution*, also has information relevant to Chapter 21. Be sure that you clear up any questions that you have from Chapter 20, *Plate Tectonics*, before going on to Chapter 21.

Website and CD Preview

http://www.whfreeman.com/presssiever

- The *Continental Collision* animation on your CD takes about 10 seconds to run and will give you an overview of the geologic circumstances that over the last 50 million years have produced the highest mountains on Earth.

DURING LECTURE

One goal for lecture should be to leave class with a good set of answers to the *Preview Questions*. To avoid getting lost in details keep the big picture in mind: Chapter 21 tells the story of continent shaping. Continental deformation is driven by plate activity and there is a clear pattern of geological features associated with each type of plate boundary. Events associated with active boundaries thicken the crust. Crust thickening results in isostatic adjustment. Isostatic adjustment results in mountain-forming uplift.

AFTER LECTURE

The perfect time to review your notes is right after lecture. The checklist below contains both general review tips and specific suggestions for this chapter.

NOTE REVIEW CHECKLIST

- ✔ All notes legible? (Rewrite so they read easily.)
- ✔ Important points clearly identified? You should now have headers in your notes that tie to each of the questions in the *BEFORE LECTURE/Chapter Preview.*
- ✔ Holes (missing material) filled in from memory?
- ✔ Areas where you don't remember what was said marked for a follow-up session with your instructor, tutor, or study partner?
- ✔ Possible test questions indicated in the margin (TQ)?
- ✔ Additional visual material?
- ✔ Reworked notes into a form that is efficient for your learning style?
- ✔ Created a brief "big picture" overview of this lecture (using a sketch or written outline)? *Hint: Complete Exercise 1 in Practice Exercises and Study Questions for Chapter 21.*

Intensive Study Session

Schedule at least one hour after lecture for intensive study. Use this time to master key concepts. Use the Website and CD Activities and Tools suggested below, along with the *Practice Exercises* and *Study Questions*, to insure you master this chapter. Do as many of these as you have time for during your scheduled study session. Pay particular attention to exercises recommended by your instructor during lecture.

> **STUDY TIP**
> Practice Exercises 1 and 2 should be the focus for this study session. Exercise 1 will help you integrate what you have learned throughout the course about how plate movement and the rock cycle drive geological processes such as mountain building.

Website and CD Activities and Tools

http://www.whfreeman.com/presssiever

Complete the Q & A at the Website. Pay particular attention to the explanations for the answers. Also at the Website are Flashcards to help you learn new terms.

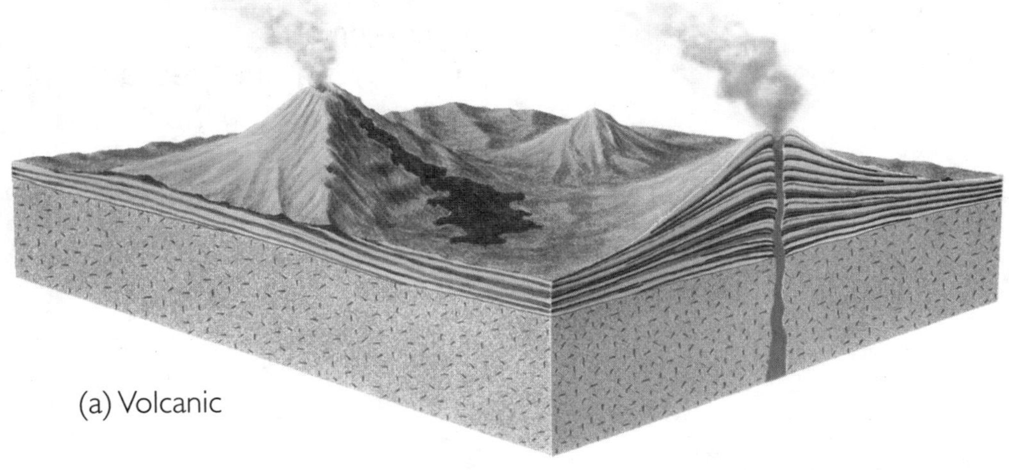

Figure 21.10 (a) / Mountains formed by volcanic action, such as the Cascades, extending from northern California through western Oregon and Washington into British Columbia, are associated with convergent plate boundaries.

PRACTICE EXERCISES AND STUDY QUESTIONS

(Answers and explanations are at the end of this chapter.)

Exercise 1: How do horizontal movements of lithospheric plates result in vertical uplift and the formation of a mountain range?

Complete the flow chart below by filling in the blanks.

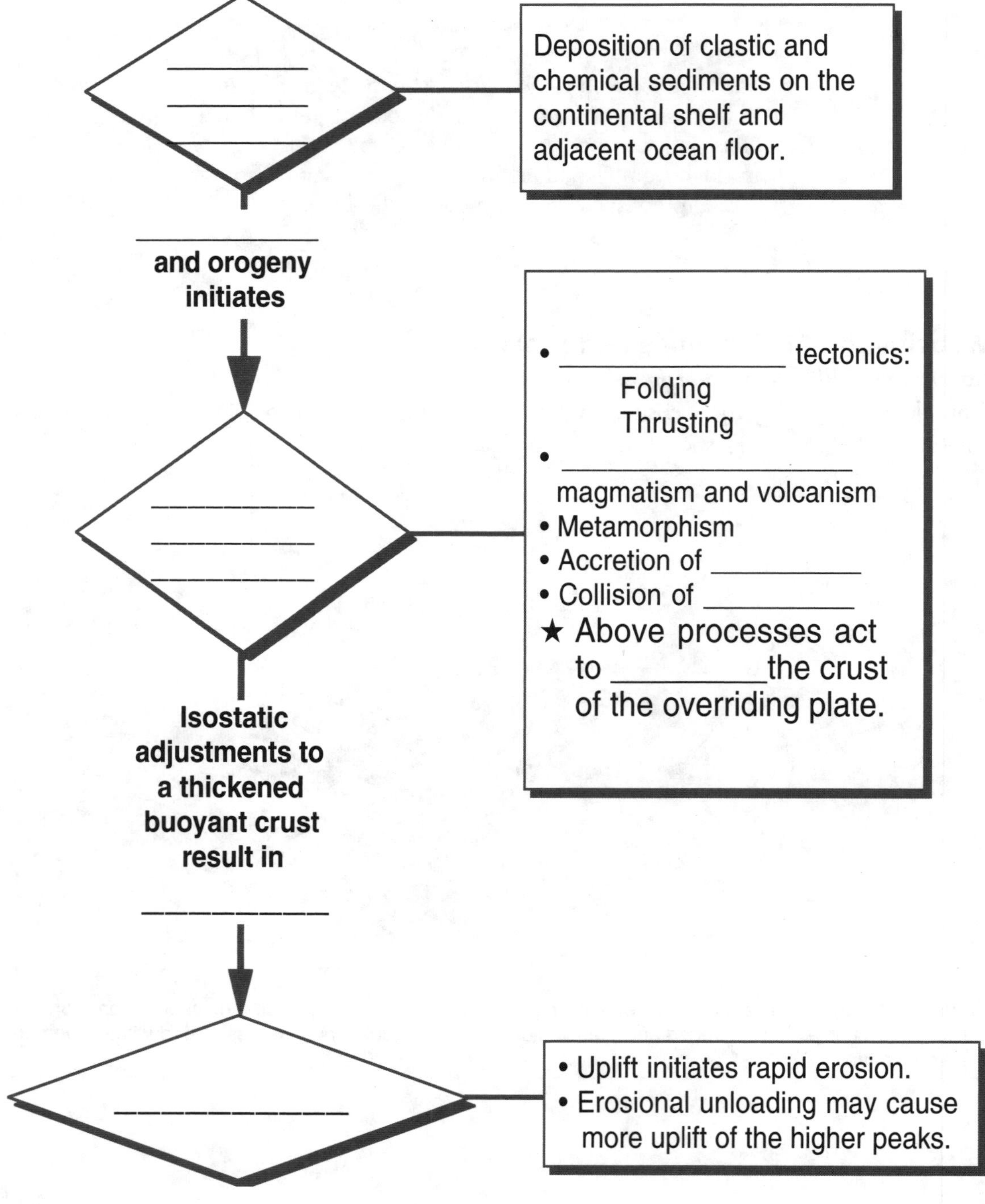

Exercise 2: The Ocean Crust vs. Continental Crust

Using Chapters 17, 19, 20, and 21, complete the table below by filling in the blank boxes.

CHARACTERISTICS	OCEAN CRUST	CONTINENTAL CRUST
COMPOSITION		
ROCK TYPE(S)		very herterogeneous, predominantly granitic and gneissic with a cover of sediments
DENSITY	3.0 g/cm^3	2.7 g/cm^3
THICKNESS	10 km	
AGE		The ages of continental crust spans 4 billion years.
TOPOGRAPHIC FEATURES		
STRUCTURE/ ARCHITECTURE	A model for the structure of the ocean crust is the ophiolite suite: deep-sea sediments, basaltic pillow lavas and dikes, and gabbro. (Note: peridotites are part of the mantle lithosphere, not the ocean crust.)	The architecture of the continents is complex. It consists of preexisting cratons, accreted microplates, island arcs, volcanic arcs, suture zones, ophiolite suites and belts representing ancient orogenic zones. Sediments cover basement rock in the interior platform of the continent.
ORIGIN		orogenic processes and accretion of preexisting crustal blocks along convergent plate boundaries

1. The massive, interior regions of continents that have been stable for extensive periods of time are called
   A. mountains
   B. plateaus
   C. cratons
   D. plains

2. Cratons like the interior of North America are characterized by all of following EXCEPT
   A. thrust sheets of geologically young passive margin sediments
   B. granitic and high-grade metamorphic basement rocks
   C. regional epeirogenic movements
   D. sedimentary basins

3. Deformation and crustal accumulation occurs on continents at
   A. convergent plate margins
   B. transform margins
   C. hot spots
   D. divergent plate margins

4. Which of the following is not associated with orogeny?
   A. thrust faulting
   B. intrusion of plutons
   C. passive continental margin
   D. metamorphism

5. The oldest rocks found on the continents are about
   A. 200 million years old
   B. 1 billion years old
   C. 4 billion years old
   D. 4.5 billion years old

6. The Basin and Range province of western North America is characterized by
   A. fault-block mountain ranges formed by crustal extension
   B. a fold and thrust fault belt formed by collision of microplates
   C. abundant volcanic centers with strataform volcanoes
   D. thrust sheets of shallow marine sedimentary layers that look like a titled stack of shingles

7. Geologists interpret both the Himalayan and the Appalachian Mountains as products of
   A. upwelling in the mantle
   B. a continent/continent collision
   C. an ocean/continent convergence
   D. doming due to a chain of hot spots

8. Cratonic and other continental interior rocks are typically ____________________ those rocks found at active continental margins.
   A. much younger than
   B. older than
   C. younger than
   D. the same age as

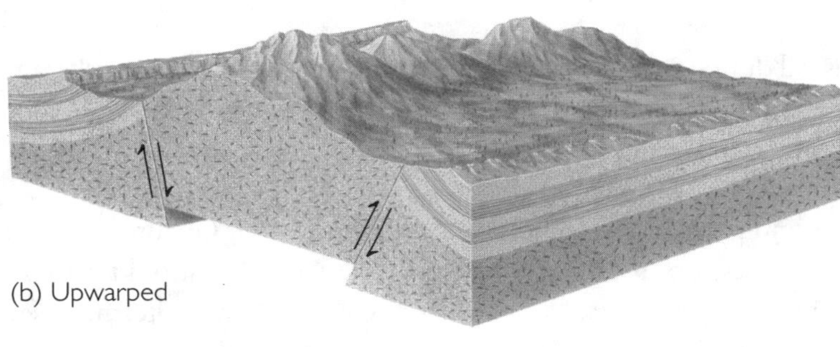

Figure 21.10 (b)
Upwarped mountains with reverse faults, such as the Front Range of the Rocky Mountains in Colorado, are an example of the rejuvenation of an orogenic belt in which renewed uplift and accelerated stream erosion sharpens the topography of the mountain range.

EXAM PREP

Many exams are in multiple choice format. Here are some tips to maximize your performance on multiple choice questions. Keep these in mind or review briefly just before you take each of your finals.

TEST-TAKING
Top Ten Tips for Taking a Multiple Choice Exam

Number 10: Answer the questions you know first.
Mark items where you get stuck. Come back to harder questions later. Often, you will find the answer you are looking for embedded in another, easier question.

Number 9: First, try to answer the item without looking at the options.

Number 8: Eliminate the distracters.
Treat each alternative as a true-false item. If "false", eliminate it.

Number 7: Use common sense.
Reasoning is more reliable than memory.

Number 6: Underline key words in the stem.
This is good to try when you are stuck. It may help you focus on what question is really being asked.

Number 5: If two alternatives look similar it is likely that one of them is correct.

Number 4: Answer all questions.
Unless points are being subtracted for wrong answers (rare) it pays to guess when you're not sure. Research indicates that items with the most words in the middle of the list are often the correct items. But be cautious—your professor may have read the research too!

Number 3: Do not change answers.
Particularly when you are guessing, your first guess is often correct. Change answers only when you have a clear reason for doing so.

Number 2: If the first item is correct, check the last.
If it says "all (or none) of the above," you obviously need to read the other alternatives carefully. Missing an "all of the above" item is one of the most common errors on a multiple choice exam. It is easy to read carelessly when you are anxious.

Number 1: READ THE DIRECTIONS BEFORE YOU BEGIN!

EXAM PREP

Materials in this section will be useful during your preparation for final exams. For optimal final exam performance, review and use the *Final Exam Prep Worksheet* (see Appendix: *How to Study Geology* for details). The basic idea is a systematic review of material divided into short study sessions.

The following *Chapter Summary* and *Practice Exam* should simplify review for Chapter 1. Read the *Chapter Summary* to begin your session. It provides a helpful overview that should get you back into the material. Next, try the *Practice Exam*. Take it just as you would a midterm: to see how you stand in regard to mastery of this chapter. After you answer the questions, score them. Finally, and most important of all: review any question that you missed. Identify and correct the misconception that resulted in missing the item.

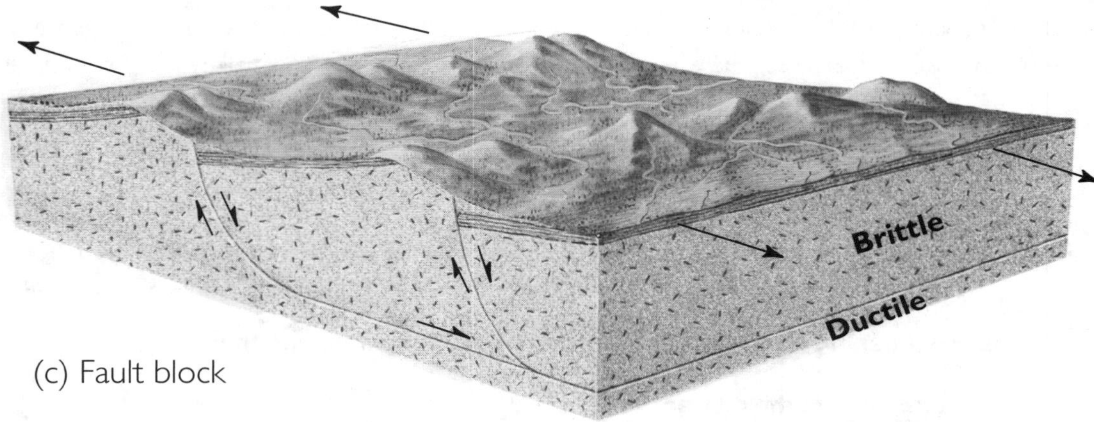

Figure 21.10 (c) Mountain ranges formed from tilted fault blocks and bounded by normal faults, with valleys between ranges, such as the mountains of the Basin and Range province. These are formed as a result of extension, where tectonic forces are pulling the crust apart.

CHAPTER 21 SUMMARY

- Orogeny, or mountain building, involves the processes of folding, faulting, magmatism, and metamorphism, typically associated with convergent plate boundaries. Orogeny and the evolution of continents are intimately related.
- Most of the continental crust can be divided into belts that have been deformed during various geologic periods. The North American continent today contains a large central stable shield and platform that have been relatively undisturbed by orogenic events since the Precambrian. Surrounding the stable interior are younger orogenic belts — the mountainous Cordillera and Appalachians, which were deformed during plate convergence at various times in the Paleozoic, Mesozoic, and Cenozoic eras.
- Plate convergence initiates orogeny and the associated crustal deformation and other processes that build mountains. If plate convergence initiates along a preexisting passive continental margin, sediments that have collected over time along that margin will be involved in the deformation and metamorphism. Folding, thrusting, emplacement of plutons, extrusion of volcanic rock, and collisions with microplates or full-sized continents all act to thicken the crust in active orogenic zones. Thicker continental crust leads to uplift and mountains.

- Continents can grow by a succession of orogenies over geologic time in several ways: by the buildup of sediments on continental shelves, by the addition of batholiths and volcanic rocks derived by melting in the mantle in subduction zones, and by microplate accretion.
- Regional movements of the crust, like postglacial isostatic rebound, that are simply up-and-down displacements without severe deformation are called epeirogenic events. In contrast, episodes of intense deformation, such as the extensive and complex folding and faulting in the Appalachians, Alps and Himalayas, are called orogenic events.

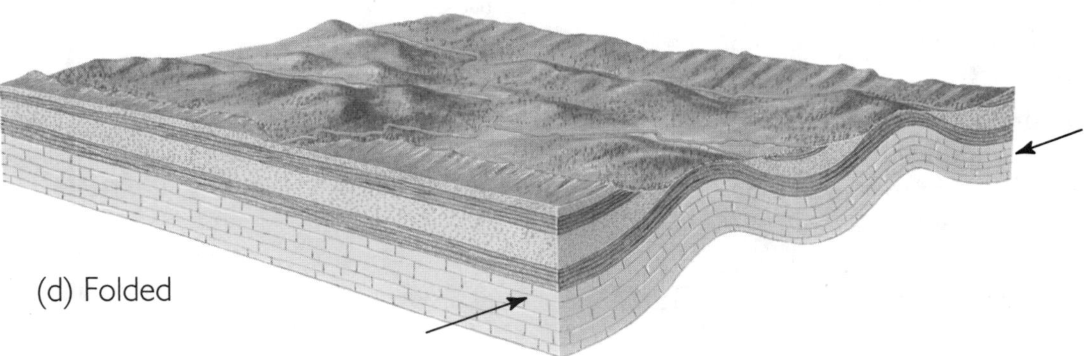

Figure 21.10 (d) / Mountains resulting from folded layers of rock, such as the Appalachian belt, occur in association with convergent plate boundaries. Folding and thrusting can be particularly intense in collision plate boundaries, such as in the Himalaya Mountains between India and Asia.

Website and CD Activities and Tools

http://www.whfreeman.com/presssiever

The *Photo Gallery* for Chapters 20 and 21 on your CD contains a virtual tour of the major mountain belts (orogenic systems) on Earth today. Be sure to read the captions along with viewing the images. Start your *Photo Gallery* tour by picking an image from a place you have visited or would like to visit.

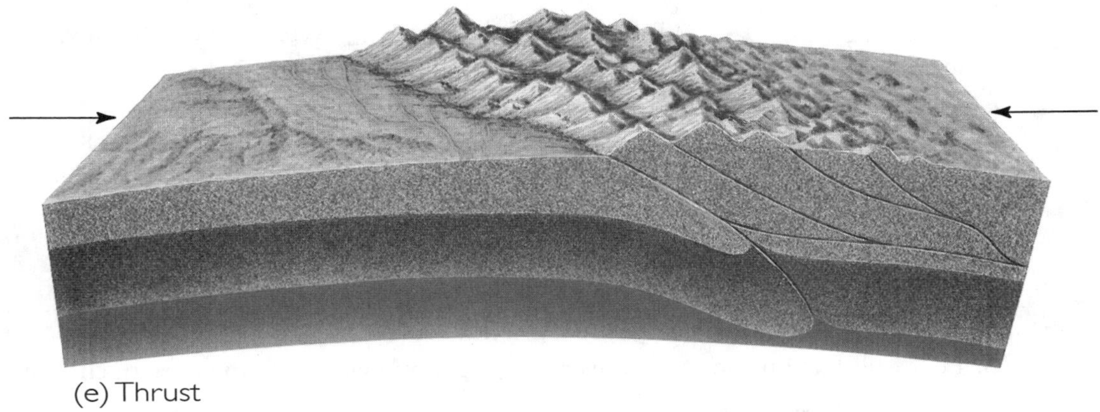

Figure 21.10 (e) / Mountains formed from stacked or overlapping thrust faults, such as the Himalayas and the southern Appalachians, are generated by the strong compressive stresses associated with ocean / continental convergence and continental / continental collision zones.

PRACTICE EXAM QUESTIONS FOR CHAPTER 21

(Answers and explanations are at the end of this chapter.)

1. Interior continental shields, or cratons, are traversed by ancient orogenic belts, but younger orogenic belts are different from these because
   A. their sediments have been subjected to extensive regional metamorphism
   B. they have deep roots of light continental rock
   C. their deformation and uplift is the result of crustal plate movements
   D. they consist almost entirely of volcanic rocks

Hint: Refer to Figure 19.6.

2. A large portion of the Cordilleran orogenic belt may consist of
   A. microplates
   B. hot spot volcanism
   C. ophiolite suites
   D. an ancient craton

Hint Refer to Figure 20.21 and 20.22.

3. Orogeny is taking place today
   A. along the east coast of North America
   B. along the west coast of South America
   C. in the center of North America
   D. in the Canadian Shield

4. Which of the following statements about orogenic systems is NOT valid?
   A. Orogeny is initiated by rifting as extension begins to open up a new ocean basin.
   B. Orogeny is initiated by subduction and the evolution of an active convergent margin.
   C. Large volumes of granite are intruded during orogenies.
   D. Folding and thrusting of preexisting rocks contributes to crustal thickening in the orogen.

5. The ______________________________ were produced by convergent plate boundary processes, including collision, during the Paleozoic.
   A. Cascade Mountains of northwestern North America
   B. Rocky Mountains of Colorado
   C. Appalachian Mountains along the eastern margin of North America
   D. Cordillera of western North, Central, and South America

6. The Cordilleran orogeny was initiated by formation of
   A. a continental rift in western North America
   B. a new subduction zone and convergent plate boundary along the western edge of North America
   C. an impact of a comet-size object in the Pacific Ocean adjacent to the western edge of North America
   D. a line of mantle or hot spots which causes doming and rifting through out western North America

7. The western Cordillera of North America is topographically higher than the Appalachians mainly because
   A. orogeny occurred more recently in the Cordillera
   B. granite batholiths were intruded in the Cordilleran belt
   C. the Appalachians eroded faster since they consist mostly of soft sedimentary rocks
   D. the Appalachians never reach the elevation of the Cordillera since collisions do not generate high mountains

8. Your assignment as a geologist is to map out ancient orogenic zones in what is now the stable interior of a continent. Recent epeirogenic uplift has resulted in good exposures of the ancient basement rocks. Which of the following features would NOT be evidence of an ancient orogenic zone?
   A. a mélange
   B. an ophiolite suite
   C. many granitic plutons with all about the same radiometric age
   D. widespread and relatively thick accumulations of coral-rich limestone, sandstones and shales

9. Why are the Sierra Nevada Mountains so much higher than the continental surfaces west and east of them?
   A. The crust is likely thicker beneath the Sierra Nevada.
   B. The crust beneath the Sierra Nevada must be very thin and hot.
   C. The crust beneath the Sierra Nevada is probably more dense than that to the east and west.
   D. the Sierra Nevada are most likely part of an ancient spreading center, which is no longer active.

Hint: Refer to Figure 19.6.

10. Mountains are both the source and product of sediments because
    A. most sediments are subducted with the ocean lithosphere and thereby contribute to subduction zone magmatism
    B. most sediments shed off mountains will end up on the margin of a continent and eventually accreted to the edge of continental crust by orogenic processes associated with convergent plate margins
    C. the melting of sediments results in mafic igneous rocks characteristic of continental crust
    D. most sediments end up on the deep ocean floor where they sit for billions of years

11. In the context of plate tectonics, a reasonable sequence of events for an orogenic "cycle" is
    A. hot spots / subduction / rifting / orogeny
    B. rifting / passive margin / subduction / orogeny / uplift
    C. rifting / collision / subsidence / erosion / uplift
    D. transform faulting / uplift / volcanism / orogeny

12. When the ancient supercontinent Pangaea broke up, it was NOT marked by the occurrence of
    A. magnetic anomaly patterns on the Atlantic Ocean floor
    B. Triassic rift valleys
    C. mafic intrusions and basaltic lava flows
    D. folding and thrust faulting of Paleozoic passive margin sediments

Hint: Refer to Figure 21.12 and Chapter 20.

13. Crustal landmasses experiencing erosion ____________ due to isostatic compensation.
    A. break into tilted blocks of crust
    B. may rise
    C. may sink
    D. remain at a constant elevation

14. The Earth's oldest continental crust can be found
    A. in active orogenic zones
    B. on the ocean floor
    C. in continental shield regions
    D. along the margins of the continents

15. The growth of continents occurs at
    A. hot spots
    B. subduction zones
    C. rift zones
    D. transform faults

16. Where could you go today on Earth to find an orogenic system with strong similarities to the Cordillera of North America? The study of this active orogenic system would provide you with a better understanding of the geologic history of western North America.
    A. Andes Mountains
    B. Himalayas
    C. Appalachians
    D. East Africa Rift

ANSWERS AND EXPLANATIONS FOR EXERCISES AND QUESTIONS

After Lecture

Exercise 1: How do horizontal movements of lithospheric plates result in vertical uplift and the formation of a mountain range?

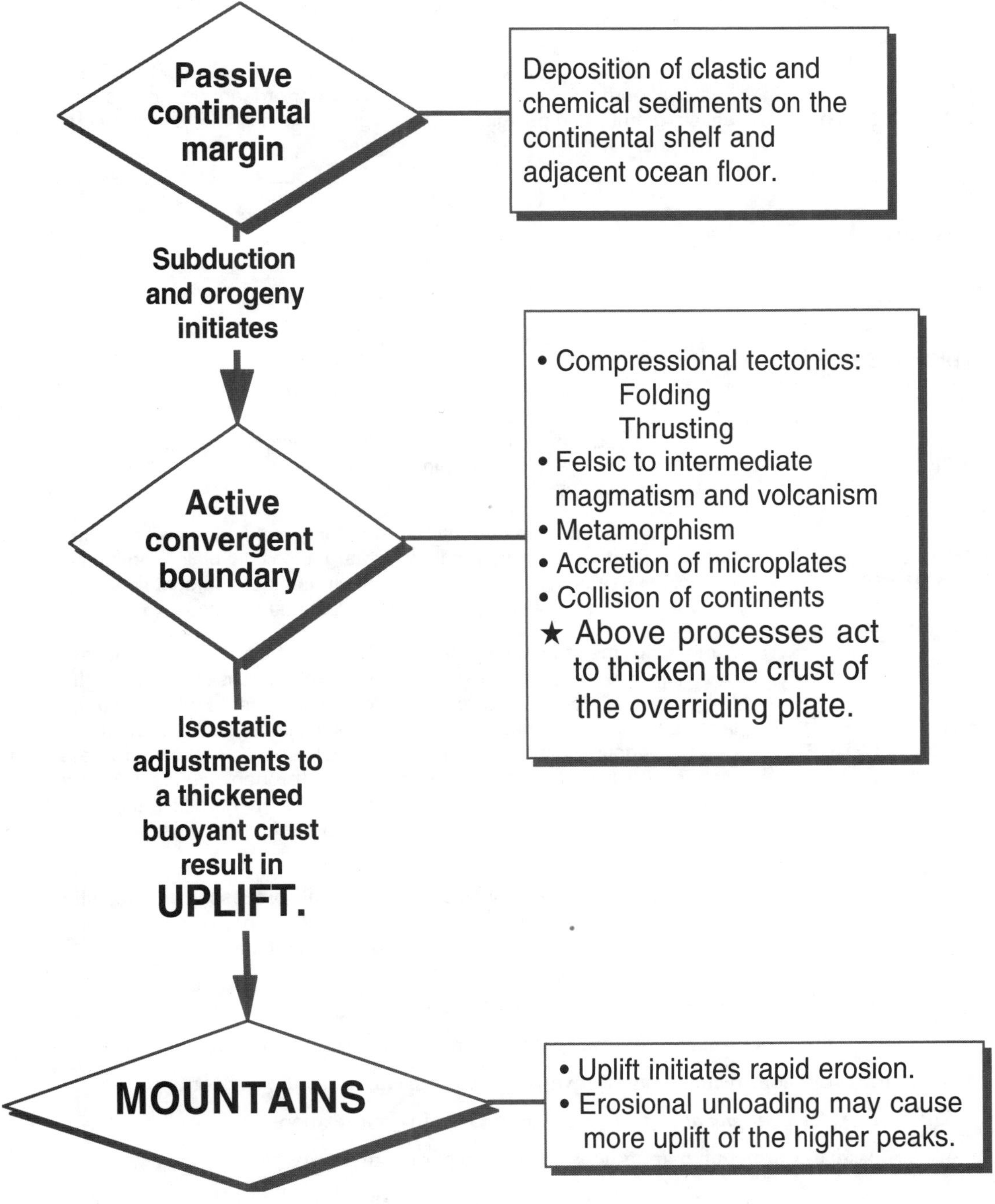

Exercise 2: The Ocean Crust vs. Continental Crust

CHARACTERISTICS	OCEAN CRUST	CONTINENTAL CRUST
COMPOSITION	mafic	felsic to intermediate
ROCK TYPE(S)	homogeneous, basalt, gabbro and pelagic sediments	very herterogeneous, predominantly granitic and gneissic with a cover of sediments
DENSITY	3.0 g/cm³	2.7 g/cm³
THICKNESS	10 km	15 - 65 km
AGE	175 million years or younger, with older fragments caught up within continents	The ages of continental crust span 4 billion years.
TOPOGRAPHIC FEATURES	abyssal ocean floor, seamounts, ridges/spreading centers, trenches, oceanic plateaus	craton, shield, platform, continental margin, coastal plain, continental shelf and slope, mountain belts, sedimentary basins
STRUCTURE/ ARCHITECTURE	A model for the structure of the ocean crust is the ophiolite suite: deep-sea sediments, basaltic pillow lavas and dikes, and gabbro. (Note: peridotites are part of the mantle lithosphere, not the ocean crust.)	The architecture of the continents is complex. It consists of preexisting cratons, accreted microplates, island arcs, volcanic arcs, suture zones, ophiolite suites and belts representing ancient orogenic zones. Sediments cover basement rock in the interior platform of the continent.
ORIGIN	mafic magmatism and volcanism at the ocean ridge system	orogenic processes and accretion of preexisting crustal blocks along convergent plate boundaries

1. C. Refer to *The Stable Interior* section in your textbook.
2. A. Thrust sheets of geologically young passive margin sediments are characteristic of active or recently active continental margins involved in convergence with another plate.
3. A. Refer to *Some Regional Tectonic Structures* section in your textbook.
4. C. A passive continental margin lies some distance from any active plate boundary.
5. C. The oldest known continental rock is radiometrically dated at 4 billion years old. It is the Acasta gneiss from the Slave Province in Canada (refer to the *Photo Gallery* of metamorphic rocks at the website or on your CD for an image of this rock).

6. A. Refer to Figure 21.10 (c).
7. B. Refer to *Case History: The Appalachians*, Figures 21.6 and 21.7, and Box 21.1, *Interpreting Earth and Its System*.
8. B. Refer to Figure 21.2.

Exam Prep

1. B. Refer to Figure 19.6 and *Orogenic Belts* section in your textbook.
2. A. Refer to Figure 20.22 and *Case History: The North American Cordilleran* section in your textbook.
3. B. The Andes Mountains along the west coast of South America are an active orogenic system associated with the subduction of ocean crust beneath the South American continent. There is an extensive collection of photos with captions about the Andes in your CD's *Photo Gallery*.
4. A. Orogeny refers to the collection of processes, typically active at convergent plate boundaries, that form mountains. Rifting leads to the formation of a new ocean basin.
5. C. Refer to the *Case History: The Appalachians* section in your textbook.
6. B. See again the *Case History: The North American Cordilleran* section in your textbook.
7. A. The Cordillera is a younger mountain belt compared to the Appalachians. In fact, portions of the Cordillera are still involved in orogenic processes. Portions of the western North American crust sit higher than the Appalachians because the Cordilleran crust is thicker and hotter, and tectonic processes remain very active.
8. D. Widespread and relatively thick accumulations of coral-rich limestones, sandstones, and shale are characteristic of a passive continental margin or a sedimentary basin within a continental platform.
9. A. Thicker continental crust typically stands higher. Hotter crust will also sit higher because heat lowers the density of the rock material.
10. B. In a sense, mountains are made from mountains. Most of the sediments eroded from previous orogenic systems ultimately end up along a continental margin where they eventually get deformed and uplifted by the evolution of an active convergent boundary. Some sediments are probably subducted, but most are deformed and entangled in thrusting and metamorphism during orogeny. Some sediments may be carried to depths where they melt. Melts derived from sediments are typically felsic, not mafic, and would crystallize within the crust to form a granitic rock. Except for regions adjacent to the continental margin, most ocean floor is surprisingly lacking in sediment. This is partly because the ocean floor is young and continually being recycled by plate tectonics and because sedimentation rates of deep marine sediments are very slow.
11. B. Rifting with the break up of a continent forms a new passive margin on which accumulate sediments. Eventually, with the development of subduction adjacent to the passive margin, orogeny is initiated. Orogenic processes thicken the overriding crust, which then begins to rise due to a buoyant balance between the mountain root and the denser mantle in which it sits. Can you think of a good reason why subduction is likely to eventually occur adjacent to an old passive continental margin?

12. D. The folding and faulting of Paleozoic passive margin sediments occurred as Pangaea was being assembled by plate tectonic processes.
13. B. Not only the visible part of the mountain, but the buoyant root extending into the mantle needs to be "eroded" in order for a mountain to be leveled. As erosion removes rock from the surface, isostatic adjustments bring rocks within the mountain root progressively towards the surface. This is why some of the deepest crustal rocks are found in old eroded orogenic belts.
14. C. Refer to Figure 21.2.
15. B. Refer to the *Some Regional Tectonic Structures* section in your textbook.
16. A. The Andes Mountains along the west coast of South America are in a more youthful stage of orogeny than the North America Cordillera and exhibit a great number of similarities to past geologic circumstances in western North America. Today, geologists actively study the Andes as a way of time traveling back into the earlier Cenozoic history of the North American Cordillera.

Chapter 22

Energy and Material Resources from the Earth

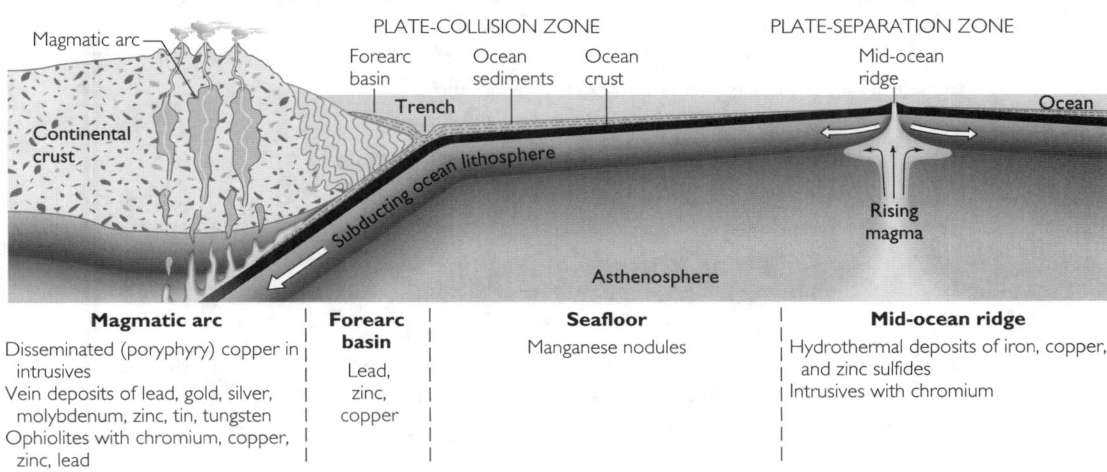

Figure 22.26 / The role of plate boundaries in the accumulation of mineral deposits. Ocean sediment and crust are enriched in metals by hydrothermal ore deposition along a mid-ocean ridge. Rising magma in the subduction zone is the source of ores that form the metal-bearing provinces of a magmatic belt such as the Cordillera of North and South America. The melting of subducting sediment and crust may contribute to ore constituents. Mineral-bearing oceanic fragments (ophiolites) accrete to the continent in the collision zone.

BEFORE LECTURE

Before you attend lecture be sure to spend some time previewing the chapter. For an efficient preview review the questions in *Chapter Preview* below.

CHAPTER PREVIEW

- **What is the origin of oil and natural gas?**
  Brief answer: Oil and natural gas form from organic matter deposited in marine sediments. Petroleum resources will be significantly depleted within about a century.

- **What is the origin of coal and how big a resource is it?**
  Brief answer: Coal is formed by the compaction and mild metamorphism of buried wetland vegetation. We have only used about 2.5 percent of the world's mineable coal.

- **What are some of the environmental concerns connected with the use of fossil fuels?**
  Brief answer: Mine reclamation, pollution, acid rain, and carbon dioxide emissions are major environmental concerns that need to be addressed.

(cont'd on next page)

CHAPTER PREVIEW (cont'd)

- **What are other alternative sources of energy?**
  Brief answer: Nuclear power can be a major energy source but only if its costs do not keep escalating and the public can be assured of its safety. Energy sources like geothermal, hydroelectric, and solar are limited presently.

- **What is an economical mineral deposit?**
  Brief answer: Mineral deposits become ore deposits when they are rich and valuable enough to mine economically.

- **How do ore deposits of metal-bearing minerals form?**
  Brief answer: Hydrothermal, metamorphic, chemical and mechanical weathering and deposition can enrich metal-bearing minerals to form economical deposits.

- **How are natural resources related to plate tectonics?**
  Brief answer: Many metal ore deposits are formed by magmatic and hydrothermal processes which are closely linked to both modern and ancient plate boundaries.

Vital Information from Other Chapters

- The concentration of energy and material resources involves a wide variety of geologic processes including magmatic, metamorphic, weathering, depositional and plate tectonic. Use the previous chapters as a resource to answer questions that may come up while reading Chapter 22.

Our entire society rests upon and is dependent upon—our water,
our land, our forest, and our minerals. How we use these resources
influences our health, security, economy, and well being.
—John F. Kennedy, February 23, 1961

DURING LECTURE

This should be an especially interesting lecture. Earth's natural resources are described in terms of origin, abundance and geological origin. You will not be surprised to learn that plate tectonics plays a role in ore forming processes. There are many important social issues connected to Chapter 22. Your instructor may make use of discussion/debate to address some of these. Try to capture important social issues and arguments in your notes.

AFTER LECTURE

The perfect time to review your notes is right after lecture. The following checklist contains both general review tips and specific suggestions for this chapter.

NOTE REVIEW CHECKLIST

✔ All notes legible? (Rewrite so they read easily.)

✔ Important points clearly identified? You should now have headers in your notes that tie to each of the questions in the *BEFORE LECTURE/Chapter Preview.*

✔ Holes (missing material) filled in from memory?

✔ Areas where you don't remember what was said marked for a follow-up session with your instructor, tutor, or study partner?

✔ Possible test questions indicated in the margin (TQ)?

✔ Additional visual material.

✔ Reworked notes into a form that is efficient for your learning style?

✔ Created a brief "big picture" overview of this lecture (using a sketch or written outline)? *After this lecture write your own brief answer to each Study Question.*

Intensive Study Session

Use the Website and CD Activities and Tools suggested below, along with the *Practice Exercises* and *Study Questions,* to insure you master this chapter. Do as many as you have time for during your scheduled study session. Pay particular attention to exercises recommended by your instructor.

Website and CD Activities and Tools

http://www.whfreeman.com/presssiever

Complete the Q & A at the Website. Pay particular attention to the explanations for the answers. Also at the Website are Flashcards to help you learn new terms.

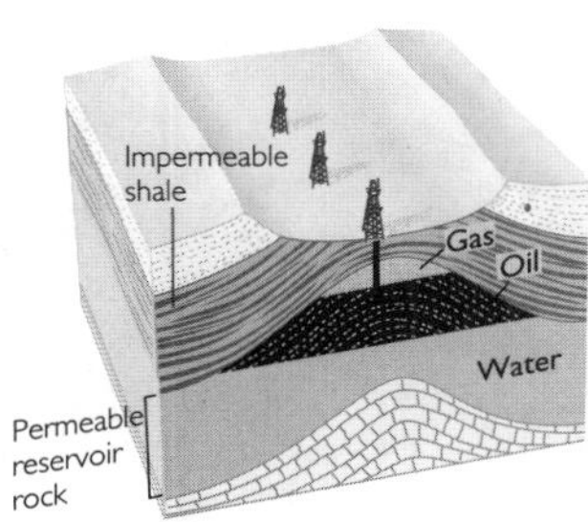

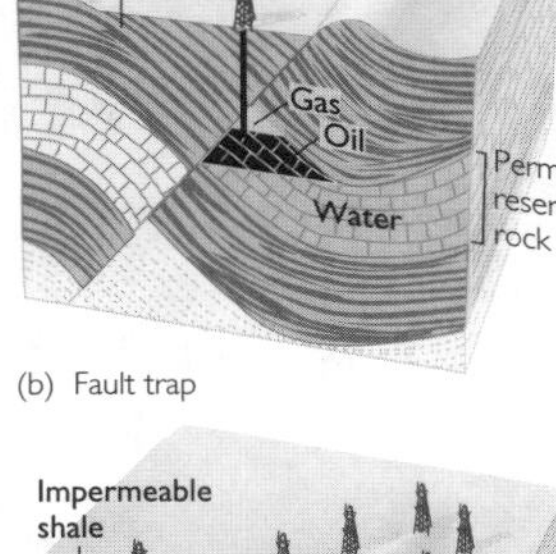

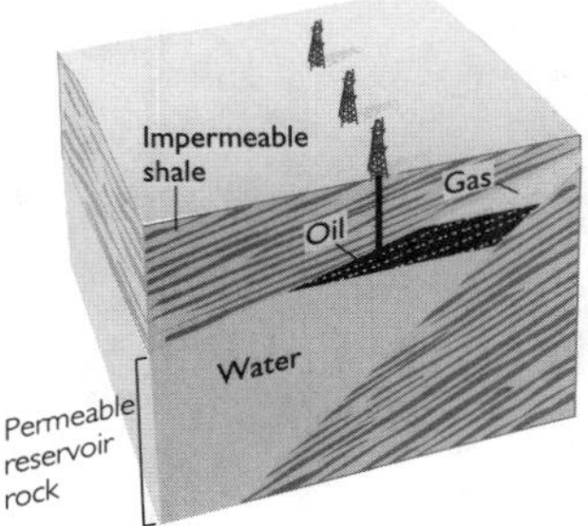

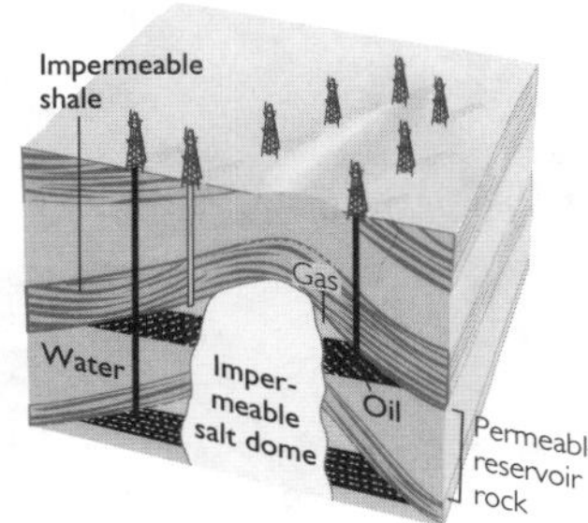

Figure 22.4
Oil traps:
(a) Anticlinal trap.
(b) Fault trap.
(c) Stratigraphic trap.
(d) Oil trapped by a salt dome. Natural gas and oil are trapped by an impermeable layer above the permeable oil-producing formation.

PRACTICE EXERCISES AND STUDY QUESTIONS

(Answers and explanations are at the end of this chapter.)

1. What is the sequence from low to high grade for the transformation of plant matter into coal?
   A. plants, peat, lignite
   B. peat, lignite, bituminous, anthracite
   C. bituminous, anthracite, peat, lignite
   D. anthracite, bituminous, lignite, peat

2. All of the following are energy sources, but which one is not a fossil fuel?
   A. natural gas
   B. coal
   C. uranium
   D. oil

3. Which of the following is not a consequence of fossil fuel consumption?
   A. mine reclamation
   B. ozone depletion in our atmosphere
   C. disposal of residual ash from the burning of coal
   D. acid rain

4. Oil and natural gas are mostly found in sedimentary rocks originally deposited
   A. in the deep ocean
   B. in river deltas and on the continental shelf
   C. in wetlands
   D. in large lakes

5. All of the following are effective oil traps except
   A. faults
   B. anticlines
   C. salt domes
   D. horizontal sedimentary and volcanic beds

   Hint: Refer to Figure 22.4.

6. When was the first oil well drilled in America?
   A. 1859
   B. 1880
   C. 1901
   D. 1940

7. The most important source of U.S. energy is
   A. coal
   B. nuclear
   C. oil
   D. hydroelectric power

   Hint: Refer to Figure 22.2.

8. How many US gallons are contained in one barrel of oil?
   A. 16 gallons
   B. 25 gallons
   C. 42 gallons
   D. 55 gallons

9. The United States ranks ________________ in oil reserves.
   A. first
   B. second
   C. eighth
   D. tenth

Hint: Refer to Figure 22.5.

10. Which country has the greatest coal reserves?
   A. China
   B. Great Britain
   C. United States
   D. former Soviet Union

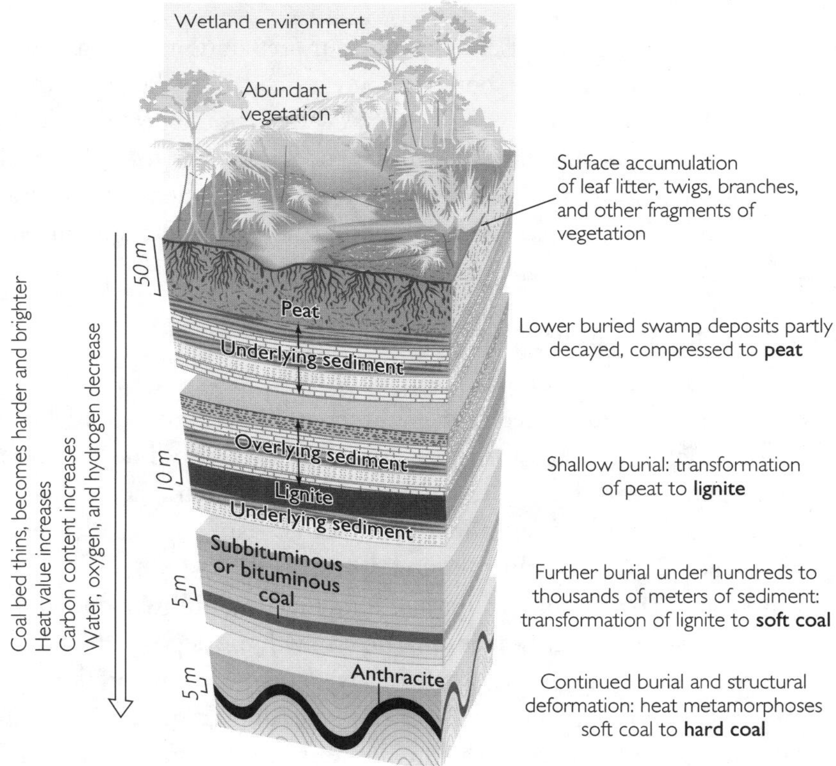

Figure 22.8 / The process by which coal beds form begins with the deposition of vegetation. Protected from complete decay and oxidation in a wetland environment, the deposit is buried and compressed into peat. Subjected to further burial, peat undergoes mild metamorphism, which transforms it successively into lignite, subbituminous and bituminous (soft) coal, and anthracite (hard) coal. As the deposit becomes more deeply buried, temperature rises, and structural deformation occurs.

FINAL EXAM PREP

Each semester in one short week you get to take an exam in each and every course, and most of those exams are comprehensive finals that cover the entire semester. Dealing with finals week successfully is a major challenge. Here are some tips that will insure you do your best work during final week.

TIPS FOR SURVIVING FINALS WEEK

- Be organized and systematic. Use the FINAL EXAM PREP WORKSHEET to help you get organized for finals. Use the EIGHT DAY STUDY PLAN for every course where the final exam will be an important factor in determining your grade.
- Stick to priorities. Say no to distractions.
- Build in moments of relaxation: Take regular short breaks, exercise, and be sure to get enough sleep.
- Be confident. By now you have built up a good set of study habits. You are a competent learner.

Materials in this section are most useful during your preparation for final exams. The following *Chapter Summary* and *Practice Exam* should simplify review still further. Read the *Chapter Summary* to begin your session. It provides a helpful overview that should get you back into the material. Next, try the *Practice Exam*. Take it just as you would a midterm: to see how you stand in regard to mastery of this chapter. After you answer the questions, score them. Finally, and most important of all: review any question that you missed. Identify and correct the misconception that resulted in missing the item.

CHAPTER 22 SUMMARY

- Oil and natural gas form from organic matter deposited in marine sediments. The organic materials are buried as the sedimentary layers grow in thickness. Heat, pressure and bacterial action transform the organic matter into fluid hydrocarbons. The fluid hydrocarbons tend to migrate out of the source rock and accumulate in geologic traps that confine the fluids within impermeable barriers. Petroleum resources will be significantly depleted within about a century.
- Coal is formed by the compaction and mild metamorphism of buried wetland vegetation. The process by which coal beds form begins with the deposition of vegetation. Protected from complete decay and oxidation in a wetland environment, the deposit is buried and compressed into peat. Subjected to further burial, peat undergoes mild metamorphism, which transforms it successively into lignite, subbituminous and bituminous (soft) coal, and anthracite (hard) coal. As the deposit becomes more deeply buried, temperature rises, and structural deformation occurs. We have only used about 2.5 percent of the world's minable coal.
- Environmental concerns associated with the use of fossil fuels include mine reclamation, pollution, acid rain, and carbon dioxide emissions.

- Alternative energy resources include nuclear, geothermal, hydroelectric and solar. Like fossil fuels, there are significant economic, technological, environmental, and political concerns associated with alternative energy resources.
- Mineral deposits become ore deposits when they are rich and valuable enough to mine economically. Hydrothermal, metamorphic, chemical and mechanical weathering and deposition can enrich metal-bearing minerals to form economic deposits. Important non-metallic mineral deposits include limestone for cement, quartz sand for glass and fiber optics, gravel for concrete, clays for ceramics, evaporites like gypsum for plaster and wallboard, plus salts and fertilizers.
- Many metal ore deposits are formed by magmatic and hydrothermal processes which are closely linked to both modern and ancient plate boundaries. Knowledge on how mineral deposits form and their association with plate boundaries has greatly facilitated the discovery of new deposits.

Website and CD Activities and Tools

http://www.whfreeman.com/presssiever

Test your knowledge of the *Fossil Fuel Cycle* by completing this Interactive Exercise at the Website. The *Photo Gallery* on your CD has a collection of images featuring energy resources and ore deposits. Test your understanding of the photos and their captions and refer to your textbook or course instructor to answer your questions about these images.

PRACTICE EXAM QUESTIONS FOR CHAPTER 22

(Answers and explanations are at the end of this chapter.)

1. Important factors contributing to the formation of coal from vegetation are
   A. heat and oxidation
   B. compaction by burial and heat
   C. biological activity and dissolution
   D. hydrothermal alteration and metamorphism

2. From what process is coal derived?
   A. the decay of marine plants and animal matter
   B. the burial, compression and heating of plant matter from wetlands
   C. from deposition and metamorphism of marine limestones
   D. transport of organic matter by rivers to their delta

Hint: Refer to Figure 22.8.

3. Acid rain forms when ______________________ from the combustion of coal and petroleum combine with rainwater.
   A. hydrogen gases
   B. sulfur dioxide gases
   C. oxygen
   D. nitrogen gases

4. Deposits of gold, diamonds and chromite found in river gravels and beach sands are classified as
   A. placers deposits
   B. hydrothermal deposits
   C. pegmatites
   D. kimberlites

5. Metallic ores
   A. form only at converging plate boundaries
   B. form only at diverging plate boundaries
   C. form at both converging and diverging plate boundaries
   D. do not have a strong association with active plate boundaries

6. The exploration division of World Amalgamated Metals (WAM) has hired you to find its next big copper deposit in the western US. On what kinds of rocks will you target your exploration?
   A. at the base of Triassic conglomerates and sandstones
   B. within and surrounding Cenozoic igneous intrusive rocks
   C. deep within Permian limestones
   D. throughout Pleistocene glacial tills

Hint: Refer to the *Disseminated Deposits* section in your textbook.

7. Kimberlites are ultramafic igneous rocks from which are mined
   A. copper, lead, and zinc
   B. nickel, copper and iron
   C. chromium and platinum
   D. diamonds

8. Evaporites are significant geologic deposits because they
   A. contain gold and silver
   B. are a major source for plaster board and chemicals
   C. are an alternative energy resource
   D. are rich in uranium and other fuels for nuclear power

9. If you were hired by Dr. Greasy's Mystery Oil Company to locate new oil and gas reservoirs in the western US, where would you focus your exploratory drilling program?
   A. where Paleozoic limestones have been metamorphosed to marble by intense magmatism
   B. where impermeable volcanic rocks cover fractured Precambrian gneiss
   C. where high angle normal faults cut Mesozoic desert sand dune deposits
   D. where crystalline rocks have been thrust over unmetamorphosed organic-rich sedimentary strata

Hint: Refer to Figure 22.4.

10. Generally, by what process are ore mineral vein deposits formed?
    A. Ore minerals are precipitated within fractures and joints when hot metal-bearing fluids are quickly cooled.
    B. Ore minerals are concentrated in veins and channel sediments by rivers and streams.
    C. Ore minerals are concentrated within mudcracks and fractures through surface evaporation.
    D. Ore minerals are concentrated in magma chambers through magmatic differentiation and fractional crystallization.

Hint: Refer to Figure 22.18.

11. Predict which well would give the best potential for oil and gas production.
    A. Well A
    B. Well B
    C. Well C
    D. Well D

Hint: Refer to Figure 22.4.

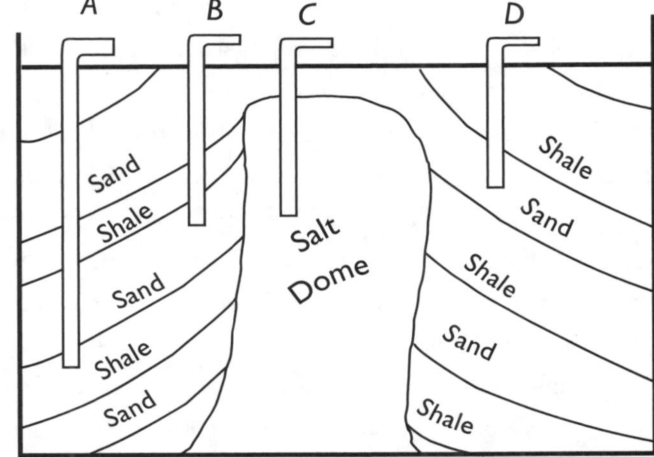

ANSWERS AND EXPLANATIONS FOR EXERCISES AND QUESTIONS

After Lecture

1. B. Refer to Figure 22.8.
2. C. Uranium is not a fossil fuel. Refer to Figure 22.3 for how energy is stored in fossil fuels.
3. B. Ozone depletion in our atmosphere is largely influenced by the release of chlorofluorohydrocarbons (CFCs), synthetic chorine/fluorine compounds used widely in aerosols and refrigerants throughout much of the last century.
4. B. Oil and gas form from organic matter derived from marine organism that thrive in shallow coastal waters.
5. D. Refer to Figure 22.4.
6. A. In 1859 the first oil well in America was drilled in northwestern Pennsylvania. The oil from this shallow well was sold for dubious medicinal purposes.
7. C. Refer to Figure 22.2.
8. C. There are 42 gallons to a barrel of oil, of which about half can be refined into gasoline for automobiles. From the rest comes jet fuel, diesel and other fuels, solvents, lubricants, greases, and asphalt.

9. C. The US ranks eighth in oil reserves. See the section, *The World Distribution of Oil and Natural Gas*.
10. D. The former Soviet Union has the greatest coal reserves. Other leading producers are the United States and China.

Exam Prep

1. B. Refer to Figure 22.8.
2. B. Refer to Figure 22.8.
3. B. Sulfur dioxide combines with rainwater to form sulfuric acid, a major component of acid rain. For this reason, low sulfur coals are environmentally more favorable to burn.
4. A. Refer to the *Sedimentary Mineral Deposits* section in your textbook.
5. C. Metallic ore deposits have a strong association with all active tectonic plate boundaries. Refer to Figure 22.26.
6. B. The formation of copper ore deposits is strongly linked to hydrothermal and magmatic processes (refer to the *Disseminated Deposits* section in your textbook).
7. D. Kimberlites are the host rock for diamonds. Because of their resistance to weathering, diamonds also occur as placer deposits. Diamonds can only be formed at the extreme pressures of the upper mantle, thus their association with Kimberlite, a type of peridotite.
8. B. Refer to the *Nonmetallic Sedimentary Deposits* section in your textbook.
9. D. Metamorphism associated with magma intrusion would probably completely decompose fluid hydrocarbons. Oil and gas would not be associated with volcanic rocks and gneisses. Only very unusual circumstances might result in the fractured gneiss being a reservoir rock for oil and gas. Mesozoic sand dune deposits might make a good reservoir rock, but are definitely not source rocks for the hydrocarbons. Oil and gas are derived from the transformation of abundant organic matter deposited in shallow marine sediments. Low angle thrust faults do produce geologic traps for oil and gas in the western United States.
10. A. Refer to the *Hydrothermal Deposits* section in your textbook.
11. B. Well B is most likely to produce oil because it is drilled into a tilted sand layer which is confined by the salt dome. Hydrocarbons are likely to accumulate in sand because of its high porosity and permeability. Because salt deposits are essentially non-porous and impermeable, the dome would seal off the sand layer and prevent the fluid hydrocarbons from escaping to the surface. Well A is drilled in shale which can be a good source rock for hydrocarbons, but its low permeability makes it a very poor reservoir rock. Well D is drilled into a sand layer that is not sealed off by the salt dome. Therefore, hydrocarbons would migrate along the Well D sand layer to the surface and be lost.

We Americans think we are pretty good!
We want to build a house, we cut down some trees.
We want to build a fire, we dig a little coal.
But when we run out of all these things, then we will find out just how good we really are.
—*Will Rogers*

Chapter 23
Earth Systems, Cycles, and Human Impacts

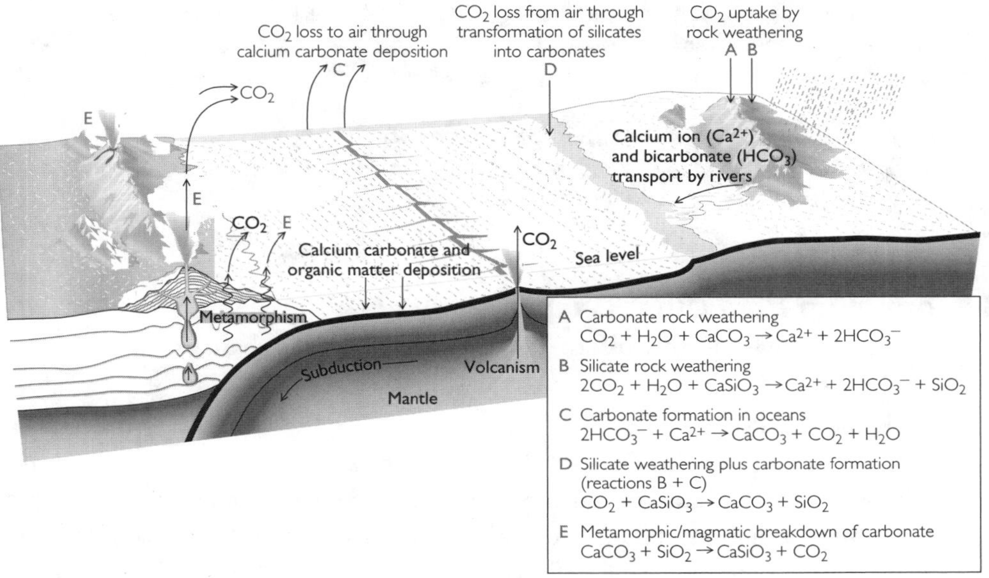

Figure 23.9 / The geochemical carbon cycle, showing reservoirs and fluxes. The fluxes of carbon dioxide into the atmosphere are balanced by fluxes out of the atmosphere. Similarly, the fluxes of carbonate and organic carbon into the Earth's interior are balanced by fluxes from the interior via volcanoes.

BEFORE LECTURE

Before you attend lecture be sure to spend some time previewing the chapter. For an efficient preview use the questions below.

CHAPTER PREVIEW

- **What is the origin of the atmosphere and hydrosphere?**
  Brief answer: The primary source of water and gases on the Earth's surface is volcanic gases, which outgasses from the planet's interior over geologic time. A notable exception is oxygen gas, which was added by photosynthetic organisms after they evolved.

- **When did life appear and how did it evolve?**
  Brief answer: The earliest fossils of bacteria are about 3.5 billion years old. Chemical evolution to form the building blocks of life preceded biological evolution.

- **How does life influence Earth's geochemical cycles?**
  Brief answer: The evolution of photosynthesizing organisms transformed the Earth's surface environments by adding oxygen which now makes up 21% of the atmosphere. Life processes have a diverse impact on the carbon cycle.

(cont'd on next page)

CHAPTER PREVIEW (cont'd)

- **What is the greenhouse effect?**
  Brief answer: Carbon dioxide and other trace atmospheric gases are transparent to sunlight, but absorb heat (IR radiation) which warms Earth's surface environments.

- **What are geochemical cycles?**
  Brief answer: Geochemical cycles trace the flux of Earth's elements like carbon from one reservoir to another. Understanding the carbon cycle is important because of its strong link to life processes and climate change.

- **What are important issues in global change today?**
  Brief answer: Climate change, ozone depletion, acid rain, and human population growth are issues of vital concern today.

Vital Information from Other Chapters

- Chapter 23 draws together information from many of the chapters.

DURING LECTURE

- One goal for lecture should be to leave class with a good set of answers to the preview questions
- Chapter 23 tells the story of Earth Systems and how we humans interact with them. Systems such as the atmosphere, hydrosphere, calcium cycle, and carbon cycle act together to maintain a balanced biosphere. This chapter deals with the many social/environmental issues created by humans as we interact with our biosphere.
- Focus on understanding the dynamics of each system.
- Again, there may be discussion/debate activities. Previewing the chapter will prepare you to take part in these activities.

AFTER LECTURE

The perfect time to review your notes is right after lecture. The following checklist contains both general review tips and specific suggestions for this chapter.

NOTE REVIEW CHECKLIST

- ✔ All notes legible? (Rewrite so they read easily.)
- ✔ Important points clearly identified? You should now have headers in your notes that tie to each of the questions in the *BEFORE LECTURE/Chapter Preview.*
- ✔ Holes (missing material) filled in from memory?
- ✔ Areas where you don't remember what was said marked for a follow-up session with your instructor, tutor, or study partner?
- ✔ Possible test questions indicated in the margin (TQ)?

(cont'd)

✔ Additional visual material. *Suggestions for Chapter 23: Table 23.2 Earth's Evolution, Figure 23.6 Calcium Cycle, and Figure 23.9 Geochemical Carbon Cycle should be added to your notes.*

✔ Reworked notes into a form that is efficient for your learning style?

✔ Created a brief "big picture" overview of this lecture (using a sketch or written outline)? *Hint: This is a good chapter to write a brief position paper on an issue which concerns you. Ask yourself what Earth system information in Chapter 23 is relevant to the issue. Try to develop a position that is based in reason and consistent with existing by science.*

Intensive Study Session

After each lecture you need to thoroughly master the concepts covered. You need to do this before you attend the subsequent lecture. The ideas of geology are like a stack of boxes. Each new idea rests on all ideas (boxes) stacked beneath it.

Schedule at least one hour after lecture for intensive study. Use this time to master key concepts. By now you know well that mastery is not gained by just reading your text. To master geology you must ask yourself questions and answer them. Use the Website and CD Activities and Tools suggested below, along with the *Practice Exercises* and *Study Questions*, to insure you master this chapter. Do as many of these as you have time for during your scheduled study session. Pay particular attention to exercises recommended by your instructor during lecture.

Website and CD Activities and Tools

http://www.whfreeman.com/presssiever

Complete the Q & A at the Website. Pay particular attention to the explanations for the answers. Also at the Website are Flashcards to help you learn new terms.

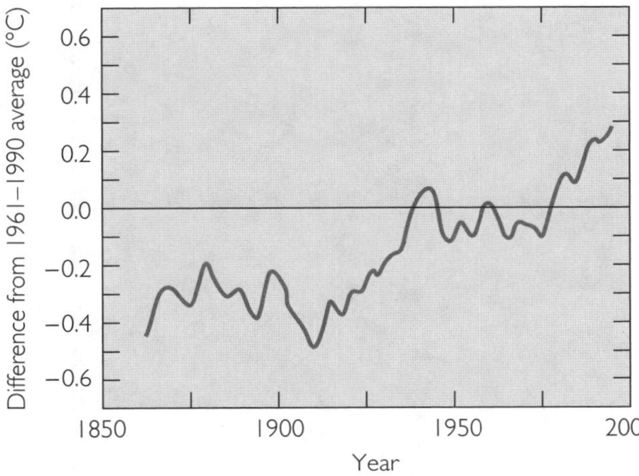

Figure 23.13
The trend in global warming. Although global temperatures fluctuate from year to year, the trend is clearly upward, owing to the rise in carbon dioxide and other greenhouse gases. [*UK Meteorological Office, University of East Anglia*]

PRACTICE EXERCISES AND STUDY QUESTIONS

(Answers and explanations are at the end of this chapter.)

Exercise 1: Characteristics of El Niño Events

Briefly describe five impacts of El Niño events:

1. ______________________________

2. ______________________________

3. ______________________________

4. ______________________________

5. ______________________________

Exercise 2: Evidence for Bolide Impact Events

Briefly describe five independent lines of evidence supporting a late Cretaceous bolide impact event on Earth.

1. ______________________________

2. ______________________________

3. ______________________________

4. ______________________________

5. ______________________________

Exercise 3: The Carbon Cycle

Complete the table below by filling in the blank boxes.

Hint: Refer to both Figure 23.9 and sections of text in Chapter 23.

CARBON FLUXES	BRIEF DESCRIPTION OF FLUX	DIRECTION OF FLUX	CLIMATIC IMPLICATIONS
VOLCANISM			
SEDIMENTATION	Calcium and carbonate ions combine to produce calcium carbonate that precipitates on the ocean floor, which is above the carbonate compensation depth.		Climate cools - CO_2 is drawn out of the oceans and atmosphere and into the crust. Because the oceans and atmosphere are in equilibrium, a drawn down of CO_2 in the oceans also results in reduced CO_2 in the atmosphere.
METAMORPHISM		CO_2 flows from the crust into the oceans and atmosphere.	
CHEMICAL WEATHERING	CO_2 in rainwater combines with silicate minerals to form calcium carbonate	CO_2 flows from the atmosphere and oceans into the crust.	
LIFE PROCESSES	Carbon is fixed in living organisms which ultimately may contribute to organic matter in sediments, e.g., fossil fuels.	CO_2 flows from the atmosphere and oceans into soils and sediments (the crust).	
HUMAN ACTIVITIES - COMBUSTION OF FUEL	The burning of fossil fuels releases CO_2 into the atmosphere.		

1. Earth's present atmosphere was formed from
   A. volcanic degassing and photosynthesis
   B. planetary formation
   C. weathering processes
   D. cometary impact

2. Which of the following gasses is most abundant in the Earth's present atmosphere?
   A. nitrogen
   B. carbon dioxide
   C. water
   D. oxygen

3. The potential for chemical weathering is ____________________ by acid rain.
   A. not effected
   B. decreased
   C. increased
   D. neutralized

4. Why is carbon dioxide considered a greenhouse gas?
   A. it absorbs heat
   B. it reflects radioactivity
   C. it absorbs UV light
   D. it reflects sunlight

5. Earth's global temperature trend is clearly
   A. upward over the last few decades
   B. downward over the last few decades
   C. unchanged over the last few decades
   D. variable, but there has been no overall change

Hint: Refer to Figure 23.13.

6. The occurrence of acid rain is most influenced by
   A. the release of radioisotopes by nuclear power plants
   B. the burning of high sulfur coals
   C. the burning of low sulfur coals
   D. the weathering of feldspars

7. Earth's present atmosphere has accumulated oxygen through
   A. the outgassing of the interior of the Earth by volcanoes
   B. the dissociation of water, forming oxygen and hydrogen
   C. photosynthesis and the burial of unoxidized organic carbon
   D. the weathering of igneous rocks that form oxides

8. The fossil record shows that the greatest loss in the diversity of marine life occurred on Earth during the
   A. Devonian
   B. Cretaceous
   C. Jurassic
   D. Permian

Hint: Refer to Figure 23.10.

EXAM PREP

The following *Chapter Summary* and *Practice Exam* should simplify review still further. Read the *Chapter Summary* to begin. It provides a helpful overview that should get you back into the material. Next, try the *Practice Exam*. Take it just as you would a midterm: to see how you stand in regard to mastery of this chapter. After you answer the questions, score them. Finally, and most important of all: review any question that you missed. Identify and correct the misconception that resulted in missing the item.

CHAPTER 23 SUMMARY

- The primary source of water and gases on the Earth's surface is volcanic gases, which outgases from the planet's interior over geologic time. A notable exception is oxygen gas, which was added by photosynthetic organisms after they evolved.
- The earliest fossils of bacteria are about 3.5 billion years old. Chemical evolution produced the building blocks of life and preceded biological evolution.
- The evolution of photosynthesizing organisms transformed the Earth's surface environments by adding oxygen, which now makes up 21% of the atmosphere. Life processes have a diverse impact on the carbon cycle.
- Much of the radiant energy from the Sun passes through the atmosphere and is absorbed by Earth's surface. The warmed surface radiates heat back to the atmosphere as infrared radiation. Carbon dioxide and other trace atmospheric gases are transparent to sunlight, but absorb heat (infrared radiation). In this way, the atmosphere acts like the glass in a greenhouse—allowing solar radiant energy to pass through, but trapping heat.
- Geochemical cycles trace the flux of Earth's elements like carbon from one reservoir to another. Understanding the carbon cycle is important because of its strong link to life processes, climate change, and plate tectonic processes.
- Climate change, extinctions, impacts, ozone depletion, acid rain, and human population growth are issues of vital concern today.

Website and CD Activities and Tools

http://www.whfreeman.com/presssiever

Test your knowledge of The Carbon Cycle by completing that Interactive Exercise at the Website.

PRACTICE EXAM QUESTIONS FOR CHAPTER 23

(Answers and explanations are at the end of this chapter.)

1. Why do scientists suspect CFCs are the source of ozone depletion in the stratosphere?
   A. CFCs contain chlorine, which reacts vigorously with ozone, while measurements of ozone in the stratosphere show it decreasing at the same time that CFCs are increasing.
   B. CFCs form a mixture with volcanic gasses in the lower atmosphere, which rises to the stratosphere and reacts with ozone.
   C. CFCs concentrate UV radiation, splitting apart ozone molecules.
   D. CFCs increase the albedo of the stratosphere, which reduces the solar radiation required for the production of ozone.

2. What is a potential link between plate tectonics and Earth's surface temperature?
   A. Changes in the shape of the seafloor induce sea-level changes.
   B. The flow of ocean currents are changed by tectonic movements.
   C. Continents can drift over the poles, leading to the growth of continental glaciers.
   D. All of the above

3. Ozone is a very reactive gas, so as a pollutant in the lower atmosphere ozone presents a significant health hazard. Why does ozone exist in the stratosphere?
   A. It is constantly leaking from the troposphere, where it is produced, up to the stratosphere.
   B. It is formed from the release of gasses out of the micrometeoric dusts that bombard the upper atmosphere.
   C. It is formed continuously in the stratosphere by solar radiation, and cannot mix or react with other gasses because of the thin atmosphere at that altitude.
   D. It is constantly produced from the oceans, rising through the troposphere to the stratosphere.

4. It has been suggested that the uplift of the Himalaya and the Tibetan Plateau could have contributed to or even caused a global cooling. The link between the Himalaya Mountains and climatic cooling is likely related to
   A. the collision of India with Asia triggering volcanism and increasing the CO_2 concentration in the atmosphere
   B. the uplift intensifying the monsoon and associated physical and chemical weathering, which resulted in a drawdown of carbon dioxide from the atmosphere
   C. the fact that high mountains generate more clouds, and their albedo (reflectivity) cools the Earth's surface
   D. El Niño and the North Atlantic deep water current

5. Which of the following is not associated with El Niño events?
   A. trade winds slacken or reverse direction
   B. volcanic activity
   C. change in ocean circulation patterns
   D. worldwide anomalous weather patterns

Hint: Refer to Box 23.1: *Interpreting Earth and Its System.*

6. As the oceans become warmer, __________ CO2 is released from the oceans into the atmosphere resulting in a __________ feedback.
   A. more / positive
   B. less / negative
   C. more / negative
   D. less / positive

7. The increase of the average temperature on Earth is linked to burning fossil fuels because
   A. the burning process consumes oxygen
   B. the burning process consumes CO_2
   C. the burning process generates CO_2
   D. the smoke given off by burning insulates the Earth

8. The surface temperatures on Venus, Earth and to a lesser extent Mars are all well above what can be explained by their distance from the Sun. What other factor significantly contributes to elevated surface temperatures for these inner planets?
   A. the presence of greenhouse gases, like carbon dioxide
   B. interior heat
   C. dust from windstorms and volcanoes which acts to trap heat
   D. presence of argon and nitrogen in the atmosphere

9. One hypothesis for the cause of the Late Permian mass extinctions is
   A. an asteroid impact
   B. climate change induced by volcanic eruptions from a superplume
   C. UV radiation from a very energetic outburst from the Sun
   D. multiple El Niño events stressed life on land and in the oceans

10. If you were looking for patterns in the fossil record to support the role of global climate change as a factor in large-scale extinctions, what might you expect to see?
   A. Latitudinal patterns in extinctions could exist, where tropical organisms are decimated, while temperate organisms experience only modest losses.
   B. Mass extinctions of land animals would be found, with little impact on marine animals.
   C. Mass extinctions of marine organisms would by seen, with little impact on land organisms.
   D. No pattern would exist because mass extinctions rule out climate change, since climate change is too gradual over geologic time.

ANSWERS AND EXPLANATIONS FOR EXERCISES AND QUESTIONS

After Lecture

Exercise 1: Characteristics of El Niño Events

1. extreme weather changes worldwide
2. flooding in some regions of the world due to an increase in storms
3. prolonged droughts in other regions of the world
4. forest fires due to extreme drought
5. landslides as a result of severe storms and rainfall
6. disrupted ecosystems, such as the collapse of Peruvian fisheries
7. spread of waterborne diseases
8. tremendous economic loss due to storm damage, crop failure, and other reasons listed above

Exercise 2: Bolide Impact Events

1. anomalous amounts of iridium in sediments at the K/T boundary
2. microtektites—spherical particles of glass
3. shock-metamorphosed quartz grains
4. the Chicxulub crater on the Yucatan Peninsula in Mexico
5. chaotic sedimentary deposits that are interpreted to be formed by the impact-generated tsunamis

Additional impact evidence is described in Box 23.2.

Exercise 3: The Carbon Cycle

CARBON FLUXES	BRIEF DESCRIPTION OF FLUX	DIRECTION OF FLUX	CLIMATIC IMPLICATIONS
VOLCANISM	Volcanoes outgas CO_2 from the crust and mantle.	CO_2 flows from the mantle and crust into the atmosphere and oceans.	Climate warms – adding CO_2 will enhance the greenhouse effect.
SEDIMENTATION	Calcium and carbonate ions combine to produce calcium carbonate that precipitates on the ocean floor, which is above the carbonate compensation depth.	CO_2 flows from the oceans and atmosphere into the crust.	Climate cools - CO_2 is drawn out of the oceans and atmosphere and into the crust. Because the oceans and atmosphere are in equilibrium, a drawn down of CO_2 in the oceans also results in reduced CO_2 in the atmosphere.
METAMORPHISM	CO_2 is liberated when carbonate minerals are replaced by silicate minerals	CO_2 flows from the crust into the oceans and atmosphere.	Climate warms – adding CO_2 will enhance the greenhouse effect.
CHEMICAL WEATHERING	CO_2 in rainwater combines with silicate minerals to form calcium carbonate	CO_2 flows from the atmosphere and oceans into the crust.	Climate cools - CO_2 is drawn out of the atmosphere and oceans and into the crust. The uplift of high plateaus and mountains may enhance this flux.
LIFE PROCESSES	Carbon is fixed in living organisms which ultimately may contribute to organic matter in sediments, e.g., fossil fuels.	CO_2 flows from the atmosphere and oceans into soils and sediments (the crust).	Climate cools - CO_2 is drawn out of the atmosphere and oceans and into the crust.
HUMAN ACTIVITIES - COMBUSTION OF FUEL	The burning of fossil fuels releases CO_2 into the atmosphere.	CO_2 flows from the crust into the oceans and atmosphere.	Climate warms – adding CO_2 will enhance the greenhouse effect.

1. A. Volcanic degassing of the Earth's interior contributed the major constituents of the oceans and atmosphere with the exception of free oxygen gas. Free oxygen gas (molecular oxygen) is not a component in volcanic gases. In fact, minor components of volcanic gases can react with free oxygen. Free oxygen gas is continually being added to the Earth's atmosphere and oceans by photosynthetic organisms.
2. A. The Earth's atmosphere is about 75% nitrogen gas.
3. C. Acids are powerful agents of chemical weathering.
4. A. Carbon dioxide is transparent to visible light and absorbs heat (IR radiation).
5. A. Refer to Figure 23.13.
6. B. Sulfur dioxide generated by burning fossil fuels high in sulfur content is the major source of acid in rain. The sulfur dioxide combines with rainfall which is already slightly acidic from the dissolved carbon dioxide to produce sulfuric acid.
7. C. Photosynthesis produces free oxygen gas and the burial of unoxidized organic matter reduces the rate at which the oxygen is used up.
8. D. Refer to Figure 23.10.

Exam Prep

1. A. Refer to the *CFCs and Ozone Depletion* section in your textbook.
2. D. Refer to the *Links between Tectonics and Climate Change* section in your textbook.
3. C. Refer to the *CFCs and Ozone Depletion* section in your textbook.
4. B. Refer to the *Links between Tectonics and Climate Change* section in your textbook.
5. B. Volcanic eruptions are not associated with El Niño events (refer to Box 23.1).
6. A. As the oceans become warmer, carbon dioxide solubility decreases. (Remember from Chapter 17, *Oceans*: carbon dioxide solubility increases as water temperature decreases.) Therefore, more carbon dioxide will be released, which is a positive feedback. A positive feedback adds carbon dioxide to the atmosphere; a negative feedback subtracts it.
7. C. Carbon dioxide is a greenhouse gas.
8. A. The greenhouse effect significantly influences surface temperatures for planets with atmospheres.
9. B. Refer to the *Superplume Eruptions and the Permian-Triassic Extinctions* section in your textbook.
10. A. Latitudinal patterns in extinctions would be strong evidence that climate played a role in causing the extinction events.

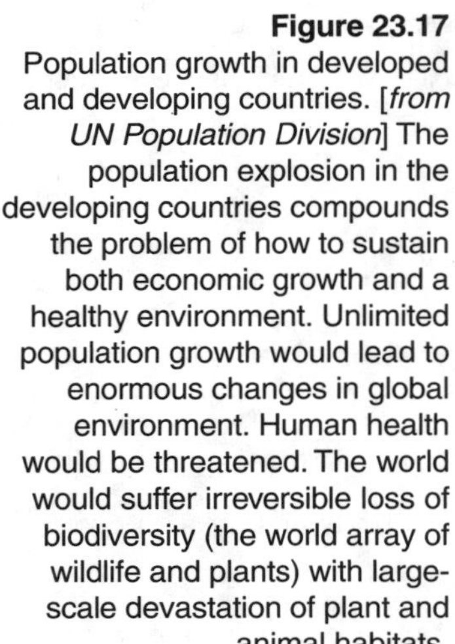

Figure 23.17
Population growth in developed and developing countries. [*from UN Population Division*] The population explosion in the developing countries compounds the problem of how to sustain both economic growth and a healthy environment. Unlimited population growth would lead to enormous changes in global environment. Human health would be threatened. The world would suffer irreversible loss of biodiversity (the world array of wildlife and plants) with large-scale devastation of plant and animal habitats.

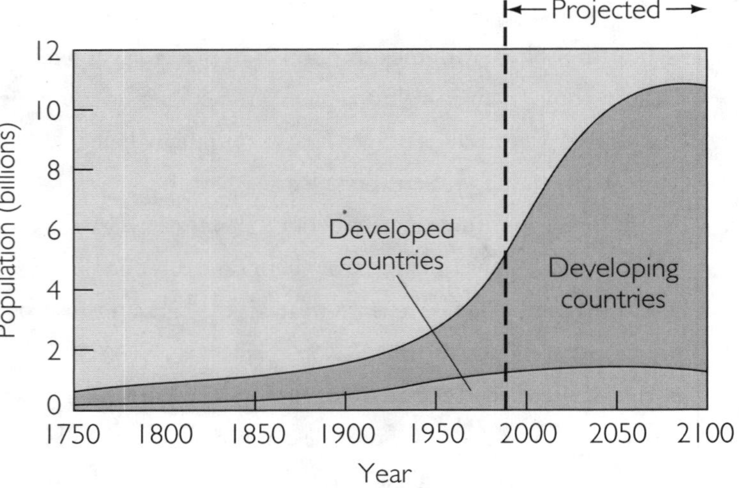

How hard to realize that every camp of men or beast has this glorious starry firmament for a roof! In such places standing alone on the mountain top it is easy to realize that . . . we all dwell in a house of one room—the world with the firmament for its roof—and are sailing the celestial spaces without leaving any track.

—John Muir, 1890

Appendix:
How to Study Geology

In this Appendix we have rolled up most of the strategies for studying geology that were introduced in this guide. One way to use the Appendix is to read it straight through. Alternatively, you can use it as a reference, to look up particular strategies, or to review forms such as the *Eight-Day Study Plan* that you may want to copy and use. We begin with a section on *How to Use Your Geology Textbook* and then procede with a review of study strategies. Strategies are organized in terms of when you should use them: *Before Lecture*, *During Lecture*, *After Lecture*, *Exam Prep*.

Meet Press and Siever: How to Use Your Geology Textbook

As you begin your study of geology, first take a minute or two to get acquainted with your text. First of all, why do you think it was written? Here is what your text authors Frank Press and Raymond Siever have to say about why they wrote this book:

"Geology fascinates and excites us. We wrote *Understanding Earth* to help you discover for yourselves how interesting geology is in its own right and how important an understanding of geology has become for making decisions of public policy. What can we do to protect people and property from natural disasters such as volcanoes, earthquakes, and landslides? How can we use the resources of Earth—coal and oil, minerals, water, and air—in ways that minimize damage to the environment? In the end, understanding Earth helps us understand how to preserve life on Earth."

Press and Siever add that people tend to enjoy what they do well. They designed their text to help you do well in your geology course. Many aids to learning are built into the text.

Clues and Tools: "What's Important?"

One of the toughest things about taking an introductory course is that you don't know the subject matter! That means that not only do you have a lot to learn but you have a lot to learn about *how* to learn it. Where should you focus your attention? What skills and concepts should receive the bulk of your study energy? Aids to help you see what material is important are built into every page of *Understanding Earth*, but you have to know where to look for these aids and how to use them. Here is a short list of learning aids in your new text and a few preliminary thoughts about how to use them to master geology:

Chapter Outline

Press and Siever begin each chapter with an outline of what will be covered. To use this tool you need to approach it like a detective: looking actively for clues. Look at the outline for Chapter 1. It probably won't surprise you that the first section in a geology text would be "The Scientific

Method." But look further. Would you expect a section about: "The Origin of Our System of Planets"? Why do you suppose a geology text would need to talk about how the planets are formed? If you don't know, read with that question in mind.

Chapter Summary

At the end of each chapter a brief summary emphasizes the most important ideas of the chapter.

> **TIME-SAVER TIP**
> Before reading a chapter of the text read the Chapter Summary first, referring as you do so to the Chapter Outline. This will provide an overview, or organizer, that will greatly accelerate your reading and understanding of material.

Key Terms

Key terms within the text are printed in bold. You can further speed up your reading time by being vigilant for key terms in bold. When you see such a bolded term you know it is the most important concept in that paragraph. Focus on understanding that concept. Read with purpose.

Pictures

Geology is a very visual science. Pictures in the text are essential. Use the pictures as a "virtual field trip" to help you learn what particular rocks and formations look like. Refer to these as you read.

> **STUDY TIP**
> Having a tough time getting yourself started with a study session? Use the pictures to motivate yourself. Start a chapter study session by scanning the pictures and captions to find material that interests you. Begin your reading with that material. This approach works like fire building. You generate a small spark of interest, then fan it by bringing in new material

Figures (Flow Charts and Tables)

Figures are even more important than pictures. Flow charts such as the Rock Cycle in Chapter 3 present the key concepts of each chapter. Pay careful attention to the arrows in flow charts and ask yourself what drives the process. If you are a visual learner you may even want to study the figures and sketches before you start to read. Then read text on an as-needed basis to clarify the figure.

Color

Color is used in the figures to provide you with clues about how each process works and what is involved. Look at the Rock Cycle (Figure 3.1. in Chapter 3). What color is used to describe processes that utilize heat? Water? Cooling? Such cues can support learning at a subliminal level of awareness, particularly for visual learners. But they work even better if you pay close attention to them.

Clues to Rock Texture

Clues to rock texture are also cleverly built into the text. Look at Figure 3.1 again. Why do you think igneous rocks are depicted as "dots and specks of varying size"? Why are sediments and sedimentary rock depicted with "horizontal markings"? Why is metamorphic rock depicted with a series of "distorted and folded bars"? You will see as you study the next several chapters that each of these textures is descriptive not only of how the rocks look but of the processes that produce them. Pay close attention to these visual learning clues. They will help you learn geology.

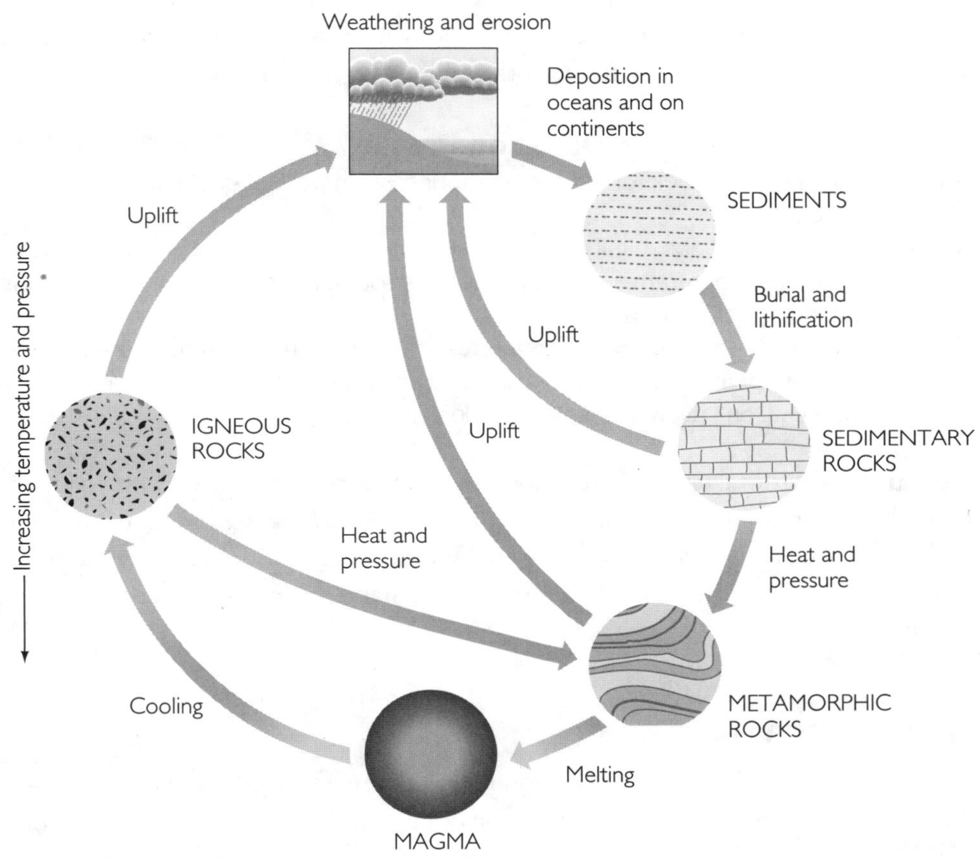

The Rock Cycle figure makes great use of patterns and colors.

BEFORE LECTURE

Chapter Preview

Why preview? Introductory courses can be difficult. There are lots of new terms, new ideas, and new skills during a first semester geology course. But which ideas are "important"? How do we focus our effort?

Picture a house under construction. Imagine further that the contractor was very careless and neglected to construct whole sections of the frame. Finishing those areas lacking a frame would be impossible. You can't tack siding onto thin air! This metaphor describes your geology lecture. You are the contractor. You need to arrive at class with an overview of the lecture already in mind.

An overview means you have already identified what geological processes will be explained and what key questions the lecture will answer. Lacking a frame of key questions is like building a house with no supporting structure. You will have nothing on which to hang the lecturer's main points (information). Without the main points the details are meaningless. As the lecture progresses you are likely to feel increasingly confused and bored. By the time the lecturer gets to the most important material you may be completely lost. Research suggests that the average note taker gets less than 15 percent of what is covered in the last segment of a lecture into their notes.

Where do you get the overview? The answer is surprisingly simple. You just need to spend a few minutes before lecture previewing the chapter that will be covered. Previewing is the method by which you generate and master a framework for listening. Here's how to do it.

Three Question Preview Method

Step 1: Skim the text. Look for the three or four most important questions the material will cover.

Step 2: List the questions. Limit yourself to just three or four so you will be sure to remember your framework during lecture. Be sure you identify the questions that are most important.

Step 3: Answer the questions. Spend the rest of your pre-lecture study time finding preliminary answers in your text. Read just enough to get a general idea of the answer. Don't allow yourself to get lost in details.

Before you go to lecture always be sure to spend some time previewing. We have made previewing easier by identifying key questions for each chapter (see *Chapter Preview* questions). These questions constitute the basic framework of the chapter. Working with the *Chapter Preview* before lecture and committing it to memory should help you understand the lecture better and get an excellent set of notes.

DURING LECTURE

Your basic goal during lecture is to get a good set of notes. Those notes should contain in-depth answers to the *Chapter Preview* questions. To avoid getting lost in details, keep the big picture in mind. Often this is best accomplished by having a copy of the *Chapter Preview* questions in front of you or, better yet, by developing preliminary answers to the questions and commiting them to memory before the lecture.

Another way to keep the big picture in mind is to bring a copy of the key figure or flow chart to lecture to refer to. In many chapters we suggest which figure you should have handy.

During some classes your lecturer may show rock formations and may pass around specimen rocks. You will get more out of these if you sit close to the front row where you can see the sample rocks as your lecturer discusses them. Remember to focus carefully on clues your lecturer provides for recognizing a particular sample in the field. Focus in particular on texture of the samples and learn to recognize differences. For example, you will learn the difference between a fine textured volcanic rock and a coarse textured plutonic rock.

As you listen to lecture, identify questions you need to ask to understand the material. Try to formulate at least one good question you can ask during every lecture.

The most important task during lecture is of course taking a good set of notes. Note taking is not easy. It is a skill that improves with practice. Here are a few tips that will help insure you take an excellent set of notes.

NOTE-TAKING CHECKLIST

- ✔ Organize your notes in a 3-ring binder so that you can easily reorganize them.
- ✔ Save space in your notes to add important visual material (flow charts, simple sketches, comparison charts) after class. To make this easy employ a double page note taking format (take notes on the right-hand page; save the left-facing page as a "sketch page").
- ✔ Date each day's notes so you can find material later.
- ✔ Take notes in a format that makes the main topics and concepts easy to identify. Some students accomplish this by taking notes in outline format, but many other approaches are possible. Visual learners sometimes find it helpful to highlight main points (after class) with a standard color. Another good approach would be to use the questions we provide under *BEFORE LECTURE/Chapter Preview* as headers. These can be added during class or during your after class Review Session.
- ✔ Keep the preview questions in front of you during lecture. Be sure to leave class with good answers to each of the three preview questions.
- ✔ Indicate areas where you need to do follow-up work with your text, instructor or tutor with a question mark in the margin.
- ✔ Indicate possible test questions by putting a "TQ" (Test Question) in the left margin where you can easily see it.
- ✔ Write assignments in the left column where you can easily find them. After class enter the due date in your Day Timer or personal calendar.

AFTER LECTURE

Review Your Notes

Right after lecture, while the material is still fresh in your mind, is the perfect time to review your notes. Review to be sure you understood the key points and wrote them down in a form that will be readable later.

Don't postpone this activity. Immediately after lecture, if there have been no interruptions, much of what was said will still be in your short-term memory. If you missed something you will still remember it and be able to put it into your notes. But short-term memory is just that—wait even one day and you will have forgotten 80 percent or more of the important points.

The basic idea in reviewing your notes is to fill in what you missed and add helpful visual material from the text. Use the following *Note Review Checklist* as a guide.

NOTE REVIEW CHECKLIST

- ✔ All notes legible? (Rewrite so they read easily.)
- ✔ Important points clearly identified? You should now have headers in your notes that tie to each of the questions in the *BEFORE LECTURE/Chapter Preview.*
- ✔ Holes (missing material) filled in from memory?
- ✔ Areas where you don't remember what was said marked for a follow-up session with your instructor, tutor, or study partner?
- ✔ Possible test questions indicated in the margin (TQ)?
- ✔ Additional visual material?
- ✔ Reworked notes into a form that is efficient for your learning style?
- ✔ Created a brief "big picture" overview of this lecture (using a sketch or written outline)?

Intensive Study Session

You should schedule at least one hour after each lecture for intensive study. This can occur anytime before the next lecture (short-term memory is no longer a problem since you have followed the *Note Review Checklist* and have a good set of notes to work with).

Why do you need an intensive study session? You learn geology much as you would build a house. Before each lecture, construct a frame of questions. During lecture, attach details and ideas to the frame. After lecture, master those ideas during an intensive study session. Now you have constructed the first story. But geology is a skyscraper with 23 floors (one for each chapter). Each chapter supports those above it. If you don't completely master this chapter, the next will be more difficult.

Schedule at least one hour after each lecture for intensive study. Devote this time to mastering key concepts. Mastery is not gained by just reading the text. Mastery occurs as the result of asking yourself questions (and answering them). To help you, we provide *Practice Exercises* and *Study Questions* for every chapter of the text. This interactive learning material is specifically designed to help you reach mastery level quickly. The greater the number of these exercises you can work into your intensive study sessions early in the course, the easier subsequent chapters will be. Plan to spend the majority of your time, say 70 percent, working on these exercises.

That means you will spend only about 30 percent of your study time reading the text. Use your text as a reference book. Read it as needed: to answer questions and master material. When you read, read efficiently: read with purpose, read to find the answer to a question you are working on.

You can make the 30 percent of the study time you spend reading the text even more efficient by skillful text marking.

How to Mark and Annotate Your Text

Marking your text as you read is a vital part of learning the material. Good marking is good thinking. First, you think about what is important. Then you mark it. Press and Siever have already marked **key terms** in bold letters. To this bold lettering you will want to add underlines, numbers, and annotations in the margin. Here are some brief suggestions for how and when to use each kind of mark

USEFUL MARKINGS FOR GEOLOGY: HOW-TO TIPS	
Underline important points. Underline to highlight the key ideas.	(1) Read before you mark. To avoid overusing underlining, set your pencil down on the table as you begin to read a paragraph. Don' t let yourself pick up the pencil until you have read the entire paragraph. Then underline what you consider the key point. (2) Be selective. Underline brief, but meaningful **phrases** or **key words** that will trigger your memory. Mark just enough so you can **review without re-reading** the paragraph. (3) Use color marking if it helps you remember.
Annotate the margin. Write a brief summary statement of **key geologic processes** in your own words in the text margin. Use these annotations to facilitate exam review.	(1) Use your own words. Putting ideas into your own words is a powerful learning strategy. (2) Be neat. This takes time. But it will pay off later when you review, because your annotation will be easily perceived. (3) Organize your annotations into categories. Grouping ideas into categories or bulleted lists makes them easier to remember. (4) Use annotations during exam review. Avoid rereading the text word for word. You don' t have time.
Other useful text marks:	(1) **Circled numbers** in the margin indicate sequences such as the processes of the Rock Cycle. (2) **Question mark?** Put a question mark in the margin to remind yourself to ask your instructor about a point you do not understand. (3) **Asterisks ***** Use asterisks to mark ideas of special importance. Use asterisks sparingly. Save them for the two or three most vital ideas in the entire chapter.

EXAM PREP

Materials in this section are most useful during your preparation for midterm and final exams. For optimal performance, midterm preparation should begin about eight days prior to the exam (see the *Eight-Day Study Plan*). The basic idea is a systematic review of material divided into short study sessions.

TIPS FOR PREPARING FOR GEOLOGY EXAMS

- ✔ Use the clues your instructor has provided in lecture about what is important. Even when a department agrees on a common core of material (rare) each instructor carves out a course that is unique, has a particular character or flavor, and distinct areas of emphasis. Your instructor is the ultimate guide in regard to the question "What is important".
- ✔ Be sure you know the format of the exam. Multiple choice? True false? Essay? Thought problems? What?
- ✔ Review your notes for material marked TQ, (Test Question).
- ✔ Ask your instructor if exams are available from the previous semester. Review these to check the format of questions, see what areas of content are stressed, and what types of problem solving are included. Don't make the mistake of assuming the same questions will be asked this semester.
- ✔ Be sure to attend review sessions if these are offered. Attending a review session will raise your midterm score.
- ✔ If your class has tutors, preceptors, supplemental instruction leaders or other peer helpers who have taken the class, ask for their suggestions about preparing for the midterm.
- ✔ Once you are clear about the nature of the midterm exam, begin your review. Conduct review in an orderly systematic manner that insures focused review of all the important material. The *Eight-Day Study Plan* is a good model for orderly review.

Eight-Day Study Plan

Adapted with permission from the University Learning Center, University of Arizona.
(Make several copies of this Study Plan and use one for every exam you take.)

Here is a guide you can use to prepare for your midterm. Everyone develops their own approach to preparing for exams, so feel free to adapt these ideas to your particular needs and situation. The basic idea is to conduct your preparation in a systematic fashion with focus on the most important material. Our plan accomplishes this by dividing the material equally and suggesting how to incorporate the *Exam Prep* materials provided in this study guide for each text chapter. You begin the plan eight days prior to the exam.

Day 8: Get organized!
Step 1: Clarify the task. Determine what sort of midterm you will take by briefly answering the following questions:

1. This exam will cover (list each chapter to be covered):

   ____

   ____

2. Material and kinds of skills to be particularly emphasized (list chapters/ideas/skills your instructor said would be particularly important):

   ____

   ____

3. The test-question format(s) will be (check all that apply):

   ___ Multiple Choice
   ___ True / False
   ___ Essay
   ___ Thought Problems
   ___ Other (specify)

3. Review session is scheduled for (date) ________________ . Be sure to attend.

Step 2 Divide the material you must review into four equal parts: A, B, C, D.

Day 7 Begin your review. Review all material in Part A.

DO THE FOLLOWING FOR EACH CHAPTER IN PART A:

1. **Chapter Summary:** To get yourself started, read the *Chapter Summary* (*Exam Prep* section of this guide) for the chapter you want to review.
2. **Practice Exam Questions:** Answer *Practice Exam Questions* (*Exam Prep* section of this guide) to see where you are with the material. Force yourself to answer all questions for the chapter without referring to the answer key. Correct only after you have tried all items. Be sure to review carefully any items you missed. Correct the misconception that resulted in error.
3. **Class Notes:** Review your class notes and annotations you made in the text margin by asking yourself questions.
4. **Focus on Visual Materials and Key Figures:** This may also be a good time to re-do some of the *Practice Exercises* in this study guide. Many Practice *Exercises* are designed to help you master the visual concepts of geology. Review visual material in your notes. Test yourself by seeing if you can reconstruct key figures from memory.
5. **Self Test:** Spend as high a proportion of your study time as possible asking yourself questions.

Study is not reading. Study is asking yourself questions.
—A good friend who earned lots of A's in college

Day 6: Review Part B. Repeat instructions for Day 7, this time reviewing Part B. If you have problems with material see your instructor at the next open office hour.

Day 5: Part C. Repeat instructions for Day 7, this time reviewing Part C. If you have problems with material see your instructor at the next open office hour.

Day 4: Part D. Repeat instructions for Day 7, this time reviewing Part D. If you have problems with material see your instructor at the next open office hour.

Day 3: Review all parts – A, B, C, D – fully. Prioritize your time. Focus on important material that will be covered. Work hardest where you are least sure of your self. If you have problems with material see your instructor at the next open office hour.

Day 2: Review all parts – A, B, C, D – fully. Prioritize your time. If you have problems with material see your instructor at the next open office hour.

Night Before: Be sure you get the amount of sleep you need to be alert and perform at your best. You don't need to cram. Just stay focused.

Zero Hour: You have prepared well. Allow yourself to be confident. Stay focused and confident during the exam. Use your best test-taking strategies.

FINAL EXAM WEEK

Each semester in one short week you get to take an exam in each and every course. Most of those exams are comprehensive finals that cover the entire semester. Dealing with finals week successfully is a major challenge. Here are some tips that will insure you do your best work during final week.

TIPS FOR SURVIVING FINALS WEEK

- Be organized and systematic. Use the FINAL EXAM PREP WORKSHEET to help you get organized for finals. Use the EIGHT DAY STUDY PLAN for every course where the final exam will be an important factor in determining your grade.
- Stick to priorities. Say no to distractions.
- Build in moments of relaxation: Take regular short breaks, exercise, and be sure to get enough sleep.
- Be confident. By now you have built up a good set of study habits. You are a competent learner.

Its time to get ready for final exam week. As with any big project, devoting some time up front to getting organized will pay big dividends. Use the following worksheet as a guide to help you get organized. Modify the suggestions to fit your personal situation and needs.

FINAL EXAM PREP
WORKSHEET

(to be completed three weeks prior to exam)

Adapted with permission from the University Learning Center, University of Arizona.

1. **COURSE SHEETS:** In your notebook, set up four or five separate sheets of paper—one sheet for each course you are taking. At the top of each sheet list the course and the grade you presently have (be realistic, not hopeful).

2. **DATE:** List the date of the final exam under each course name.

3. **COMPREHENSIVE FINALS:** Mark with a "C": each course with a comprehensive final.

4. **EXAM FORMAT:** Identify the format of the exam (multiple choice, essay, etc.) under the date of the final for each course.

5. **TASK:** Identify the levels of thinking expected. Hint: Previous midterms are your ultimate resource on this question. List all kinds of questions. Estimate what percent of total points will be devoted to each kind of thinking.
    - Application to Real World Situations
    - Problem Solving
    - Critical Thinking
    - Understanding Principles
    - Memory of Basic Facts

6. **RANK FINALS IN ORDER OF IMPORTANCE:** In the upper right-hand corner of each sheet, rank in order the most critical and important final to the least important final—the final that will make the least difference in your grade. (Be aware of how much impact your final exam has on your overall class grade.)

7. **LIST WHAT THE TEST WILL COVER:** For each course on each sheet, list everything the test will cover—remember which exams are comprehensive.
   - Handouts??
   - Chapters?? (which ones??)
   - Lectures??
   - Discussions??
   - Other??

   Check your syllabi to be sure you have not left anything out.

8. Draw a line beneath this list. Then list what you still have left to do for that particular course.
   - Which chapters do you still have to preview?
   - Which lecture notes do you need to review and update?
   - Which *Practice Exercises* and *Study Questions* do you need to complete?
   - Which labs do you still have to hand in?
   - What papers do you still have to write?

9. Draw another line. Now list the test preparation strategies you will use to study for the exam—study groups or study patterns, self-questioning using the annotations, mapping, charting, questions and answers, concept cards, going over old tests and quizzes, making up your own problems.

10. Now fill in the calendar, identifying exams, finals and when papers are due. Each day you need to do something from No. 8, but you will also need to study and review for the finals at least two hours a day. Be sure to use all of your available time—weekends, waiting time, etc.

Work Toward These Goals:

- Finish all work under No. 8. Review *Practice Exercises* and *Study Questions*, etc. one week prior to your first final. Review all lecture notes by asking yourself the questions out loud or by having someone quiz you five days prior to your first final (allow two to three hours).
- Divide the work that remains so that you do an *Eight-Day Study Plan* for each course that you assigned a high priority in No. 6.
- Remove the distractions from your life. This is not the week to be captured by TV or other addictions. Stick to your priorities. Tell friends and family that you need to focus all your energy on your finals until they are over.
- Avoid burnout. Build into your schedule time for adequate sleep, relaxation and exercise.

TEST-TAKING TIPS

Test Taking and Learning Styles

VISUAL LEARNERS

- Use written directions.
- When you get stuck on an item, close your eyes and picture flow charts, pictures, field experiences or text.

AUDITORY LEARNERS

- Pay attention to verbal directions.
- Repeat written directions quietly to yourself (moving your lips is often enough).
- When you get stuck, remember your lecturer's voice covering this section.

KINESTHETIC LEARNERS

There are a variety of things kinesthetic learners find helpful when they get stuck on a test item. Try some of these:

- When you get stuck, move in your chair or tap your foot to trigger memory.
- Feel yourself doing a lab procedure.
- Sketch a flow chart to unlock memory of a process.
- Stuck on a geology problem? Sketch what is being described to get you started.

Many exams are in multiple choice format. Here are some tips to maximize your performance on multiple choice questions.

TEST-TAKING

Top Ten Tips for Taking a Multiple Choice Exam

Number 10: Answer the questions you know first.

Mark items where you get stuck. Come back to harder questions later. Often, you will find the answer you are looking for embedded in another, easier question.

Number 9: First, try to answer the item without looking at the options.

Number 8: Eliminate the distracters.

Treat each alternative as a true-false item. If "false", eliminate it.

Number 7: Use common sense.

Reasoning is more reliable than memory.

Number 6: Underline key words in the stem.

This is good to try when you are stuck. It may help you focus on what question is really being asked.

Number 5: If two alternatives look similar it is likely that one of them is correct.

(cont'd on next page)

TEST-TAKING (cont'd)

Number 4: Answer all questions.

Unless points are being subtracted for wrong answers (rare) it pays to guess when you're not sure. Research indicates that items with the most words in the middle of the list are often the correct items. But be cautious—your professor may have read the research too!

Number 3: Do not change answers.

Particularly when you are guessing, your first guess is often correct. Change answers only when you have a clear reason for doing so.

Number 2: If the first item is correct, check the last.

If it says "all (or none) of the above," you obviously need to read the other alternatives carefully. Missing an "all of the above" item is one of the most common errors on a multiple choice exam. It is easy to read carelessly when you are anxious.

Number 1: READ THE DIRECTIONS BEFORE YOU BEGIN!